Camouflage

Camouflage

Edited by:

Shiv Sanjeevi Sripathi

Delve
PUBLISHING

www.delvepublishing.com

Camouflage

Shiv Sanjeevi Sripathi

Delve Publishing
224 Shoreacres Road
Burlington, ON L7L 2H2
Canada
www.delvepublishing.com

Email: orders@arclereducation.com

ISBN: 978-1-77407-697-2 (Hardcover)

DECLARATION

Some content or chapters in this book are open access copyright free published research work, which is published under Creative Commons License and are indicated with the citation. We are thankful to the publishers and authors of the content and chapters as without them this book wouldn’t have been possible.

ABOUT THE EDITOR

ShivSanjeevi Sripathi completed his Masters in Biotechnology from Mumbai University in 2008. He was awarded for academic excellence in both his Bachelors and Masters for securing second rank in Mumbai University in 2006 and first rank in his college: Kishinchand chellaram College. For his Masters he secured first rank in his college KET's V.G.Vaze College. He qualified CSIR and NET and and TOEFL in September 2008. He then worked on a stem cell project at the Specialized Centre for Cell Based Therapy (SCCT),KEM Hospital at Mumbai on a project entitled, : Isolation & detection of stem cells from Human Umbilical cord/ amniotic membrane" following which he worked at Junior Research Fellow at Microbiology & Cell Biology Department, Indian Institute of Science, Bangalore from August 2009 to November 2011. The field of work involved cloning of cell wall genes and transcription factors in *E.coli & M.smegmatis*. As a writer, he has authored and co-authored 25 books with Delve and Arcler International Press. He loves to read and share on interesting aspects of life sciences in books. In his free time he loves to travel and explore ancient customs and traditions.

TABLE OF CONTENTS

LIST OF CONTRIBUTORS

Shivsanjeevi Sripathi

Prerna Pandey

Suzanne Amador Kane
Physics & Astronomy Department, Haverford College, Haverford, Pennsylvania, United States of America

Yuchao Wang
Physics & Astronomy Department, Haverford College, Haverford, Pennsylvania, United States of America

Rui Fang
Physics & Astronomy Department, Haverford College, Haverford, Pennsylvania, United States of America

Yabin Lu
Physics & Astronomy Department, Haverford College, Haverford, Pennsylvania, United States of America

Roslyn Dakin
Migratory Bird Center, Smithsonian Conservation Biology Institute, National Zoological Park, Washington DC, United States of America

Bettina von Helversen
Max Planck Institute for Human Development, Center for Adaptive Behavior and Cognition, Berlin, Germany,
Department of Psychology, University of Basel, Basel, Switzerland, University of Sussex, United Kingdom,

Lael J. Schooler
Max Planck Institute for Human Development, Center for Adaptive Behavior and Cognition, Berlin, Germany,

Uwe Czienskowski
Max Planck Institute for Human Development, Center for Adaptive Behavior and Cognition, Berlin, Germany,

R. Gasparini
Department of Health Sciences, University of Genoa, Via Pastore 1, 16132 Genoa, Italy

D. Panatto
Department of Health Sciences, University of Genoa, Via Pastore 1, 16132 Genoa, Italy

N. L. Bragazzi
Department of Health Sciences, University of Genoa, Via Pastore 1, 16132 Genoa, Italy

P. L. Lai
Department of Health Sciences, University of Genoa, Via Pastore 1, 16132 Genoa, Italy

A. Bechini
Department of Health Sciences, University of Florence, Viale G.B. Morgagni 48, 50134 Florence, Italy

M. Levi
Department of Health Sciences, University of Florence, Viale G.B. Morgagni 48, 50134 Florence, Italy

P. Durando
Department of Health Sciences, University of Genoa, Via Pastore 1, 16132 Genoa, Italy

D. Amicizia
Department of Health Sciences, University of Genoa, Via Pastore 1, 16132 Genoa, Italy

Adriana Gutiérrez-Hoya
Laboratorio de Oncología Molecular, FES Zaragoza, Universidad Nacional Autónoma de México, Batalla 5 de mayo s/n Col. Ejército de Oriente, CP 09230 Ciudad de México, Mexico
Cátedras CONACyT, CONACYT, Avenida Insurgentes Sur 1582, Benito Juárez, Crédito Constructor, 03940 Ciudad de México, Mexico

Octavio Zerecero-Carreón
Laboratorio de Oncología Molecular, FES Zaragoza, Universidad Nacional Autónoma de México, Batalla 5 de mayo s/n Col. Ejército de Oriente, CP 09230 Ciudad de México, Mexico

Arturo Valle-Mendiola
Laboratorio de Oncología Molecular, FES Zaragoza, Universidad Nacional Autónoma de México, Batalla 5 de mayo s/n Col. Ejército de Oriente, CP 09230 Ciudad de México, Mexico

Martha Moreno-Lafont
Departamento de Inmunología, Escuela Nacional de Ciencias Biológicas, Instituto Politécnico Nacional, Prolongación de Carpio y Plan de Ayala S/N. Colonia Santos Tomás, Miguel Hidalgo, 11340 Ciudad de México, Mexico

Rubén López-Santiago
Departamento de Inmunología, Escuela Nacional de Ciencias Biológicas, Instituto Politécnico Nacional, Prolongación de Carpio y Plan de Ayala S/N. Colonia Santos Tomás, Miguel Hidalgo, 11340 Ciudad de México, Mexico

Benny Weiss-Steider
Laboratorio de Oncología Molecular, FES Zaragoza, Universidad Nacional Autónoma de México, Batalla 5 de mayo s/n Col. Ejército de Oriente, CP 09230 Ciudad de México, Mexico

Isabel Soto-Cruz
Laboratorio de Oncología Molecular, FES Zaragoza, Universidad Nacional Autónoma de México, Batalla 5 de mayo s/n Col. Ejército de Oriente, CP 09230 Ciudad de México, Mexico

Luís Felipe Toledo
Museu de Zoologia "Prof. Dr. Adão José Cardoso", Instituto de Biologia, Universidade Estadual de Campinas, Rua Albert Einstein s/n, CP 6109, 13083-863 Campinas, SP, Brazil

Célio F. B. Haddad
Departamento de Zoologia, Instituto de Biociências, Universidade Estadual Paulista, CP 199, 13506-970 Rio Claro, SP, Brazil

Kate L. A. Marshall
Behavioural Ecology Group, Department of Zoology, University of Cambridge, Cambridge, United Kingdom,

Kate E. Philpot
Centre for Ecology and Conservation, College of Life and Environmental Sciences, University of Exeter, Penryn Campus, Penryn, Cornwall, United Kingdom,

Isabel Damas-Moreira
CIBIO Research Centre in Biodiversity and Genetic Resources, InBIO, Universidade do Porto, Campus Agrário de Vairão, Rua Padre Armando Quintas, Vairão, Vila do Conde, Portugal, University of Sussex, UNITED KINGDOM,

Martin Stevens
Centre for Ecology and Conservation, College of Life and Environmental Sciences, University of Exeter, Penryn Campus, Penryn, Cornwall, United Kingdom,

Xueqiang Gu
College of Mechatronic Engineering and Automation, National University of Defense Technology, Changsha, Hunan 410073, China

Lincheng Shen
College of Mechatronic Engineering and Automation, National University of Defense Technology, Changsha, Hunan 410073, China

Jing Chen
College of Mechatronic Engineering and Automation, National University of Defense Technology, Changsha, Hunan 410073, China

Yu Zhang
College of Mechatronic Engineering and Automation, National University of Defense Technology, Changsha, Hunan 410073, China

Wanpeng Zhang
College of Mechatronic Engineering and Automation, National University of Defense Technology, Changsha, Hunan 410073, China

LIST OF ABBREVIATIONS

CSF	Contrast sensitivity function
JND	Just noticeable distances
ROI	Regions of interest
BBB	Blood-brain barrier
CEACAMs	Carcinoembryonic antigen cell adhesion molecules
CAMPs	Cationic antimicrobial peptides
CREEs	Correia repeat-enclosed elements
CRs	Correia repeats
HSPG	Heparan sulphate proteoglycan
HTTs	High-throughput technologies
HAMPs	Host-associated molecular patterns
IM	Inner membrane
IS	Insertion sequence
LPS	Lipopolysaccharide
MS	Mass spectrometry
MAC	Membrane attack complex
MDA	Meningococcal disease-associated
MMEs	Minimal mobile elements
MAPK	Mitogen-activated protein kinase
MWNTs	Multiwalled carbon nanotubes
NKs	Natural killer
NIMEs	Neisserial intergenic mosaic elements
OM	Outer membrane
OMP	Outer membrane protein
PAMPs	Pathogen-associated molecular patterns
PRRs	Pattern recognition receptors
RNS	Reactive nitrogen species
ROS	Reactive oxygen species

SSM	Slipped strand mispairing
CP	Themeningococcal capsule
UTRs	Untranslated regions
WGS	Whole-genome sequencing
ADP	Adenosine diphosphate
AMP	Adenosine monophosphate
CSC	Cancer stem cells
EGFR	Epidermal growth factor receptor
SCCHN	Squamous cell carcinoma of the head and neck
PCA	Principle component analysis
CF	Coordination function
CV	Coordination variables
DFT	Differential flatness theory
ECAVs	Electronic combat air vehicles
GPM	Gauss pseudospectral method
IP	Initial point
LG	Legendre-Gauss
NFZ	No-fly zone
NLP	Nonlinear programming
OCPs	Optimal control problems
UAV	Unmanned aerial vehicle
VMC	Virtual motion camouflage
VTD	Virtual time domain

PREFACE

An insect with a backpack! Read ahead to know more!

We have heard the phrase "master of disguise" right from childhood. While we disguise or camouflage our "emotions" or "scars", plants and animals camouflage themselves for protection. Darwin's concepts of the "struggle for existence" and "survival of the fittest" have led to living organisms evolve several approaches to increase their survival.

Remember a chameleon quickly adjusting its colors?

Fish that resemble ocean floors?

Snakes that effortlessly blend with a pile of dried leaves?

Crabs that decorate themselves with algae?

Well, there are also octopus species that imitate certain fish. There are birds that roll their eggs in the soil to help match with the background thus lowering detection. Mammals such as zebras show dazzle camouflage-the presence of patterns on their body creates a blur when they quickly move across the visual field. This makes it difficult to trace their exact speed or path. These are all examples of camouflage-either to capture prey to feed or to protect oneself from getting preyed.

This book is a concise presentation of camouflage encompassing the plant, animal and human worlds. Plants were initially considered defenseless and senseless creatures. However, certain examples of mottled leaves called disruptive coloration prevent easy recognition by color blind herbivores. A plant called Elaeocarpus hookerianus in New Zealand is mottled brown to match its surroundings. Similarly, several examples of camouflage in the animal kingdom have been discussed.

There is an interesting example of the assassin bug- this insect is a backpacker! In order to lower its identification, the insect coats itself with dust, soil and corpses of its own prey!

The book goes on to present an example of camouflage in humans: a disease called Autism spectrum disorder (ASD) is incident in both boys and girls. However, boys are detected with more ease than girls. This is attributed to "camouflaging" in girls with ASD making the diagnosis complicated. Camouflage is an interesting aspect of biology. While most of us humans have learned to camouflage our emotions or intentions for our survival, plants and animals camouflage themselves to meet the same common thread unifying all living organisms-Survival.

Editors

1

Understanding Camouflage

Shivsanjeevi Sripathi

ShivSanjeevi Sripathi completed his Masters in Biotechnology from Mumbai University in 2008. He was awarded for academic excellence in both his Bachelors and Masters for securing second rank in Mumbai University in 2006 and first rank in his college: Kishinchand chellaram College. For his Masters he secured first rank in his college KET's V.G.Vaze College. He qualified CSIR and NET and and TOEFL in September 2008. He then worked on a stem cell project at the Specialized Centre for Cell Based Therapy (SCCT),KEM Hospital at Mumbai on a project entitled, : Isolation & detection of stem cells from Human Umbilical cord/ amniotic membrane" following which he worked at Junior Research Fellow at Microbiology & Cell Biology Department, Indian Institute of Science, Bangalore from August 2009 to November 2011. The field of work involved cloning of cell wall genes and transcription factors in *E.coli & M.smegmatis.* As a writer, he has authored and co-authored 25 books with Delve and Arcler International Press. He loves to read and share on interesting aspects of life sciences in books. In his free time he loves to travel and explore ancient customs and traditions.

1.1 CONCEPT

Detection or recognition can be avoided with camouflage such as body coloration that is evident in the animal kingdom. Apart from vision, other examples include morphological structures or environmental material. These strategies range from marine isopods that breakup their body appearance to many moths that possess markings or colors to their surroundings. Certain cephalopods and chameleons can alter their colors to match that of their backgrounds. There are fish that look like dead leaves and insects that look like bird droppings. Shadows are hidden by certain animals by bioluminescence while butterflies can mimic leaves. Some crabs can decorate themselves using items from their environment.

Thus, we can see that camouflage is a broad field with single aim-avoiding enemies or predation-the greatest selection pressures in biology!

(Stevens and Merilaita, 2009)

1.2 AN ONION FOR SURVIVAL

Erasmus Darwin in 1794 discussed animal camouflage as "The colors of many animals seem adapted to their purposes of concealing themselves, either to avoid danger or to spring upon their prey".

The 'survivability onion' is used to explain defensive strategies using the common military framework. It refers to avoiding detection, identification, engaging, acquiring, penetration or getting affected. This concept can be extended to biology by avoiding predators or the detection of prey.

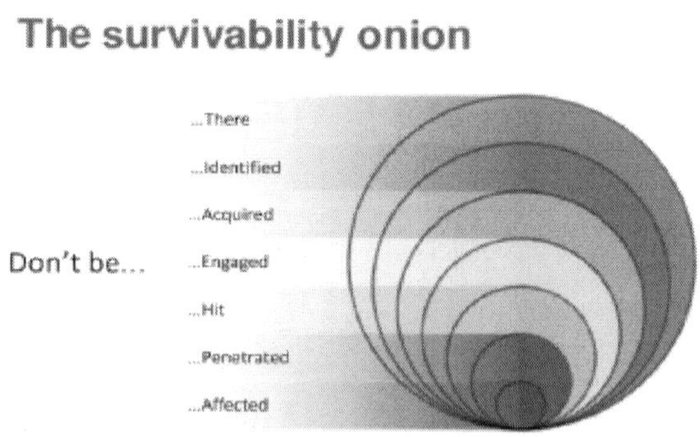

Figure 1.1: Survivability onion (Cuthill, 2019)

The first course is moving away, vigilance and avoiding danger to achieve safety. In the case, a predator cannot be avoided another option is crypsis:

this refers to 'hiding in plain sight'. In the event that detection cannot be avoided then strategies to avoid identification or promote misidentification is the next line of defense. In the event that identification occurs then selection by the predator or danger sign can be avoided by preventing attack or selection: what is referred to as 'target acquisition'.

When the predator discovers or attacks, another option is 'deimatic' display or startling the predator by targeting its reflex actions. The use of unpredictable ('protean') escape patterns can prevent immediate capture. The outer layer of the onion refers to preventing injury or death or deflection of attack.

(Cuthill, 2019)

1.3 TERMS USED IN CAMOUFLAGE BIOLOGY

The various systems used by animals to conceal themselves have been addressed using several terms:

- Camouflage refers to all strategies involved in concealment as well as preventing crypsis(detection).
- Background matching refers to making an appearance to match the background in terms of color, lightness and pattern.
- Self-shadow concealment involves the creation of shadows when countershading cancels out shadow formation due to directional light.
- Obliterative shading is when the three-dimensional form is abolished by the use of countershading.
- Disruptive coloration refers to preventing the recognition of an object by obliterating its true shape and form using false edges and boundaries.
- Flicker-fusion camouflage refers to preventing the recognition of a moving animal by the use of stripes that blur and match the color of the surroundings.
- Distractive markings as the name suggests, distract a predator or observer from phenotypes or appearances that allow identification of an animal.
- Masquerade refers to an animal resembling an uninteresting object, such as a leaf or a stick to prevent recognition.

- Motion dazzle makes the process of understanding the speed or path of an animal difficult hence, lowering the chance of capture.
- Motion camouflage is adopting movements that lower the probability of movement detection.

1.4 CEPHALOPODS: ADEPTS AT CHANGING

While the majority of animals undergo movement to reach the appropriate background and lighting and later display camouflage, cephalopods — octopus, cuttlefish and squids directly alter their colors. These colors can match several patterns ranging from sand to seagrass beds to kelp to rocks to coral reefs. An example was published by Hanlon (2007) where the response of *Octopus vulgaris* in the Cayman Islands was recorded. The animal showed a rapid change from camouflaged to fully conspicuous in 2.02 seconds.

Figure 1.2: *Octopus vulgaris* when it spotted a diver (predator). The direct neural control of the skin resulted in the rapid second-level change from camouflaged to conspicuous. Video clip available at http://www.mbl.edu/mrc/ hanlon/ video.html (Hanlon,2007)

The direct neural control of chromatophores (muscle-ended pigment sacs in the skin) allows for such quickness in color change. Each chromatophore organ muscle is innervated by motor neurons originating from the brain and reaching the muscle without any synapse. Due to the presence of changeable skin papillae, the camouflage can be observed at three-dimensional effects.

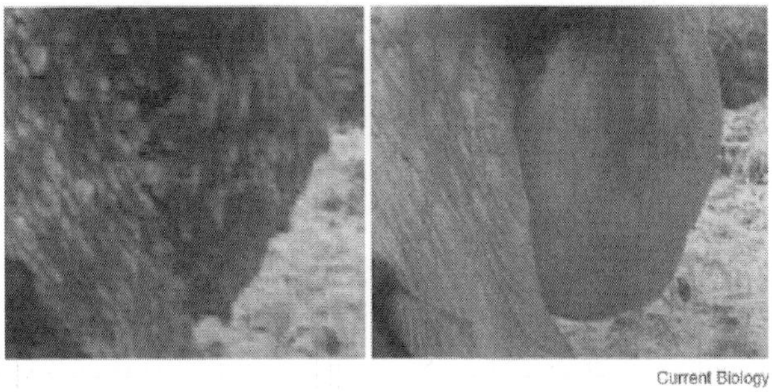

Figure 1.3: Observing the changes in changeable skin papillae (Hanlon, 2007)

Camouflage is the first strategy adopted by cephalopods especially by staying motionless and showing appropriate patterns. Yet, some octopus species can maintain their camouflage even when they are moving slowly. There are *Thaumoctopus mimicus* in Indonesia and *Octopus defilippi* in the Caribbean that mimic flounders as they traverse sand plains without any significant features. The camouflage is displayed the moment the animal stops to resemble a fish rather than an octopus!

Figure 1.4: *Bothus mancus* (a flounder) (Hanlon, 2007)

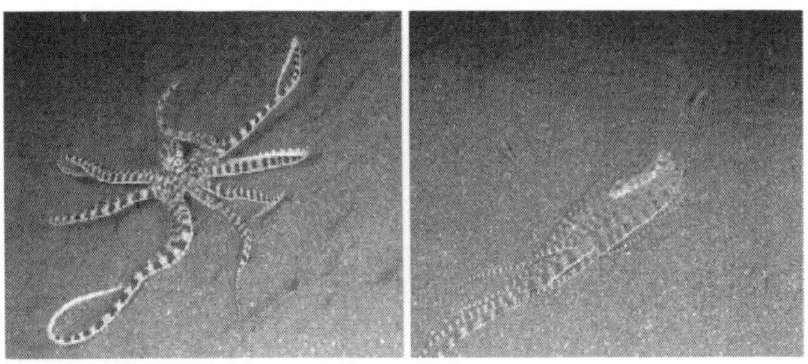

Figure 1.5: *Thaumoctopus mimicus* (left) that adopts the shape as well as the swimming speed of a flounder resembling Figure 1.4 (Hanlon, 2007)

Certain patterns such as light and dark patches or repeated patches are referred to as mottle patterns. These patterns serve to resemble the surroundings-background matching. There are disruptive patterns that have orientations and patches at high contrast. These patterns provide false edges and patterns to conceal the true outline of an animal. These are against uniform patterns that have minimal changes in contrast.

These 3 patterns were tested using European cuttlefish, *Sepia officinalis* that shows camouflage given that it is a choice prey of many ocean hunters and is also adapted to lab surroundings. In order to assess how a cephalopod sees its surroundings and adapts to it quickly, uniform, mottle or disruptive body patterns were tested in the lab.

While the matching of a cephalopod's color to the surroundings is laudable, these animals are colorblind! The presence of only one visual pigment was confirmed using the large checkerboard assay. The uniform, mottle or disruptive camouflage patterns could be elicited by small to large high-contrast checkerboards. Based on the scale of the checker, the mottle and disruptive pattern backgrounds are constructed. Checks made of contrasting yellow and blue shades were made to match the animal's 492 nm (green) pigment. The animals could not distinguish the yellow/blue checkerboard as they showed uniform coloration.

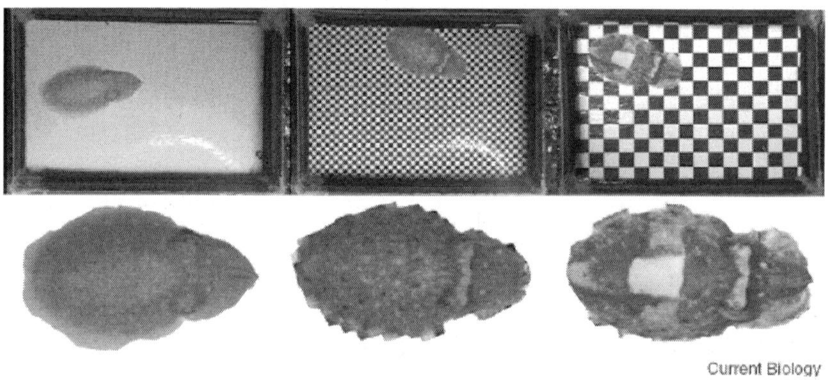

Figure 1.6: Eliciting uniform, mottle or disruptive camouflage using checkerboards. The figure shows the variation in shapes and contrasts in disruptive mottling (Hanlon, 2007)

(Hanlon, 2007)

1.5 CHANGING COLORS LIKE CHAMELEONS

The dangers of predators serve as a constant source of threat to animals that then evolve different systems for different predators. This is due to differences in the sensory systems among different predators. Thus, different predators can elicit different responses or defensive behavior or appearances. A research article published in 2008 tested whether dwarf chameleons (*Bradypodion taeniabronchum*) exhibited different colors to different predators: a bird and a snake.

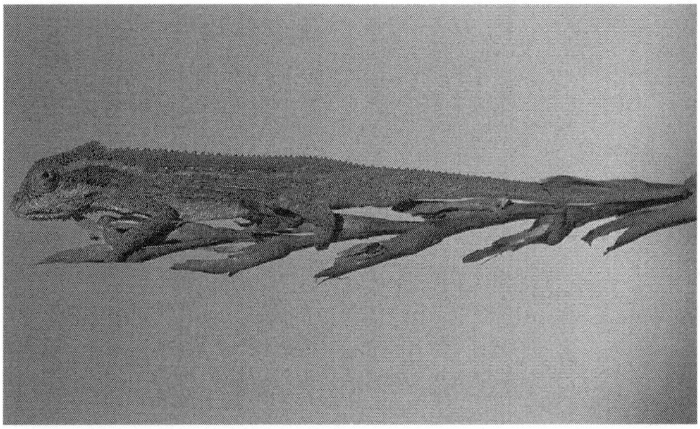

Figure 1.7: *Bradypodion taeniabronchum*

Source: *https://commons.wikimedia.org/wiki/File:Bradypodion_taeniabronchum1.jpg*

As the main anti-predator strategy in chameleons is changing color rapidly against a stationary background, these animals were chosen for the study. Their primary predators are the fiscal shrike, *Laniuscollaris* that impales chameleons on thorns and a snake called the boomslang, *Dispholidustypus*. The former being a bird is a tetrachromat with 4 visual pigments over the latter that is a trichromat (like humans) with 3 visual pigments.

Field trials were conducted using 8 chameleons of both sexes by placing them on a natural branch. Predator models were presented to them: stuffed fiscal shrike (*L. collaris*) and a resin boomslang snake (*D. typus*). Reflectance measurements were done as soon as the animals displayed anti-predator behavioral responses. Visual modeling and statistics revealed that the background matching in chameleons was more when exposed to birds than snakes!

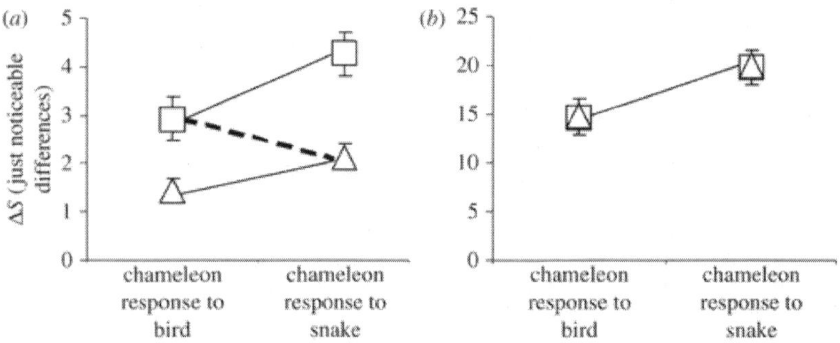

Figure 1.8: Differences in the perceptual distance, Δ*S*, between two colors in units of 'just noticeable differences' (JNDs) (Stuart-Fox and Moussalli, 2008)

This increased background matching in the case of birds can have several reasons. One is increased camouflage in chameleons to match the higher color discrimination of birds. Second is the abundance of avian predators over that of snakes. Though the achromatic (brightness) discrimination of birds and snakes are similar, birds view their prey from above against a low illumination background while snakes view from below against a high illumination background.

Thus, animal camouflage is not just about changing colors randomly but careful and planned response of animals to definite situations.

(Stuart-Fox and Moussalli, 2008)

1.6 SPIDERS: CHANGING THEIR BODY AND WEB COLORS

Another animal that displays camouflage is the spider. *Misumena vatia* and *Thomisus onustus* of the family Thomisidae have received attention since the 19th century. The spider called *Misumenavatia* can change its color from white to yellow and back that is dictated by the color of prey and the background color. The spiders match their background exceptionally well to target their prey.

The pigments behind these colors are ommochromes from tryptophan mainly seen in arthropods.

Figure 1.9: One crab spider *M. vatia* is shown: (pale yellow female) The camouflage is dependent on the location of the animal on the plant

(*a*,*b*) the different colors detection of the animal more difficult,

(*c*) the green leaves may shine through the translucent legs

(*d*) Guanine reflects the yellowness of the corolla to effective camouflage (Thery and Casas, 2009)

Spiders also camouflage themselves using detritus decorations. For example, the intensity of predation of the spider was lowered when detritus was added to the webs of *Cyclosafililineata* and *Cyclosamorretes. Early research conducted by Bruce and team in 2005 examined* achromatic and chromatic contrasts of web decorations. These decorations are silk structures added to the webs whose functions were yet to be ascertained. The visual properties of spiders of Araneidae and Uloboridae were chosen-4 *Argiope* species that make decorative patterns such as cruciate, linear and discoid while the uloborid, *Zosis geniculatus* makes an irregular mat.

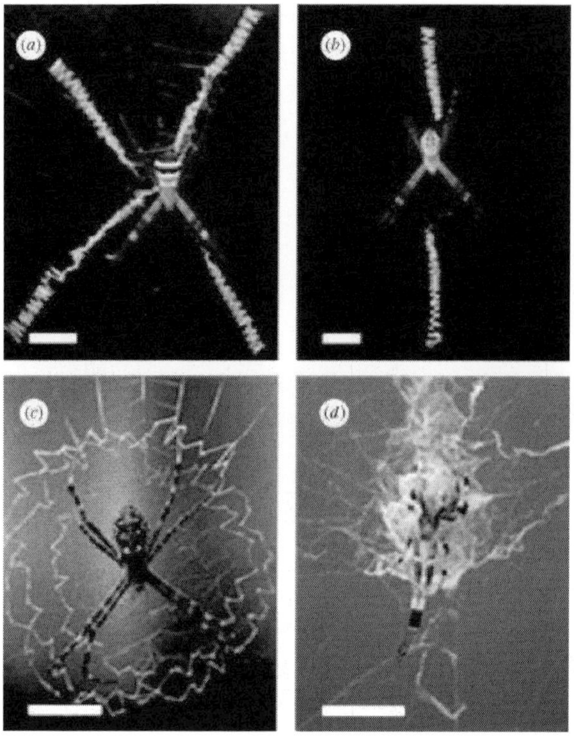

Figure 1.10: shows the animals studied:

(a): cruciate decoration by *A. aetherea*

(b) linear decoration by *A. picta*

(c) discoid decoration by *A. mascordi*

(d) irregular mat decoration by *Zosis geniculatus*

(Bruce et al, 2005)

The results showed that the conspicuousness of the spiders against green vegetation or their spiders was increased by these web decorations for both predators and prey. The camouflage was seen in the case of the discoid decoration by *Argiope mascordi* .

(Thery and Casas, 2009)

1.7 AVOIDING THE HUMAN EYE

The visual detection of an animal is lowered by blending an animal's patterning into the color and texture of the background. This is referred to as background matching. Other animals show disruptive coloration where the misleading shape of an animal is conveyed by patterns on its surface. Both these systems of camouflage are seen in animals. One type of disruptive coloration is edge enhancement where various colors on different parts of the body are separated by salient and high-contrast edges.

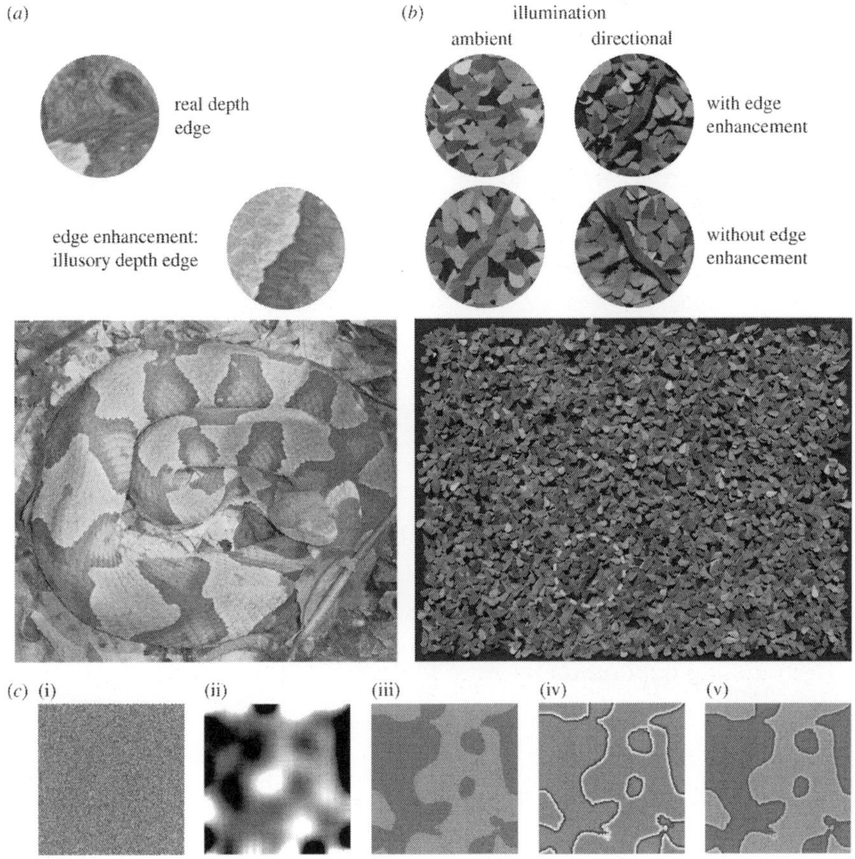

Figure 1.11: shows edge enhancement and background matching in the copperhead snake (Adams et al, 2019)

The result of edge enhancement is the development of fake depth and hence prevents the detection of separating parts from each other and their background-segmentation. Another example of edge enhancement is seen in the Australian Limnodynastes frog that blends with its surroundings.

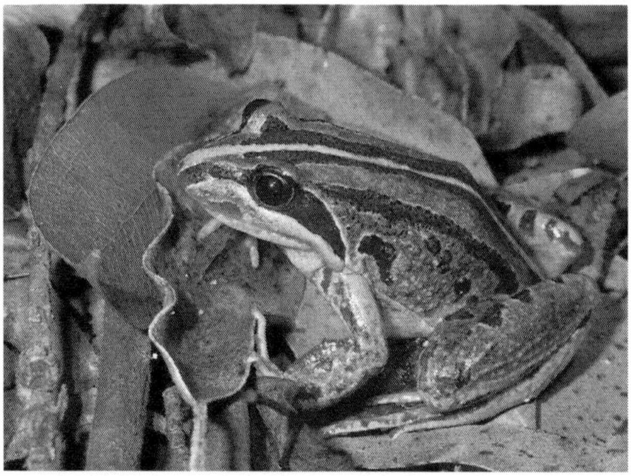

Figure 1.12: The Australian Limnodynastes frog

Source: *https://australianmuseum.net.au/learn/animals/frogs/striped-marsh-frog/*

Research has shown that depth changes at enhanced edges and edge-enhanced patterns are challenging for humans. They find such targets at a slower pace. Hence, background matching that hides the boundaries of an animal and edge enhancement that fakes the contours complement each other: ultimately to delay detection or predation by us!

(Adams et al, 2019)

1.8 CAMOUFLAGE- AN ARRAY OF SYSTEMS IN ANIMALS

Kallima inachus or the Indian leaf mimic butterfly was discovered by Wallace at the Malay Archipelago was found to sit on dead leaves and twigs but was not observed to sit on green vegetation. Other animals too were found to select appropriate surroundings that are suitable for camouflage such as ground-nesting birds laying their eggs matching suitable locations.

One approach to camouflage is a rapid change in color as seen with chameleons and cephalopods and even fish discussed above. However, there is a mismatch when the background changes and the animal too has to switch its color. Another option is the resting or sitting of an animal in an area where its color matches that of its surroundings.

(Stevens and Ruxton, 2019)

Figure 1.13: the various aspects of animals matching appropriate backgrounds (Stevens and Ruxton, 2019)

(A) *Kallima* leaf-mimicking butterfly resting on dead vegetation

(B) Ghost crab (*Ocypode* sp.) sits on sand that matches its color

(C) Bronze-winged courser (*Rhinoptilus chalcopterus*) lay eggs on backgrounds that match with the eggs

(D) The plumage of fiery-necked nightjars (*Caprimulgus pectoralis*) match nesting backgrounds

(E) Aegean wall lizards (*Podarcis erhardii*) rest on sites that match their body colors

(F) Many moths adjust body positions/orientations to match their surroundings (seen here is a tree bark)

(G) Caterpillars adopt postures to resemble the shape of twigs and bird droppings

(H) Grasshoppers adopt positions to align themselves with the background

Another aspect of camouflage is the role of behavior such as shown below in Figure 1.14: (Stevens and Ruxton, 2019)

(A) Some insects match the background by mimicking twigs or objects

(B) Lowering the shadow by eyed-hawk moth caterpillar (*Smerinthusocellata*) via body

(C) Decorating itself using materials in the environment by the long-legged spider crab (*Macropodia rostrata*)

(D) Motion dazzle by striped animals, potentially zebra (*Equus*)

(E, F) Some species like Kittlitz's plovers (*Charadrius pecuarius*) camouflage eggs by modifying the nesting (E, natural clutch; F, uncovered)

Figure 1.14

(Stevens and Ruxton, 2019)

1.8.1 Evaluating the environment

- Dark soil or mixed sand was chosen by American toads (*Bufo americanus*) than plain soil to lower the risk of predation by snakes.

- Tortoises and some lizards prefer microhabitats made of rocks and stones to hide themselves
- A bird called the common potoo (*Nyctibius griseus*) chooses sites that match its color and appearance. It often resembles tree branches as shown below in Figure 1.15:

Figure 1.15

Source: *https://www.hbw.com/ibc/species/common-potoo-nyctibius-griseus?page=1*

- Snakes such as the western terrestrial garter snake (*Thamnophis elegans*) bask at sites that offer better background matching than random sites to lower predation by birds or mammals.
- The eggs of Japanese quail (*Coturnix japonica*) range from light-colored to match light-colored surroundings while dark eggs with a prominence of dark markings were laid at darker soils.
- As plovers and coursers leave a nest when there is danger, the focus of camouflage is on matching the eggs with their backgrounds. However, as nightjars tightly guard their nests, the focus is on their plumage! Thus, behavior determines camouflage in some species.
- The tadpoles of *Bokermannohyla alvarengai*: the tree frog resemble several materials found in waters to lower the risk of predation as shown in Figure 1.16:

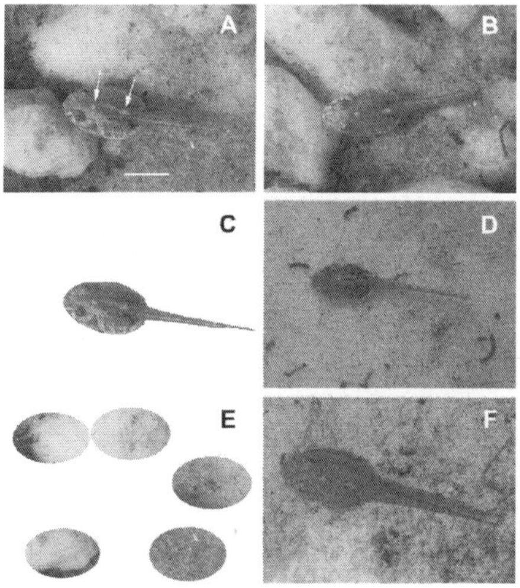

Figure 1.16

Source: *https://www.researchgate.net/figure/Pictures-of-Bokermanno-hyla-alvarengai-tadpoles-showing-an-example-of-tadpole-position-A_fig1_225026754*

- Early thorn moth *Selenia dentaria* caterpillars resemble sticks: the selection of stick size is that which is at par with their own body size. Larger caterpillars choose longer sticks as shown below in Figure 1.17:

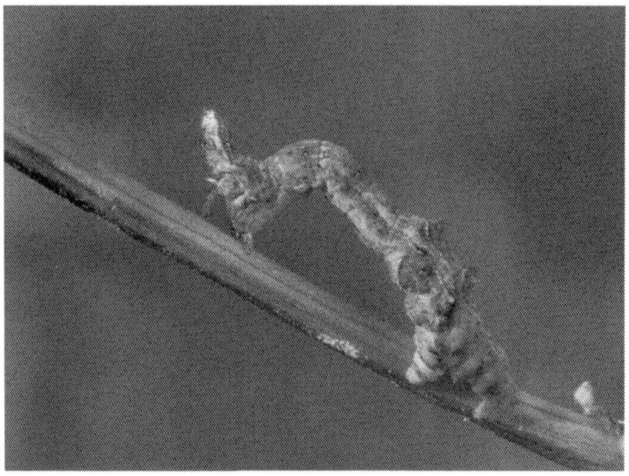

Figure 1.17

Source: *https://commons.wikimedia.org/wiki/File:70.237_BF1917_Early_Thorn,_Selenia_dentaria,_larva_(3950419088).jpg*

- Grasshopper species can change their color over days and weeks to match their background. For example, azure sand grasshoppers (*Sphingonotus azurescens*) that live in street pavements (urban settings) occupy the spaces between paving stones to lower detection. The behavior was also interesting with those grasshoppers with better camouflage allowing predators to come close before they escaped.

Figure 1.18: shows how the animal is adept at matching backgrounds:

Source: *https://www.youtube.com/watch?v=5U2T8fTDxG8*

- Camouflage can also be determines by the microhabitat. The prawn *Hippolyteo bliquimanus* has two main morphs – a homogenous type relying on background matching that resembles algae while a striped form becomes transparent. While the former changes its color across algal backgrounds, the latter shows mobile behavior.

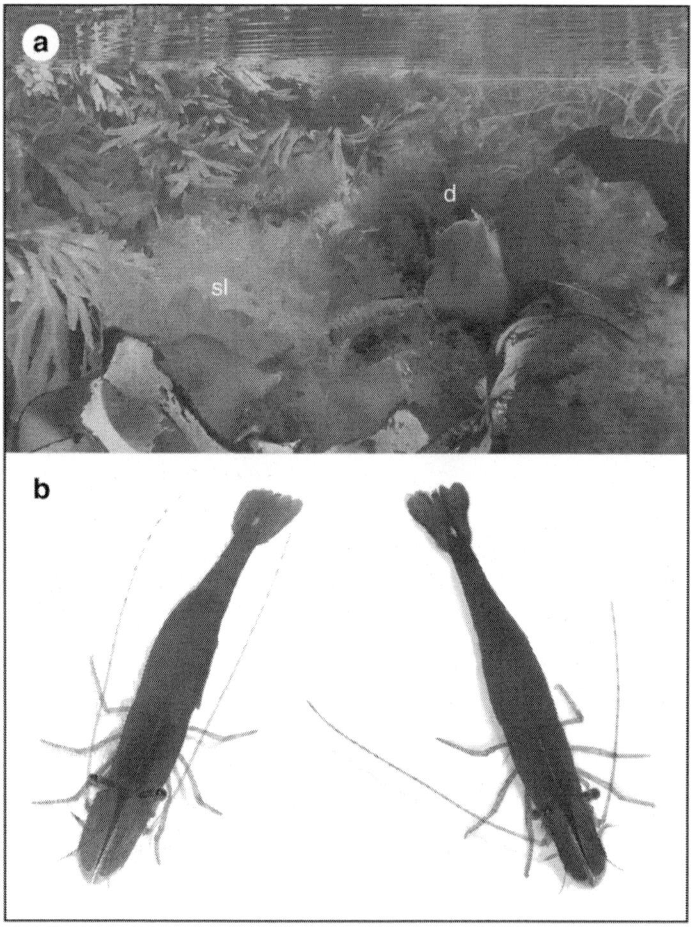

Figure 1.19: the prawns closely resemble the seaweeds (Green, S.D., Duarte, R.C., Kellett, E. *et al.* Colour change and behavioural choice facilitate chameleon prawn camouflage against different seaweed backgrounds. *Commun-Biol* 2**,** 230 (2019) doi:10.1038/s42003-019-0465-8)

- Experiments, where the moths: *Hypomecisro boraria* and *Jankowskia fuscaria* were released at night, revealed that the animals moved their position after landing to match the backgrounds. The detection was found to be lowered due to changed body orientation and the selection of a new microhabitat.

(Stevens and Ruxton, 2019)

1.8.2 Modifying the Background

A decorator is defined as 'an organism that (by means of specialist behavior and/or morphology that has been favored by selection for that purpose) accumulates and retains environmental material that becomes attached to the exterior of the decorator'.

- 75% of crabs of the superfamily Majoidea are decorators. These animals place environmental materials on hooks on their body. These decorations are moreover their rostrum and thus, help to conceal these large and sedentary predators from attack.

Figure 1.20: is an example:

Source: *https://www.newscientist.com/article/dn27702-zoologger-decorator-crabs-accessorise-to-avoid-being-eaten/*

- Recognition is lowered by an interesting mechanism seen in assassin bugs. They have a coat of dust and soil followed by a 'backpack' of prey corpses on their backs plus fecal material. Figure 1.21 shows the "backpack" of the assasin:

Figure 1.21

Source: *http://www.saferbrand.com/advice/insect-library/insect-education/assassin-bug*

- Behavior is seen in the bird called the blue-footed booby (*Sula nebouxii*) that involves matching the appearance of its eggs with that of the surroundings by rolling the eggs on the soil. The coating with soil increases the match of the eggs to the surrounding substrate.

Figure 1.22: shows this phenomenon:

Source: *https://www.sciencenews.org/blog/wild-things/blue-footed-boobies-dirty-their-eggs-hide-them-predators*

- Cosmetic coloration or altered appearance is seen with certain birds. This involved staining the feathers by the birds using self-made oils or even environmental substances. The start of the breeding season sees the rock ptarmigan (*Lapogus mutus*) as completely white birds. With the melting of the snow, females molt and form brown patterned feathers to offer camouflage. However, males coat themselves with soil and form brown feathers at later stages!

Figure 1.23: shows the birds:

Source: *https://www.inaturalist.org/guide_taxa/384101*

- The Kentish plovers (*Charadrius alexandrines*) cover their eggs with nest materials before leaving the nest. This allows for camouflage as well as thermoregulation as shown below in Figure 1.24:

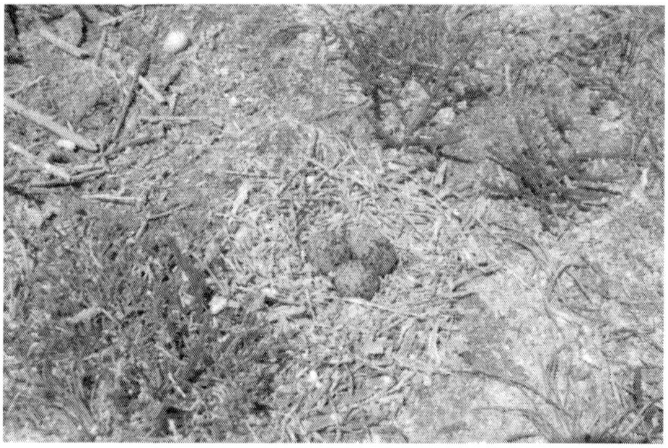

Figure 1.24

Source: *https://www.sciencedirect.com/science/article/pii/S1018364716300076*

Similarly, little grebes (*Tachybaptus ruficollis*) and great crested grebes (*Podiceps cristatus*) display similar behaviors with their floating nests made of vegetation.

(Stevens and Ruxton, 2019)

1.9 CAMOUFLAGE WHILE MOVING

- A rapid movement of a patterned object across the visual field of a viewer can blur the appearance of the object. Male adders (*Vipera berus*) with zigzag morph had better survival against the melanistic morph while the opposite effect was reported for females.

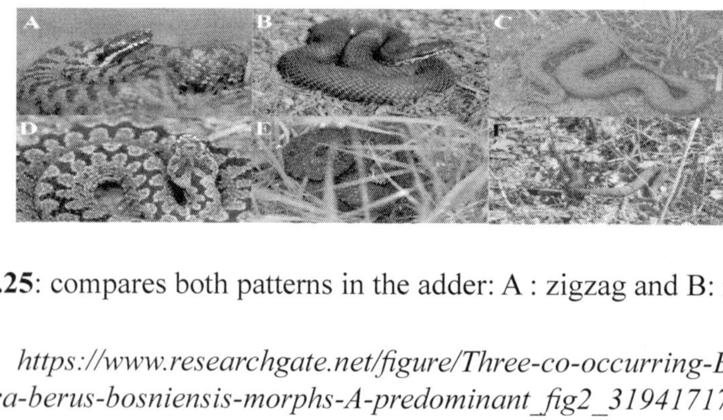

Figure 1.25: compares both patterns in the adder: A : zigzag and B: melanistic

Source: *https://www.researchgate.net/figure/Three-co-occurring-Balkan-Adder-Vipera-berus-bosniensis-morphs-A-predominant_fig2_319417176*

- Motion dazzle is the phenomenon involving challenges in estimating the speed or trajectory of a moving object. The presence of patterns can cause motion dazzle in the case of zebras as shown below in Figure 1.26:

Figure 1.26

Source: *https://www.inverse.com/article/6375-the-science-of-motion-dazzle*

And even certain lizards as the example of Scincid lizard (*Plestiodon japonicus*) with striped body coloration is shown in Figure 1.27:

Figure 1.27

Source: *https://indiabioscience.org/news/2016/how-or-why-the-lizard-got-his-stripes*

An interesting example was presented by Pike (2015) where the success of a bird predator was lowered by interference coloration. Interference coloration involves variation in the perceived color and brightness with the angle of illumination or observation. The research was conducted where the role of interference coloration was tested using virtual insects targeted by Japanese quail (*Coturnix japonica*).

It was observed that the presence of interference-like colors in prey required more number of pecks by the birds. Thus, the widespread occurrence of interference coloration is due to the role of this phenomenon in anti-predator benefits.

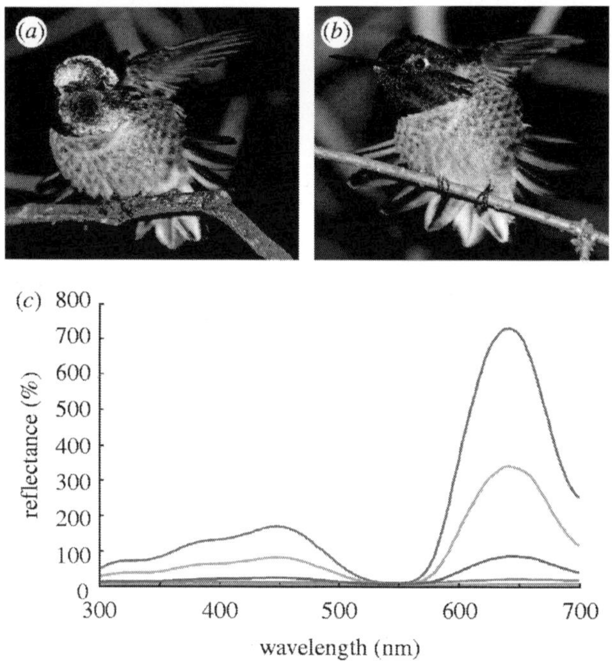

Figure 1.28: shows an example of differences in color perception by an observer of Anna's hummingbird (*Calypte anna*). The change in the shades and geometry can be seen between the angles of viewing at (a) and (b).

Source: Doucet SM & Meadows MG. (2009) Iridescence: a functional perspective. ***J. R. Soc. Interface*** 6, S115–S132.

Certain animals like the stick insect make movements to match the background motion- such as the swaying of its body to match the movement of vegetation with the wind.

Figure 1.29: A stick insect blending with its surroundings

Source: https://www.sutori.com/item/roz-learns-about-camouflage-from-a-stick-insect-she-realizes-she-can-use-camou

Table 1.1: Below summarizes different strategies employed by animals for camouflage (Stevens and Ruxton, 2019)

Mechanism	Example	Role	Details
Background choice	Ground-nesting birds choosing nest sites	Microhabitat selection	A resting place is chosen
	Grasshoppers	Habitat selection	Improved camouflage at their habitat
	Moths	Patch selection	Selection of a specific patch or area within a habitat for camouflage
Body position	Moths	Resting site orientation	Specific body positioning and orientation against the background for increased matching
	Caterpillars	Changing posture	The animals alter their bending to resemble uninteresting objects such as bird droppings
Decoration	Crabs	Modifying appearance	Animals attach environmental materials to lower detection

Modify the environment	Birds	Modifying local surroundings	Materials are added to eggs to lower detection
	Birds	Hiding	Hiding of nests or eggs
Movement	Snakes	Flicker-fusion	Blurring of a moving animal due to patterns on its body
	Zebras	Motion dazzle	Speed and path estimation becomes challenging due to high-contrast markings
	Stick insect	Match environmental motion	The motion of sticks in the wind is imitated by the insect.

1.10 CONCLUSIONS

While men are considered to be masters of camouflage in terms of "hiding" their true emotions, in the animal world, camouflage becomes necessary protection. Several concepts of camouflage ranging from motion dazzle while in movement to background matching have been discussed. Few experiments conducted along with their interesting results have also been presented. Some animals modify their own appearance using embellishments while others hide their nests or resting spots. Others mimic the motion present in the background. In a changing world, animals cope up with changes by changing their own appearance or environment to emerge successful in the struggle for existence.

1.11 REFERENCES

1. Stevens M and Merilaita S (2009) Animal Camouflage: Current Issues and New Perspectives. Phil Trans R Soc B. 364:423–427.
2. Cuthill IC (2019) Camouflage. J Zool. 308(2019):75–92
3. Hanlon RT (2007) Cephalopod Dynamic Camouflage. Curr Biol. 17:400–404. doi:10.1016/j.cub.2007.03.034
4. Stuart-Fox D, Moussalli A and Whiting MJ (2008) Predator-Specific Camouflage in Chameleons. Biol. Lett.4 326–329.
5. Théry M and Casas J (2009) The Multiple Disguises of Spiders: Web Colour And Decorations, Body Colour And Movement. Philos Trans R Soc B. 364(1516):471–480.doi:10.1098/rstb.2008.0212.
6. Bruce MJ, Heiling AM and Herberstein ME (2005) Spider Signals: Are Web Decorations Visible to Birds and Bees? Biol Lett. 1(3):299–302. doi:10.1098/rsbl.2005.0307.
7. Adams WJ, Graf EW and Anderson M (2019) Disruptive Coloration and Binocular Disparity: Breaking Camouflage. Proc R Soc B. 286:2018-2045.
8. Stevens M and Ruxton GD (2019) The Key Role of Behaviour in Animal Camouflage. Biol Rev. 94:116–134.
9. Pike TW (2015) Interference Coloration as An Anti-Predator Defence. Biol Lett. 11:20150159.

2

Plant Camouflage

Shivsanjeevi Sripathi

ShivSanjeevi Sripathi completed his Masters in Biotechnology from Mumbai University in 2008. He was awarded for academic excellence in both his Bachelors and Masters for securing second rank in Mumbai University in 2006 and first rank in his college: Kishinchand chellaram College. For his Masters he secured first rank in his college KET's V.G.Vaze College. He qualified CSIR and NET and and TOEFL in September 2008. He then worked on a stem cell project at the Specialized Centre for Cell Based Therapy (SCCT),KEM Hospital at Mumbai on a project entitled, : Isolation & detection of stem cells from Human Umbilical cord/ amniotic membrane" following which he worked at Junior Research Fellow at Microbiology & Cell Biology Department, Indian Institute of Science, Bangalore from August 2009 to November 2011. The field of work involved cloning of cell wall genes and transcription factors in *E.coli & M.smegmatis.* As a writer, he has authored and co-authored 25 books with Delve and Arcler International Press. He loves to read and share on interesting aspects of life sciences in books. In his free time he loves to travel and explore ancient customs and traditions.

2.1 INTRODUCTION

Since ancient times, all of us are made to believe in a very simple approach that plants are one of the most defenseless, inanimate living species on earth that have accepted their fate as to be eaten up by herbivores. Recently, researchers have proven this to be a myth; further confirming an evolutionary pressure plants have to face by integrating and evolving complex defense arsenals against their attackers. The fossil research has further confirmed that over the period of evolution, various traits have been adopted by plants for their protection, against herbivores. This may be associated with a physical defense like the presence of wax crystals on the epithelial cuticles; as well as a chemical defense system, with the integration of tannins, making the material less digestible.

In this regard, current chapters intend to provide an idea as to how plants have evolved with the integration of different defensive traits and what strategies have been integrated to decrease their visual presence to herbivores.

2.2 EVOLUTIONARY PERSPECTIVES

Evolutionary biologists are being able to detail various defense mechanisms incorporated by plants that have been categorized into either physical as well as a chemical in nature. Further to which, in-depth analysis in the same could evidently be explored that plants deploy various stepwise defense strategies for their protection; which can be detailed as:

- The first step would be to minimize presence in front of herbivores; if this has been accomplished successfully, the other means of protection can be subtle (Le-Yadun; 2006).
- In the second step of protection, plants employ strategies named the "onions of survivability"; which is also being identified as the incorporation of chemical defense mechanisms that can be more heavy and quantitative (Yamamura & Tsuji, 1995).
- In an advanced set, researchers are expecting that plants would have adopted towards an advanced mechanism of self-defense; in order to further prevent obvious presence in front of the attackers.

Hence, researchers all over the world are struggling to identify more strategies that have been adopted by plants; in order to decrease their visual presence. The current chapter is intended to provide detailed information related to different strategies named plant camouflage.

In this regard, an immediate question that generally peeps in our mind is, "what are plant camouflage; and how do you define them?" and the answers for the same have been listed in subsequent sections.

2.3 WHAT ARE PLANT CAMOUFLAGES?

Every living organism on earth has been blessed with a different defense mechanism, by Mother Nature. One of the common examples is of the living chameleon, which apparently changes color when he senses the attacker. A popularly known bush, "Touch Me Not" shrinks its leaves in inward motion, when touched. So these are the type of mechanisms adopted by living beings to protect themselves from attackers. Similarly, plants do. Camouflage is a type of adaptation that is being implemented by an organism to look blended with the environment; which it belongs. Although, it is a much-generalized definition; an organism actually can implement a variety of strategies to save himself from the environment.

With reference to the definition, camouflage can generally be categorized into "mimicry" and "crypsis"; however, scientists still have to greatly decide as to which biological phenomena need to be incorporated. Just to proceed further, it is important from our side to understand the exact meaning of these terms, which will further help us to justify criteria that are being assessed in the plant, further in this chapter.

2.4 HISTORICAL PERCEPTION

In 1981, Endler has proposed that six classes of phenomena, such as masquerade, Batesism, Mullerian, crypsis, convergence and polymorphism can be compiled together to explain camouflage in details (Endler, 1981); although only two are directly associated with it, such as crypsis and masquerade. It is very crucial to understand that the two phenomena that have been mentioned herewith, if selected do not affect with evolutionary dynamics of the model; however, others that are remaining can have a significant impact. At the same time, it should also be noted that the above-mentioned mechanisms are previously thought to be particularly associated with animals; however, very recently, it has been observed that even the mechanism is also applicable in the case of plants. While crypsis is resemblance towards the random sample of the environment, masquerade involves, matching with the irrelevant background object.

In this regard, the crypsis and masquerade are primarily focused on further discussion; but only pertaining to plant camouflage.

2.5 THE EVOLUTIONARY CRITERIA OF "CRYPSIS"

As discussed previously, the mechanism of crypsis is being adopted by wide categories of animals; including chameleon. But when it comes to the terrestrial animal species, it is quite straightforward that they attempt to blend with the environment; in order to minimize their presence. However, in the case of plant camouflage; it has become quite obligatory to understand how the mechanism is being displayed in plants; especially when plants themselves are in the background.

Many scientists tried to summarize the mechanism of "crypsis" by plants, including Burns (Burns, 2010). According to which, the herbivore predator should be visually orientated; at the same time, corresponding plant species should be habitat specialists. Studies have quite relevantly observed that plant species living in different habitats cannot exhibit crypsis; as they have to be in close proximity to the surrounding environment. In another language, the mechanism of crypsis cannot be adopted by tall-statured plants; but at the same time, it should be noted that higher plants have hardly anything to lose because of it.

Scientists further proposed that for adopting crypsis, the background needs to be unpalatable; otherwise it would fail to provide any further evolutionary benefit.

2.6 NEW ZEALAND: EVIDENCE OF PLANT CAMOUFLAGE

Birth is the most crucial as well as a vulnerable stage that is faced by every living being on this earth; whether through seed, egg and/or mother's womb. In this regard, archeologists have further hypothesized that during this critical period, even the selection pressure for exhibiting and/or evolving with inconspicuous traits is also high. This can further be explained with *Elaeocarpus hookerianus* an endemic species native to New Zealand, which was further being analyzed for morphometric and spectrographic traits. It was further evident through analysis that their progeny that are mottled brown in color adapt themselves with the leaf litter background through the expression of chromatic as well as achromatic background (Fadzly& Burn,

2010). The mechanism is further helpful for them to be protected from the visual perception of browsing avian, the herbivores (**Figure 2.1**).

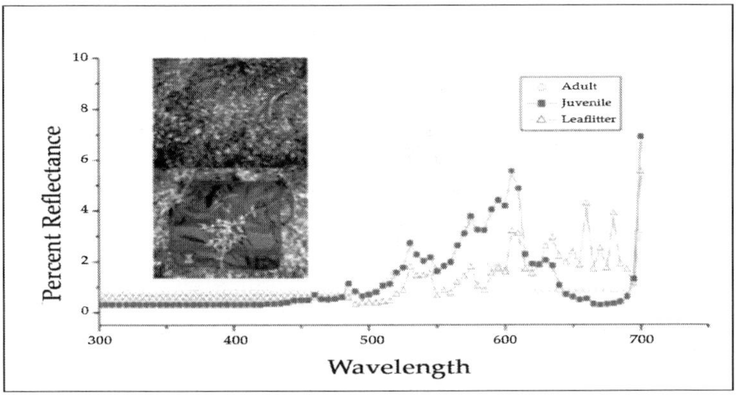

Figure 2.1: The spectrographic analysis of the leaf litter background adopted by the juvenile version of *E. hookerianus*. The mechanisms of cryptic morphologies are being explained through two different images of the same sampling captured with a different background(Greenwood & Atkinson, 1977).

One of the most potent among these herbivores was "moa", although the bird is now extinct, it was then a large, ratite bird with the height of 3m (Greenwood & Atkinson, 1977). It is further very essential to note that at this height, *E. hookerianus* regains its tree-like appearance, and become less cryptic in nature. Interestingly, this abrupt morphological change can be noted as heteroblasty. Once the plant is being in the safe zone through heteroblasty; all the energy is then concentrated towards its growth and expansion.

The analysis further shed some light on the fact that certain physiological adaptations can be incorporated by the plants, due to above ontogenetic changes; like in case of *E. hookerianus* as it matures, it can adapt to changing conditions of the environment due to its peculiar structure with divaricating branching (Day & Gould; 1997); further important for supporting the plant's ability to explore as well as exploit light alternative environment. In its juvenile version, the characteristic morphology of its leaves further facilitates easy and straight penetration of light into its interior; further responsible for more efficient gas as well as a heat exchange system.

The coast of New Zealand shares a great taxonomic similarity in the context of its floral diversity with the Chatham Islands; another archipelago, with the only difference being a complete absence of ratite birds. In spite of this difference, evidence of heteroblasty has been noted on Chatham, with the help of a study involving comparative analysis of leaf morphologies (Burns & Dawson, 2009). When the consistency of this system is analyzed, it was found to be quite different; in order to support this fact, many theories have been proposed, the promising amongst them is emphasizing on timing required for colonization. According to this theory, various species of plants require variable time to reduce their expression of heteroblasty and be normalized to the homoblastic stage.

At the same time, some other theories are also being proposed that certain other factors should also be responsible and taken into consideration; while analyzing the evolution of heteroblasty. For example, there were two species that had larger leaves during their juvenile period; and were easily visible by many herbivores (Bond et.al; 2004). However, when these species pass by the juvenile stage, the leaves become smaller; thereby reducing the mechanical stress on stalks, further important for offering biomechanical benefit to the plant.

Another complex yet obvious example of heteroblasty is *Pseudopanaxcrassifolius;* the species is originally a native to New Zealand. The analysis further confirmed that in its juvenile stage; the species protect themselves from gazing herbivores due to crypsis. However, when the plant grows or become mature, it is quite visible to other due to emergence of tissue at the end of spine-like projections (**Figure 2.2**); and is identified as a perfect mechanism of plant camouflage; wherein unique achromatic perceptive properties of the avian eye have been found to be quite useful to detect the same (Hart et.al, 1998). The same phenomenon has been observed in other genera of cacti, such as Agave, Aloe as well as Euphorbia (Lev-Yadun, 2011). However, it should further be noted that this type of phenotypic plasticity is definitely very individualistic.

Further research carried out in the same direction has confirmed that the traits of the heteroblastic phenotypes, we are observing today; is preferably due to evolutionary changes during multiple selection procedures. However, it should further be noted that definitive evidence still needs to be explored more(Lee & Gould; 2002).

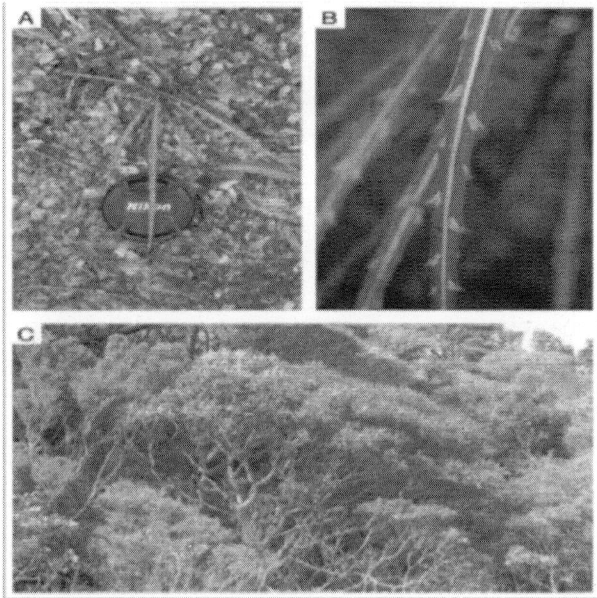

Figure 2.2: (a) the seedlings with cryptic adaptations. (b) Aposematic Juveniles (C)Adults native from New Zealand exhibiting heteroblasticity(Fadzly et.al; 2016)

Thus, for the quantitative camouflage assessments, the species with a large number of herbivores opposition should be selected. Further to this, if the results obtained from such experiments are as expected the hypothesis that visual perception of moa lead to the ecologic evolution of camouflage will be supported.

2.7 TYPES OF DECEPTIONS

There are certain species that maintain a symbiotic relationship with animals to protect themselves from grazing herbivores, like certain species of *macaranga* genus. This particular genus is surrounded by many ants (*myrmecophily*), who offer active protection to them from cattle grazing around. Further to this, the detailed spectroscopic analysis confirmed that these species also adapt to the red background; wherein they are grown (Fadzly et.al; 2016). These species are more vulnerable to herbivores, especially upon relocation. They further help themselves by avoiding predators' attention through the production of extra pigments and reduced rate of photosynthesis. At the same time, it should be noted that other physiological mechanisms can be expressed alternatively, depending upon the circumstances (Lee & Gould; 2002).

Further investigation on seedlings has been done by many researchers, including Aviezer& Lev-Yadun; who conducted research on genus *Pisum* through comparing colors of seedlings from different species. The adaptive mechanism underwent by them is primarily background matching as well as disruptive coloration. It was initially believed that herbivores that are visually oriented can hardly detect prey.

Some of the examples of species expressing disruptive coloration patterns can be noted as mottled color patterns on the leaves that were present in US flora, and since they were not being detected by their predators; their predators were previously thought to be color-blind (Givnish; 1990).

Recent investigation on the same has confirmed that this type of lead variation can be categorized under the camouflage, and was not being noted by Endler earlier, but has been named as "dazzler camouflage" (Lev-Yadun; 2014).

Thus, it should be understood that the hypothesis of camouflage cannot be understood on the basis of spectroscopic analysis as well as a simple observational pattern; but needs proper experimental manipulation for a definitive conclusion. For that matter, the exact quantification of crypsis is required that has been elaborated further.

2.8 QUANTIFIED CRYPSIS

Undoubtedly chlorophyll is a characteristic biochemical that is responsible for genotypical as well as the phenotypical variation of the plant and can significantly impact its appearance. However, there are certain plants like mycoheterotrophs; which are non-photosynthetic in nature and instead of relying on chlorophyll; further, have direct association of mycorrhizal fungi. Being non-pigmented these species can suitably adapt to the surrounding environment; further expressing the cryptic characteristic, especially by the adult version of these plants (Klooster et.al; 2009). Investigation in this regard supported the proposed hypothesis; wherein the herbivory rate was found to be substantially low in *M. odorata*, as prominent in (**figure 2.3**).

However, it should further be noted that spectroscopic analysis of the same experimental design has come up with complementary results; and has come up with an overview expressing the possibility of the co-evolutionary mechanism of both plant species as well as a corresponding herbivore. This was done on the basis of the quantitative analysis of the cryptic mechanism.

Moreover, some recent studies involving *C. benecincta* that is a characteristic low growing alpine plant, with grey leaves; for cost-benefit analysis, done on the basis of their quantification of ability to discriminate leaves by their peculiar herbivore Parnassius butterflies.

Figure 2.3: (a) *M.odarata* with crypsis (b) *M.Odorata* without crypsis(Niu et.al; 2014)

The difference in polymorphism in leaf color further facilitated a reliable foundation of cost-benefit trade-off (Niu et.al; 2014).

Conclusively, the two described case studies further supported the hypothesis of plant camouflage with the cryptic mechanism. At the same time, it should be noted that these two pieces of evidence are only discussing the benefits associated with the same, on the basis of morphological observation. In this regard, further experimentation can be advisable to support the hypothesis on the basis of genotypic expression of traits as well. Researchers also are optimistic about the technological advancements that may further lead to an in-depth analysis of these species and their association with herbivores on an ecological scale.

2.9 REFERENCES

1. von Helversen B, Schooler LJ, Czienskowski U (2013) Are Stripes Beneficial? Dazzle Camouflage Influences Perceived Speed and Hit Rates. PLoS ONE 8(4): e61173. doi:10.1371/journal.pone.0061173.
2. Kojima T, Oishi K, MatsubaraY, Uchiyama Y, Fukushima Y, Aoki N, et al. (2019) Cows Painted with Zebra-Like Striping Can Avoid Biting Fly Attack. PLoS ONE 14(10): e0223447.
3. Waage JK (1981) How the zebra got its stripes–biting flies as selective agents in the evolution of zebra colouration. J Entomol Soc South Afr. 44: 351–358.
4. Caro T, Argueta Y, Briolat ES, Bruggink J, Kasprowsky M, Lake J, et al. Benefits of zebra stripes: Behaviour of tabanid flies around zebras and horses. PLoS ONE. 2019; 14(2): e0210831.

3

Interesting Aspects and Applications of Camouflage

Shivsanjeevi Sripathi

ShivSanjeevi Sripathi completed his Masters in Biotechnology from Mumbai University in 2008. He was awarded for academic excellence in both his Bachelors and Masters for securing second rank in Mumbai University in 2006 and first rank in his college: Kishinchand chellaram College. For his Masters he secured first rank in his college KET's V.G.Vaze College. He qualified CSIR and NET and and TOEFL in September 2008. He then worked on a stem cell project at the Specialized Centre for Cell Based Therapy (SCCT),KEM Hospital at Mumbai on a project entitled, : Isolation & detection of stem cells from Human Umbilical cord/ amniotic membrane" following which he worked at Junior Research Fellow at Microbiology & Cell Biology Department, Indian Institute of Science, Bangalore from August 2009 to November 2011. The field of work involved cloning of cell wall genes and transcription factors in *E.coli & M.smegmatis.* As a writer, he has authored and co-authored 25 books with Delve and Arcler International Press. He loves to read and share on interesting aspects of life sciences in books. In his free time he loves to travel and explore ancient customs and traditions.

Let us discuss some characteristics and applications of camouflage as reported in experiments.

3.1 STRIPES AND CAMOUFLAGE

Camouflage is nothing but the capacity of patterns to hinder the possibility that the target will be acquired and usually applies to objects that are stationary. At the other end of the spectrum, the concept of dazzle has been produced to protect the target in much the same way, except that it uses patterns that are disruptive. The animal kingdom and most specifically the dorsal patterns on snakes have been the focus of the study of the dazzle effect, by researchers. With not just the animal kingdom, Dazzle has also played a role in the influence of human perception. They navy painted their warships in bright high contrast patterns in order to make them a difficult target, during the world wars, as a result of adoption of the concept of dazzle, despite the fact that there is no existing correlation that high contrast dazzle designs protect the ships from being a target, at that time. However subsequent research has demonstrated that there is a correlation in the increasing degree of acquiring a target that has high contrast dazzle patterns on it as a moving target. Despite these findings, there is no clear understanding of why this happens, why it is difficult to acquire a moving target that has dazzle patterns.

The general perception is that contrasting patterns impair the observer's ability to locate the target accurately, thereby reducing the ability to capture it. A study indicated further that the ability to perceive speed is impaired with respect to objects that are coated with dazzling patterns, as they appear to move slower than they actually are. Despite these elaborate studies indicating that there is a reduced efficiency in capturing objects that are coated with high contrast patterns, such as zigzags and checks, there are some psychological studies that are contrarian to these findings, in that they establish that there is only a minor impact on the ability to capture the targets. Perception and action are revealed as existing on the basis of different types of information, according to various studies. There was a computerized model that tracked the perceived speed with respect to the positioning and negated the correlation with target fixation and the actual hand movement. Therefore, though it has been established that dazzle patterns actually do influence retrogradely on the ability to capture a target, the impact of such perception on the actual ability to acquire targets has been demonstrated as minimal.

The principal goals set were:

- To study if there is an influence by the dazzle patterns on (a) the perceived speed of the target objects, and (b) the ability to capture said targets, and
- To study if the ability to capture the targets is impaired by this perceived speed. There was also a secondary study about specifics – whether longitudinal or vertical patterns were the most inhibitive for acquiring the targets.

While it is universally accepted that stripes are an influence on speed perception, researchers don't seem to have a consensus on which type of stripe pattern causes the greater influence, viz. vertical or horizontal. Stripes can be either parallel to the direction of movement or perpendicular to the direction of movement. Therefore, here, we will refer to the stripes that are parallel to the direction of movement as longitudinal stripes and those that are perpendicular to vertical stripes.

There is a debate on whether longitudinal stripes increase or decrease the speed perception. Gabor patches aligned with the movement direction, suggest that these longitudinal patterns make perception of the object faster, however, as a contrarian view, it has also been argued that the longitudinal stripes having a clear advantage when compared to vertical stripes in high-speed flight scenarios adopt a snake-like tendency which actually makes the object travel slower than it actually is. When the snake stays still, the stripe patterns act as a cryptic enhancer and is very difficult to detect, but when the snake moves, the same stripes, thanks to their contrast elements, act as reference points and actually enhance detection and calculation of speed. There are studies that have found a genetic correlation between the probabilities of escape and the dorsal patterns of garter snakes, however, that being said, contrarian studies have struggled to find any provable correlation between the patterns and the number of strikes on target and the assessment of seed.

Two experiments were conducted to gauge the effect of the dazzle patterns. In the first experiment, studies were made to assess how the longitudinal or vertical stripes influence the probability with which the fleeing target is hit, and in the second experiment, the studies focused on the impact of the pattern type with regard to perceived speed and analyzed the hit rate as well as the link between that and the perceived speed.

Experiment 1

The ability of people to hit objects of the same color in a computer game was the base of the investigation in the case of experiment One. The game started in such a way that the objects moved across the screen from top to bottom and left to right. Hit was accomplished by physical movement of the cursor, using a joystick, to hover over the object that was transitioning and depressing the fire keystroke. As shown in Figure 1 the cursor was positioned at location L of the width of the screen and the cursor could be moved vertically. With the ability to move the cursor vertically the objective was to hit the fire button as the object came under the cursor. Three different surface patterns were used – uni-coloured surface, longitudinal stripe patterns and vertical stripe patterns. If dazzle patterns actually have an advantage the number of times the uni-coloured objects get hit should exceed the times the same object under striped patterns gets hit.

However if the studies were to be proven right and if the patterns of dazzle do indeed have an advantage, and if the orientation of the patterns does indeed make a difference in the catchability, then the longitudinal patterns should be hit a far lesser number of times than the vertical dazzle patterns. In addition to the pattern, we imposed conditions were we varied the speed of the objects and the time of disappearance of the objects before they appeared to start moving, and studied the time to get hit and the misses. The reason for varying the speed is to assess that the pattern stayed true to the characteristic at varying speeds. The disappearance factor was added because in the natural world, the snake may be hidden in the tall grass for durations of time in its escape path from the predator. Besides, by introducing the condition of disappearance into the experiment the participant would not be able to predict the starting position of the object and would be true to the requirement to actually rely on the speed perception of the object while taking the hit.

The layout of the experiment was such that there was a span of 50 trials, wherein there was a test phase of 360 trials, with 30 trials in each condition. Task familiarization was afforded to the participants by allowing them to train with uni-coloured objects. Adaptive algorithms were deployed in the training phase, in order to get the average hit rate of 50%, by increasing the rate of speed after every hit and decreasing the rate of speed after every miss. At the end of every hit, the object would turn red and an auditory and visual cue would indicate to the participant that the trial had ended. However, in the case of a miss, the object continued its path until it vanished from the screen. One shot per trial was what was given to the participants. The space bar started a fresh trial and after every trial, the cursor would return automatically to the starting position. The error variance was kept in check by positioning 80% of the objects in the middle quarters of the screen while the 20% was when the objects appeared in the extremes of the screen.

The time interval between when the objects disappeared and when they appeared to be hit was altered, the speed of the objects was altered and the pattern of the objects was also varied in Experiment One. Using repeated analysis of variance (ANOVA) impact of pattern change in target acquisition with variation in speed was determined, using the percentage of hits as the dependent variable and speed, pattern and duration of disappearance as within subject factors. 288 experimental trials that were consigned to the middle quarters of the screen at 24 trials per condition were used in focus to minimize variance. And then contrasts between longitudinally striped objects and vertically striped objects were compared with uni-coloured ones and calculations were made.

The participants hit 62% of the objects at an average during the trials, this was determined. Among the findings was the fact that objects that had disappeared for 150 ms were hit on less often than the objects that were hit on for 450 ms. Dazzle patterns influence speed perception, which in turn influences the rate at which moving targets are acquired, this much has been demonstrated in the two experiments. While in Experiment One longitudinally striped objects were hit more than uni-coloured objects, but in Experiment 2 vertically striped objects were hit more than uni-coloured objects as against Experiment 1. And more vertically striped objects were hit than uni-coloured objects, in Experiment 2 (and not in Experiment 1).

As a result of these trials, it was proven definitively that striped objects were perceived as moving faster than uni-coloured objects, in Experiment 2. These findings are consistent with the research reporting that longitudinally patterned snakes give out a more difficult perception of speed as a result of the dazzle effect. In summary, the relation between the perception of speed and the dazzle pattern is more complex owing to a multitude of factors, such as for instance, the object being white in our study and black in the other study. The pattern's effect seemed to vary between the experiments. While there was a demonstrable effect of the impact of vertical stripes in Experiment 2, there was no effect of the same in Experiment 1. Now, this can be attributed to several factors, like say, the role of chance, or the differences in task setup procedures. While in the case of Experiment 1 the participant not only had to fire the cursor at the exact moment he/she thought that the object was under it, but the cursor also had to be positioned vertically to coincide with the moving object, however, in Experiment 2 the cursor was fixed, and the fact that the vertical stripes would have impeded the participant's perception of the position of the object would have actually reduced the hit rate when compared to the earlier experiment. Another limiting factor in the Experiment could be that there was a vast difference in the luminosity and contrast of the uni-coloured object when compared to the high contrast striped objects (80% grey and 20% black). Perceived speed can be influenced by the visual contrast and therefore differences, in contrast, could have led to the perception of differences in speed and hit rates. A proponent against this explanation is that the contrast patterns that are low were found to be moving slower than high contrast patterns and high and low contrast patterns displayed a similar disparity in that the high contrast patterns had a greater effect on the perceived speed.

Clearly, in these experiments, it has been categorically determined that the estimation of speed is influenced by the dazzle patterns, and therefore this has a direct impact on the probability of hitting the objects. But then in both the experiments, uni-coloured objects were hit more or less equally as dazzle striped objects, throwing into question the fundamental proposed benefit of the dazzle patterns themselves. While on the one hand, research has determined that the dazzle patterns actually reduce the number of times the object is hit, in both the experiments, the fact that the striped objects were hit more than the uni-coloured ones, affords two possible explanations.

One of the explanations is that the result is deterministic due to the way the objects are captured, for instance, the dorsal pattern on lizards and their flight pattern is an evolutionary result of predation from the air, as opposed to predation from the ground. In our controlled experiments, the participant hits the object by waiting until it is right below the cursor, while in real life, there is a lot of pursuit by the predator to acquire the prey, in which scenario the impact of the dazzle patterns may differently affect the ability to get the prey. The second explanation is that the path of flight could actually determine the efficacy of the impact of the dazzle patterns. In the cases of certain snakes, the impact of the dazzle pattern varied when the snake was fleeing sideways as against straight movement. Since the linear movement of the snake itself is not strictly linear and is more winding, this disparity may feature into the outcome of the experiment. To summarize, future evaluations should not be content with taking into account the types of patterns to assess the impact on the speed perception but instead should factor in the interaction of the pattern with the natural habitat as well in order to have a more accurate determination of the impact of dazzle patterns on speed perception, in the natural world.

Conclusion

A small but reliable difference was found by us in our experiments, between the impact on the speed perception between uni-coloured objects and dazzle striped objects, with striped objects being hit 2% more than uni-coloured objects.While at the outset, 2% may sound like a negligible variance, the impact on the selection process is significant when you look at it from the evolutionary perspective. The differences in the result of striped objects and uni-coloured objects found in our experiments, when compared to previous research, where the argument was made that the misrepresentation of magnitude was enough to protect the group from a roving predator, are similar.

(von Helversen et al, 2013)

3.2: MAKING COWS INTO ZEBRAS: THE BENEFITS

Bold black and white stripe patterns characterize the animal Zebra. There are many functioning hypotheses surrounding the reason for the evolution of the black and white stripes, ranging from thermoregulation to confusing predators to the avoidance of biting flies. For instance, as shown by Caro et al (2019), tabanid flies were less likely to land and bite on horsebacks that are covered with black and white coats as compared with uni-coloured ones. But then researchers have not supported the other hypotheses such as predator confusion, camouflage, heat management and social interaction. There has been further evidential research pointing out that biting flies eschew white, striped and spotted surfaces. The critical width of 5cm or narrower, and in the case of spots, 10cms or narrower diameter, have been demonstrated to prevent fewer flies from biting when compared with the effect on white surfaces.

The issue of biting flies is palpable, and significant losses can be demonstrated as they have been proved to alter the behavior of livestock, as they reduce grazing time, and cause problems due to twitching, foot-stomping, head throwing and tail flicking. Bunching also increases the chances of injury among cattle and directly impact weight gain and dairy yield in cattle. It was hypothesized that whether the painting of the bodies of the cattle in white stripes would actually reduce the impact of the flies biting the surface of the skin. To the best of scientific knowledge, this is the only study to analyze the effect of zebra-like painting of the body of cattle to assess the impact on biting flies. This is also a very ecologically positive approach in that we are able to control the impact of biting flies on cattle without using environmentally harmful pesticides and pollutants.

3.2.1 Approach

A 3X3 Latin square design was adopted for the treatment of the cows. This consisted essentially of white painted stripes (Black and White), Black Painted stripes and the control is no stripes – natural black. (Fig 3.1). Japanese cows were used for the experiment, and since Japanese cows are usually black, we used commercial waterborne white lacquer and for the B experiment, commercially waterborne black lacquer. 30 instantaneous scans per minute for 30 minutes, twice daily were performed on each cow.

Figure 3.1. Here are the examples of the stripes painted cows (Kojima et al, 2019)

(a) Black-and-white striped cattle with commercially available waterborne white lacquer,

(b) Black-striped cattle (B) with commercially waterborne black lacquer, and

(c) Cattle with no stripes.

This clearly helped observe the fly repellant behavior on every cow during the experimental period. Tail flick, ear beat, skin twitch and leg stomp were recorded. A commercial camera, the Pentax Optio RZ10 was used to take photos of biting flies and the right sides of the three cows in the experiment were observed simultaneously in bright sunlight and in shade, with photos being snapped in after every 30-minute observation. Biting flies were counted from the photos.

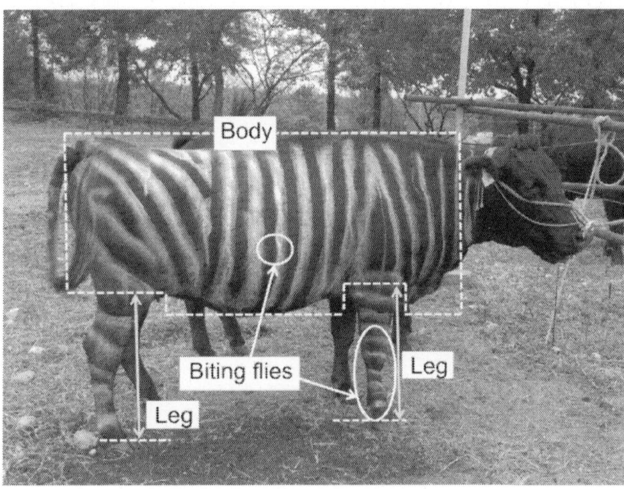

Figure 3.2. The span of the area on the body and leg of the cow used to count the number of biting flies from the photos (Kojima et al, 2019)

The total number of biting flies on the leg and body of the Black and White cows and the total sum of biting flies on the Black and White cows were almost half as those on the control and the B cows.

The most harmful arthropod pests of cattle in the territorial United States and worldwide even are the biting flies and their impact was estimated at $2211 million per year. Insecticidal control measures are often used to reduce the economic losses associated with the pest. But the problem with insecticides is that the pests evolve rapidly to become immune to the insecticides as quickly as within a decade of introduction. For instance, horn flies were found to have quickly become resistant to insecticides and the popular insecticide ear tags were abandoned due to ineffectiveness in the United States.

The number of flies repelling acts by the cow, except for the skin twitch was reduced in the Black and White Stripes experiment as part of this study. It needs to be pointed out that acts such as head throws and foot stomps are more energy consuming acts and tail flicks and skin twitches are much less energy-consuming activities and typically occur in the cases of fewer fly loads. The present experiment demonstrates clearly that the Black and White striped cows had the least amount of biting fly loads and therefor had more skin twitches when compared to non-Black and White ones. As a result of this experiment it can be extrapolated that Black and White striping of cattle is a more humane and ecologically positive way of controlling pests, and not only does it increase the yield from cattle, it also reduces the volume of infestation and has a positive impact on the general environment and on human health. Painting has also been used as an external identifier to detect livestock from a distance, and the entire technique is cheap and is totally animal-friendly.

A minor caveat is that the technique of striping is temporary and exists for a few weeks to a few months depending on the technique employed. Research needs to be commissioned to develop longer-lasting striping techniques that would last through the biting fly season – usually 3 to 4 months.

Also, in support of our study, there are earlier studies that proposed that biting flies are less likely to land on Black and White painted surfaces of livestock. Waage in 1981 determined that black and white striped cylinders rotating slowly trapped much fewer flies than cylinders that were evenly coated in black or white. Additionally, Caro et al (2019) proved that horses wearing coats with a striped pattern had far fewer tabanid flies landing on them when compared to coats of plain black or white color. As a result of this and other attempts, horse keepers have been known to zebra paint their horses and Black and White striped horse coats have actually become a retail item.

While there is the assumption that the odor emanating from the lacquer may have played a part in affecting the landing behavior of the flies on cattle bodies, the fact that there was no marked difference in the flies landing on the black strip lacquer and the control cow suggested that indeed there was a marked reduction in the number of flies that landed on a Black and White striped cow. The present study also clearly establishes the fact that there is benefit in painting zebra-like patterns on the bodies of horses despite the fact that they actually do not thwart the tabanid flies from a distance. The success seems to be premeditated on the confusion that arises on the part of the pests to perform a landing on the surface that is coated in Black and White stripes. However, there is no doubt that more study needs to be conducted to help improve our fundamental understanding of the mechanics of this confusion, in order to further support our hypothesis (Kojima et al, 2019).

3.3 REFERENCES

1. von Helversen B, Schooler LJ, Czienskowski U (2013) Are Stripes Beneficial? Dazzle Camouflage Influences Perceived Speed and Hit Rates. PLoS ONE 8(4): e61173. doi:10.1371/journal.pone.0061173.
2. Kojima T, Oishi K, MatsubaraY, Uchiyama Y, Fukushima Y, Aoki N, et al. (2019) Cows painted with zebra-like striping can avoid biting fly attack. PLoS ONE 14(10): e0223447.
3. Waage JK. How the zebra got its stripes–biting flies as selective agents in the evolution of zebra colouration. J Entomol Soc South Afr. 1981; 44: 351–358.
4. Caro T, Argueta Y, Briolat ES, Bruggink J, Kasprowsky M, Lake J, et al. Benefits of zebra stripes: Behaviour of tabanid flies around zebras and horses. PLoS ONE. 2019; 14(2): e0210831.

4

Camouflage and Language Skills

Prerna Pandey[1], Shivsanjeevi Sripathi[2]

[1]Dr Prerna Pandey, a biotechnologist with several years of wet lab research experience, has worked as a senior research fellow at International Center for Genetic Engineering and Biotechnology, New Delhi. Her field of research in PhD included isolation and molecular characterization of Geminiviruses, genome sequencing, gene annotation, as well as gene silencing using the RNA interference technology. She has published papers in peer-reviewed journals in the field and has a submitted a number of annotated Geminiviral genome sequences in the GenBank, including two novel ones. Prerna has also completed her editing and proof-reading courses from SfEP, UK and now works as a freelance scientific editor, writer and a subject matter expert. She has edited and co-authored several books on a range of topics in Life-Sciences. When Prerna is not busy with her assignments, she loves visiting historical places.

[2]ShivSanjeevi Sripathi completed his Masters in Biotechnology from Mumbai University in 2008. He was awarded for academic excellence in both his Bachelors and Masters for securing second rank in Mumbai University in 2006 and first rank in his college: Kishinchand chellaram College. For his Masters he secured first rank in his college KET's V.G.Vaze College. He qualified CSIR and NET and and TOEFL in September 2008. He then worked on a stem cell project at the Specialized Centre for Cell Based Therapy (SCCT),KEM Hospital at Mumbai on a project entitled, : Isolation & detection of stem cells from Human Umbilical cord/ amniotic membrane" following which he worked at Junior Research Fellow at Microbiology & Cell Biology Department, Indian Institute of Science, Bangalore from August 2009 to November 2011. The field of work involved cloning of cell wall genes and transcription factors in *E.coli & M.smegmatis*. As a writer, he has authored and co-authored 25 books with Delve and Arcler International Press. He loves to read and share on interesting aspects of life sciences in books. In his free time he loves to travel and explore ancient customs and traditions.

4.1 INTRODUCTION

Autism spectrum disorder (ASD) is a behaviorally defined condition predominantly found in males [Parish-Morris J et al, 2017; Loomes, Hull and Mandy, 2017). Recent research suggests that girls with ASD may "camouflage" real struggles with social communication by engaging in social mimicry and behaving in typical ways, thus complicating diagnosis (Lai et al, 2011). For example, nonverbal communication (e.g., gesture) is broadly impaired in ASD (Parish-Morris J et al, 2017; American Psychiatric Association, 2013). Using 3D motion capture, a recent study showed that girls with ASD gesture in ways that are more vibrant and noticeable than boys with ASD, despite similar struggles with social communication (Rynkiewicz et al, 2016). In this way, girls effectively modified their behavior to mask a traditional area of weakness. In the present study, we use a language-based analysis--by producing gender (male/female)-typical filled pauses during a naturalistic conversation--to explore another way in which girls with ASD achieve this social camouflage effect.

Filled pauses in the course of normal conversation, interlocutors pause, revision, and re-work on their utterances, a process termed disfluency (Parish-Morris et al, 2017), which affects how speakers are perceived by others, and elevated rates of speech disfluencies, have been linked to negative social perceptions of the speaker (Boyle, 2015). Pauses are disfluencies that occur during natural speech and can either remain unfilled (silent pauses) or be filled by words like UM, UH, like, and you know (filled pauses). UM and UH, in particular, are thought to be social pragmatic hesitation markers with communicative value (e.g., facilitating comprehension by signaling to the listener that the speaker needs more time to finish communicating their current thought, or that they would like to hold/cede the floor; Fox Tree, 2001). However, there is growing evidence to suggest that UM and UH are not the same; UH is used to signal a short delay while UM may be used to signal more significant delays. Largescale cross-linguistic studies of spoken and written language show that these two markers are used with varying frequency by people from different demographic groups (Parish-Morris et al, 2017; Wieling et al, 2016). For example, UM is used relatively more often by women, educated individuals, and younger generations than by men, less educated individuals, and older generations. In contrast, UH is used relatively more frequently by men, less educated individuals, and older generations than by women, more educated individuals, and younger generations. This pattern is robust across American English, British English,

Scottish English, Dutch, Norwegian, German, Danish, and Faroese (Wieling et al, 2016).

4.2 ASD AND FILLED PAUSES

A significant body of research suggests that children and adults with ASD produce unusually high rates of various speech disfluencies (Parish-Morris et al, 2017; MacFarlane et al, 2017), but only three studies have specifically examined UM and UH. First, 4- to 8-year-olds with ASD use UM at lower rates than children without ASD. This finding suggests that UM and UH may hinge on distinct cognitive processes, with UM more powerfully affected by the presence of autism (Parish-Morris et al, 2017; Heeman et al, 2010). Subsequent research showed that UM used by 8- to 21-year-olds normalizes when overall ASD symptoms wane off (Irvine et al, 2016), again lending support to the hypothesis that UM is associated with social competence. In a third study, language samples from children with ASD were compared to samples from children with specific language impairment, and individual variation in UM use was found to be a marker of pragmatic ability rather than a consequence of general language skill. Taken together, these studies suggest that UM and UH are used differently by children with ASD and that UM ratios [um/(um + uh)] associate with ASD symptoms (Parish-Morris et al, 2017).

A major limitation of this line of research is that gender differences in filled pause use have never been studied in individuals with ASD. If females have higher UM ratios than males (Wieling et al, 2016), slight sex ratio differences across diagnostic groups could lead to erroneous results. For example, Gorman et al. (2016) included 10 % of girls in their ASD group and 30 % of girls in their typical group. Elevated UM ratios in the typical group relative to the ASD group might thus be a consequence of sex ratio imbalances (because girls may produce higher UM ratios than boys)—and the TD group was enriched for girls. The observation, that slight differences in sex ratio by the diagnostic group could change the results of scientific studies in a variety of domains affects how we evaluate the presence and severity of ASD symptoms in girls and boys and will influence the hunt for behavioral markers of ASD more broadly.

In the case of UM and UH specifically, if girls with ASD exhibit gender normative speech patterns, it could bias perception of their ASD symptoms. The current literature is unable to address this question, primarily out of an absence of speech data from females with ASD. Apart from Gorman et al. (2016), the other studies that explored UM vs. UH in children with ASD either did not report sample sex ratios at all (Heeman et al, 2010), or included insufficient numbers of female participants with ASD to assess potential effects of gender on filler production (3 out of 24 or 12.5 %) (Parish-Morris et al, 2017; Irvine et al, 2016).

Filled pauses in the course of normal conversation, interlocutors pause, revise, and re-work their utterances, a process termed disfluency (Clark and Fox Tree, 2002). Disfluencies affect how speakers are perceived by others, and elevated rates of speech disfluencies have been linked to negative social perceptions of the speaker (Boyle, 2015). Pauses are disfluencies that occur during natural speech and can either remain unfilled (silent pauses) or be filled by words like um, uh, like, and you know (filled pauses). UM and UH, in particular, are thought to be social pragmatic hesitation markers with communicative value (e.g., facilitating comprehension by signaling to the listener that the speaker needs more time to finish communicating their current thought, or that they would like to hold/cede the floor) (Fox Tree, 2001). However, there is growing evidence to suggest that UM and UH are not the same; UH is used to signal a short delay while UM may be used to signal more significant delays.

Largescale cross-linguistic studies of spoken and written language show that these two markers are used with varying frequency by people from different demographic groups. For example, UM is used relatively more often by women, educated individuals, and younger generations than by men, less educated individuals, and older generations. In contrast, UH is used relatively more frequently by men, less educated individuals, and older generations than by women, more educated individuals, and younger generations. This pattern is robust across American English, British English, Scottish English, Dutch, Norwegian, German, Danish, and Faroese (Wieling et al, 2016).

4.3 SUBTLE LINGUISTIC MARKERS

Subtle linguistic markers, such as UM is a social pragmatic marker used more often by younger people and women as compared to older people and men, and UH is used relatively more often by older people and men than by younger people and women (Wieling et al, 2016). Unusually low UM ratios have been reported in children/adolescents with ASD (Irvine et al,

2016), and have been argued to mark difficulties with social communication (Gorman et al, 2016). The outcome showed that girls with ASD and typical girls/boys exhibited higher UM ratios than boys with ASD, while girls in both diagnostic groups suppressed UH relative to their male counterparts. Importantly, filled pause differences in boys and girls with ASD in this study were not attributable to increased social pragmatic ability in girls, as girls and boys in our sample had equivalent social communication skills and comparable autism symptom severity. The present findings suggest that UH suppression and higher UM ratios may serve as "linguistic camouflage" to normalize the way a girl with ASD sounds relative to same-aged typical peers, while elevated rates of UH (relative to UM) may cause boys with ASD to sound particularly atypical. Intentional or not, overtly typical-sounding speech in girls could have some benefits (to more easily "blend in" with classmates), while simultaneously complicating the detection of ASD, leading to missed or delayed diagnosis, and misdiagnoses (Lai et al, 2015).

4.4 UM RATIO AND GENDER DIFFERENCES

Normally, girls with ASD did not fill more pauses than boys; they filled them differently. Like typical children and children with specific language impairment, girls used UM during > 70 % of all filled pauses. Boys with ASD, in contrast, used UM during only in 55 % of filled pauses (Parish-Morris et al, 2017). Although girls are using typical-sounding speech patterns (as reflected by the UM ratio), it is not necessarily reflected in improved parental perceptions of their social communicative ability.

Why might girls with ASD fill conversational pauses with "typical-sounding" words, and suppress atypical words like UH, despite comparable social communication deficits to boys? One hypothesis is that girls are not using filled pauses as communicative tools, but rather produce higher UM ratios as a form of unconscious social mimicry or scripting (Parish-Morris et al, 2017), a girl with ASD might produce typical UM ratios to appear less atypical and improve her chances of successfully integrating during social situations. However, she may not necessarily understand the social meaning behind different filled pause types and struggle with social communication as much as boys that do not "normalize" their UM ratios. Indeed, our results suggest that girls' linguistic camouflaging is more successful in some ways than others.

Although girls with ASD produced higher UM ratios than boys with ASD due to UH suppression, their overall filled pause rate was still lower typically, resulting in incomplete camouflage, due to the nature of the sample in this study. Girls who have not yet been identified as autistic or referred for an evaluation may engage in more successful (or complete) linguistic mimicry. Another likely explanation for the gender differences reported here hinges on powerful forces of gender socialization (Parish-Morris et al, 2017).

Research on typically developing children shows that parent perceptions, play practices, and styles of interaction differ systematically gender of the child, with infant girls hearing more language, receiving more eye contact, and having more opportunities to engage in social-emotional interaction than infant boys (Cheslack-Postava, 2012), and maybe in full effect for 2 to 3 years before most ASD diagnoses are made (Parish-Morris et al, 2017), and may act as pressure on young girls to conform to "girl" expectations that include pragmatic competence. Thus, perhaps girls with ASD are responding to forces of gender socialization that expect girls to sound less pragmatically impaired, by suppressing UH and producing higher UM ratios. girls with ASD and typical girls are subject to the same gender-based influences during the first few years, thus, their language might be similarly affected. Indeed, we found that typical girls also suppress UH and produce higher UM ratios than typical boys. This is only one possible hypothesis that likely relates to the differences observed in this study and the causal pathways that lead to this type of linguistic variation in girls need to be investigated.

Consistent with prior research documenting UM and UH differences in adult men and women, gender differences in UH production are robust in children. Negative relationships have been observed between UH production and perceived social ability across both genders in ASD; this builds on several studies linking speech disfluencies to negative social perceptions of speakers (Parish-Morris et al, 2017).

4.5 IQ SCORES AND MOTOR SKILLS

In one study, boys and girls did not differ significantly on IQ, although, basic cognitive differences are unlikely to have caused elevated UH use by boys. The groups in this study had comparable verbal IQ scores, but it is still possible that boys in this study had slightly compromised or uneven language abilities relative to girls, who tend to have better language skills than boys in typical samples as well (Messinger et al, 2015). A third explanation, that poor coordination between motor and language systems results in greater UH use by boys with ASD relative to girls, maybe due to subtle motor control differences in ASD (Cook, 2016).

Elevated UH rates in boys with ASD may, therefore, indicate disrupted coordination across motor and language systems, and could provide clues about a global underlying motor-based pathobiology that partially accounts for social communication problems in ASD (Cook, 2016). However, we know little about gender-based differences in motor features of ASD; if UH is a marker of asynchrony across motor and language systems that disproportionately affects boys, and elevated UH production drives reduced UM ratios in boys, it could be that motor dysregulation leads to pragmatic problems even when the cognitive aspects of language are grossly intact. But the motor ability was not measured in this study, which may or may not vary by gender.

Combined with a recent study showing gestural camouflaging in girls with ASD (Rynkiewicz et al, 2013), the linguistic camouflage effect identified in this study suggests that subtle gender differences in behavioral domains relevant to ASD (e.g., language) could contribute to girls being missed—or misdiagnosed—due, in part to a male-focused conceptualization of ASD and male-normed diagnostic tools.

In addition to gender-specific identification and characterization tools, behavior dependent intervention outcome measures may also need to differ by gender. The goal of naturally and automatically integrating information about putative gender differences into the hunt for behavioral biomarkers could lead to creative new approaches that are more sensitive to the way ASD manifests in boys and girls. Recent large-scale research shows that gender differences in the cognitive and motor profiles of infant siblings of children with ASD are not unique to ASD risk, but rather reflect broader gender differences in the general population (Messinger et al, 2015), there is a need to compare girls with ASD to typically developing girls.

An inherent limitation of analyzing language produced during semi-structured evaluations is that clinicians are free to ask questions in any order or to skip questions completely if appropriate from a clinical perspective. Clinicians possibly differed in the number of times they asked questions vs. used open-ended comments to start conversations. These variations were limited to the extent possible, but are inherent in any research that relies on ADOS evaluations.

We did not examine the influence of co-occurring disorders or dimensions, such as anxiety, attention-deficit/ hyperactivity disorder, and executive dysfunction in ASD, on filled pause use. Further, there was no relationship between age, IQ, and filled pause type. The lack of an age effect may be due to the age range of our sample (which does not extend into adulthood). Future research with older men and women with ASD is needed to elucidate how our findings change across the lifespan.

4.6 FUTURE DIRECTIONS

Children and adolescents with ASD produce more fillers when cognitive demands are high (Irvine et al, 2016), which for individuals with social challenges, may occur more often when discussing social topics. Although others have shown that UM is used relatively less frequently by children with ASD across a variety of tasks (Heeman et al, 2010), future research should specifically compare children's responses to highly social questions and less-social questions. The discrepancy in UM and UH production between different question types that vary by social load may be even more informative than relative usage collapsed across an entire interaction. Future studies with older and larger samples of individuals with ASD, as well as human raters, are needed to test this hypothesis. Filled pauses vary by gender in ASD, but the daily consequences of these variations are unknown.

Given that early detection and early intervention are critical for maximizing functional outcomes in ASD (Kim and Lord, 2013), issues of under-referral and inadequate measurement for girls with ASD are not trivial. Suppressed UH and typical UM ratios in girls correlate with the perception of social normalcy by peers, thus helping to establish or maintain friendships. Perhaps girls with ASD suppress UH more when talking to peers than when talking with adults, mimicking typical behavior and increasing their chances of social affiliation in classrooms or on playgrounds.

4.7 SUMMARY

ASD experts make diagnostic decisions based on observable behavior and subtle differences in how a child moves or talks will influence the way they are perceived. Gender socialization or social mimicry may lead to "camouflaged" behavior in girls with ASD, which, combined with widely held gender biases about how girls and boys should behave and true biological gender-based differences, likely complicate efforts to effectively identify and treat boys and girls with ASD. Recent attempts to reduce bias by directly sampling behavior and using objective, computational measurement tools hold promise over existing parent report and clinician rating scales (Parish-Morris et al, 2017; Rynkiewiczet al, 2016), but even these new tools will likely be influenced by variables such as age, gender socialization, socio-economic status, physical and mental health, and home and cultural environment.

4.8 REFERENCES

1. Parish-Morris J, Liberman MY, Cieri C, Herrington JD, Yerys BE, Bateman L, Donaher J, Ferguson E, Pandey J, and Schultz RT. Linguistic camouflage in girls with autism spectrum disorder. ParishMorris et al. Molecular Autism (2017) 8:48 DOI 10.1186/s13229-017- 0164-6
2. Loomes R, Hull L, Mandy WPL. What is the male-to-female ratio in autism spectrum disorder? A systematic review and meta-analysis. J Am Acad Child Adolesc Psychiatry. 2017;56(6):466–74. https://doi.org/10. 1016/j.jaac.2017.03.013
3. Lai M-C, Lombardo MV, Pasco G, Ruigrok ANV, Wheelwright SJ, Sadek SA, et al. A behavioral comparison of male and female adults with high functioning autism spectrum conditions. PLoS One. 2011;6(6):e20835. https://doi.org/10.1371/journal.pone.0020835
4. American Psychiatric Association. Diagnostic and statistical manual of mental disorders, 5th edition: DSM-5. 5th ed. Washington, D.C: American Psychiatric Publishing; 2013
5. Rynkiewicz A, Schuller B, Marchi E, Piana S, Camurri A, Lassalle A, BaronCohen S. An investigation of the 'female camouflage effect' in autism using a computerized ADOS-2 and a test of sex/gender differences. Molecular Autism. 2016;7(1) https://doi.org/10.1186/s13229-016-0073-0
6. Boyle MP. Identifying correlates of self-stigma in adults who stutter: further establishing the construct validity of the Self-Stigma of Stuttering Scale (4S). J Fluen Disord. 2015;43:17–27. https://doi.org/10.1016/j.jfludis.2014.12.002
7. Fox Tree JE. Listeners' uses of um and uh in speech comprehension. Mem Cogn. 2001;29(2):320–6. https://doi.org/10.3758/BF03194926
8. Wieling M, Grieve J, Bouma G, Fruehwald J, Coleman J, Liberman M. Variation and change in the use of hesitation markers in Germanic languages. Lang Dyn Change. 2016;6(2):199–234.
9. MacFarlane H, Gorman K, Ingham R, Presmanes Hill A, Papadakis K, Kiss G, van Santen J. Quantitative analysis of disfluency in children with autism spectrum disorder or language impairment. PLoS One. 2017;12(3):e0173936. https://doi.org/10.1371/journal.pone.0173936
10. Heeman PA, Lunsford R, Selfridge E, Black L, Van Santen J. Autism and interactional aspects of dialogue. In Proceedings of the 11th Annual Meeting of the Special Interest Group on Discourse and

Dialogue. Association for Computational Linguistics. 2010. pp. 249–52. Retrieved from http://dl.acm.org/citation.cfm?id=1944551.

11. Irvine CA, Eigsti I-M, Fein DA. Uh, um, and autism: filler disfluencies as pragmatic markers in adolescents with optimal outcomes from autism spectrum disorder. J Autism Dev Disord. 2016;46(3):1061–70. https://doi.org/ 10.1007/s10803-015-2651-y
12. Gorman K, Olson L, Hill AP, Lunsford R, Heeman PA, & van Santen JPH. "Uh" and "um" in children with autism spectrum disorders or language impairment. Autism Res. 2016, n/a-n/a. https://doi.org/10.1002/aur.1578.
13. Clark HH, Fox Tree JE. Using uh and um in spontaneous speaking. Cognition. 2002;84(1):73–111
14. Lai M-C, Lombardo MV, Auyeung B, Chakrabarti B, Baron-Cohen S. Sex/ gender differences and autism: setting the scene for future research. J Am Acad Child Adolesc Psychiatry. 2015;54(1):11–24.
15. Cheslack-Postava K, Jordan-Young RM. Autism spectrum disorders: toward a gendered embodiment model. Soc Sci Med. 2012;74(11):1667–74. https:// doi.org/10.1016/j.socscimed.2011.06.013
16. Messinger DS, Young GS, Webb SJ, Ozonoff S, Bryson SE, Carter A, et al. Early sex differences are not autism-specific: a Baby Siblings Research Consortium (BSRC) study. Mol Autism. 2015;6(1):32. https://doi.org/10.1186/ s13229-015-0027-y
17. Cook J. From movement kinematics to social cognition: the case of autism. Philos Trans R Soc B. 2016;371(1693):20150372. https://doi.org/10.1098/rstb.2015.0372
18. Kim SH, Lord C. The behavioral manifestations of autism spectrum disorders. In The neuroscience of autism spectrum disorders. Elsevier. 2013. pp. 25–37. Retrieved from http://linkinghub.elsevier.com/retrieve/pii/ B9780123919243000028.

5

How Conspicuous are Peacock Eyespots and other Colorful Feathers in the eyes of Mammalian Predators?

Suzanne Amador Kane[1], Yuchao Wang[1], Rui Fang[1], Yabin Lu[1], Roslyn Dakin[2]

[1]Physics & Astronomy Department, Haverford College, Haverford, Pennsylvania, United States of America
[2]Migratory Bird Center, Smithsonian Conservation Biology Institute, National Zoological Park, Washington DC, United States of America

ABSTRACT

Colorful feathers have long been assumed to be conspicuous to predators, and hence likely to incur costs due to enhanced predation risk. However, many mammals that prey on birds have dichromatic visual systems with only two types of color-sensitive visual receptors, rather than the three and four photoreceptors characteristic of humans and most birds, respectively. Here, we use a combination of multispectral imaging, reflectance spectroscopy, color vision modelling and visual texture analysis to compare the visual signals available to conspecifics and to mammalian predators

Citation: Kane SA, Wang Y, Fang R, Lu Y, Dakin R (2019) How conspicuous are peacock eyespots and other colorful feathers in the eyes of mammalian predators? PLoS ONE 14(4): e0210924. https://doi.org/10.1371/journal.pone.0210924.

from multicolored feathers from the Indian peacock (*Pavo cristatus*), as well as red and yellow parrot feathers. We also model the effects of distance-dependent blurring due to visual acuity. When viewed by birds against green vegetation, most of the feathers studied are estimated to have color and brightness contrasts similar to values previously found for ripe fruit. On the other hand, for dichromat mammalian predators, visual contrasts for these feathers were only weakly detectable and often below detection thresholds for typical viewing distances. We also show that for dichromat mammal vision models, the peacock's train has below-detection threshold color and brightness contrasts and visual textures that match various foliage backgrounds. These findings are consistent with many feathers of similar hue to those studied here being inconspicuous, and in some cases potentially cryptic, in the eyes of common mammalian predators of adult birds. Given that birds perform many conspicuous motions and behaviors, this study suggests that mammalian predators are more likely to use other sensory modalities (e.g., motion detection, hearing, and olfaction), rather than color vision, to detect avian prey. This suggests new directions for future behavioral studies and emphasizes the importance of understanding the influence of the sensory ecology of predators in the evolution of animal coloration.

INTRODUCTION

Ever since Darwin, colorful feathers have been assumed to present salient visual signals readily detectable by their natural predators [1,2]. For this reason, sexually-selected ornaments like the iridescent eyespot feathers of the Indian peacock (*Pavo cristatus*) (Fig 1A) have been proposed to incur a cost due to increased predation [2–4]. As Zahavi argued in his paper introducing the handicap principle: "The more brilliant the plumes, the more conspicuous the male to predators" [5]. Evidence for such opposing selection pressures has been found in ornamented guppies preyed upon by fish [6]. On the contrary, susceptibility to cat predation was not found to correlate significantly with sexual dichromatism in birds [7], and a recent experimental study found that conspicuous plumage does not enhance predation by avian predators ([8] and references therein). Conspicuously-colored plumage also has been proposed to function as warning coloration for aposematism [9,10]. However, while these hypotheses are predicated on the predator being able to detect prey visual signals [11], no studies have tested whether this is true for the mammalian predators that prey on many birds. For example, the primary predators of adult peafowl are carnivorans (felids and canids, S1 Appendix), and cats are a major threat to bird populations world-wide [12]. These predators all have dichromatic visual systems; i.e., they have only two types of cone visual receptors with distinct spectral sensitivities, not the four characteristic of most birds or the three found in most humans [13]. Because dichromatic mammals lack red-green color discrimination, they are unlikely to detect many of the chromatic visual cues evident to birds and humans [13–15]. Studies of visual ecology have considered how prey appear to various types of predators (birds, insects and fish) for many types of prey, including insects and birds [16,17], fish [18], cuttlefish [19], crustaceans [20], primates [21] and lizards [11]. Two previous studies also have studied the iridescence reflectance spectra of peacock eyespots and how they are perceived by peahens (females) [22,23]. So far, no studies have compared how visual signals from peacocks and other birds appear in the vision of their mammalian predators.

Figure 1. Peacocks and the model peacock train. A) An Indian peacock displaying his erect train to a peahen (female) in the foreground and B) another individual holding his train folded while walking. C) Model peacock train assembled from a collection of eyespot feathers used to evaluate the appearance of the train viewed against vegetation.

During courtship displays, male Indian peafowl ("peacocks") attract mates by spreading, erecting and vibrating their fan-like train ornament (Fig 1A), causing it to shimmer iridescently and emit mechanical sound [24–26]. Several lines of evidence indicate that these feathers are assessed during mate choice: train-rattling performance by peacocks is obligatory for mating success [24], eye-tracking experiments have shown that train-rattling displays are effective at attracting and holding the peahen's gaze [27], and eyespot iridescence has been shown to account for approximately half of variation in male mating success [22,23,28]. Because peacocks spend the majority of their time in activities other than courtship displays even during the breeding season [24,29], any test of visual saliency must also consider the appearance of the folded train. Furthermore, because the peacock's head, neck and breast are covered by iridescent blue contour feathers [30], the visual cues generated by this body plumage are also relevant for salience to potential mates and predators.

Here, we use multispectral imaging to estimate how detectable peacock feathers are to conspecifics and dichromatic mammalian predators, as measured by color, brightness, and texture contrast relative to green background vegetation, following similar studies of prey that model camouflage against predators with a variety of visual systems [31]. We also use reflectance spectroscopy to compare the spectral reflectances of the various feather and foliage samples with each other, and with the known sensitivities of each viewing animal's photoreceptors. Our goal was to test the assumption that colorful feathers that are highly conspicuous to conspecific birds are also readily detectable by these predators. To determine how generalizable our results were to other hues of colorful plumage, we also measured reflectance spectra and multispectral images of red and yellow parrot feathers. We then used psychophysical vision models to estimate whether conspecifics and dichromatic mammalian predators can readily detect the color and brightness contrasts between feathers and green vegetation. Our analysis modeled the appearance of feathers at various distances to determine when each observing species could distinguish color patches relative to the surrounding environment. We also reviewed the literature to determine the light niches and sensory modalities used by mammals that prey on peafowl and other birds, and to understand whether an enhanced risk of predation has been documented for peacocks relative to other prey.

In addition to color cues, visual salience depends on the presence of pattern features that are perceptually discriminable from the background. To compute whether predators might detect the peacock's train using such visual texture cues, we analyzed images of live peacocks and of the model train relative to that of background vegetation using two pattern analysis methods motivated by visual processing in vertebrates [32]. Granularity analysis is a spatial filtering method that determines the contributions to image contrast of features with different sizes; this image processing technique has been used to compare pattern textures in studies of cephalopod, avian egg, fish and shore crab camouflage, as well as humans searching for objects against various backgrounds [20,32–35]. A second method, edge detection, provides a complementary measure of texture complexity by using image processing to detect sharp gradients in intensity [36].

MATERIALS AND METHODS

Feather Samples

Five Indian peafowl eyespot (Fig 2A), three blue peacock contour breast feathers (Fig 2B), four scarlet macaw (*Ara macao*) wing feathers (two red and six yellow patches total) (Fig 3A), two Amazon parrot (*Amazonica ochrocephala panamensis*) wing feathers (two red and two yellow patches total) (Fig 3B), and four red African grey parrot (*Psittacus erithacus*) tail feathers (Fig 3C) were obtained from Moonlight Feather (Ventura, CA USA) and Siskiyou Aviary (Ashland, OR USA). Because the psittacofulvin pigments in parrot feathers have reflectance spectra with similar spectral features and reflectances as found for red and yellow carotenoid pigments [37–39], our results should be representative of red and yellow feathers in general. An earlier study of variation in color measurements for feathers [40] showed that three measurements per patch on one individual per species is sufficient for quantifying the colorspace coordinates of a feather color patch to within 5% of the mean as needed for our visual signal analysis. For mounting, eyespot feathers were cut off below the outermost colored ring at the proximal end. All feather types were mounted on black matte art quality paper with a magnetic backing that adhered to the tilt stages used for spectroscopy and multispectral imaging. Feather samples were stored without compression in sealed boxes in acid-free envelopes at 75% relative humidity and ambient temperature (22 ± 2 deg C). The different peacock eyespot color patches (colored rings and central disk) are referred to using the names and two letter abbreviations indicated in Fig 2A.

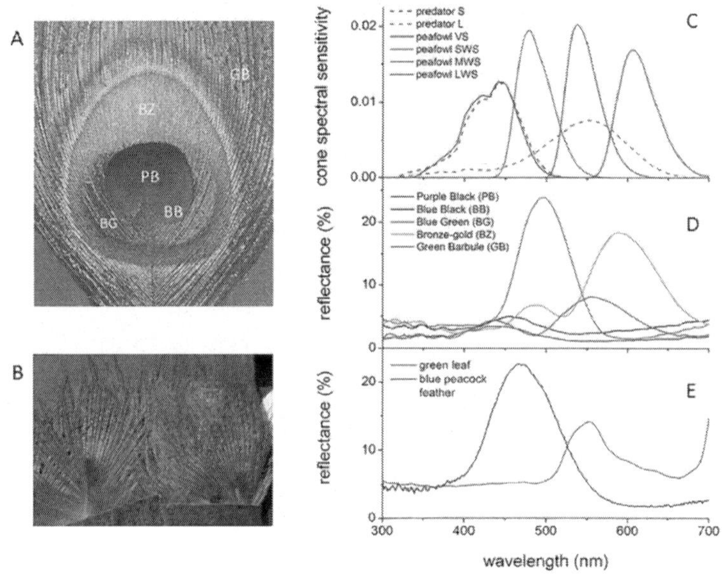

Figure 2. Peafowl feathers, cone sensitivity spectra and reflectance spectra for feathers and green vegetation. (A) An Indian peacock eyespot feather showing the color patch names used in the analysis. (B) Peacock blue breast plumage. (C) Comparison of the cone photoreceptor spectral sensitivities for the Indian peafowl and ferret, which has dichromatic color vision very similar to that of cats and dogs. All spectra are multiplied by the D65 illuminance spectrum used to model sunlight and normalized to unit area. Reflectance spectra of (D) peacock feather eyespots and (E) peacock iridescent blue body plumage and the green saucer magnolia (*Magnolia x soulangeana*) leaf used as a background for the feather sample images.

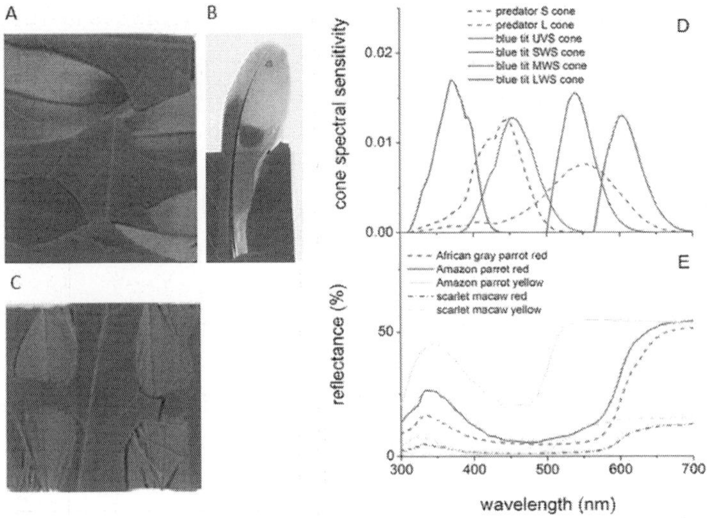

Figure 3. Parrot feather images, cone sensitivity spectra and feather reflectance spectra. (A) Scarlet macaw, (B) Amazon parrot and (C) African grey parrot feather samples. (D) Comparison of the cone photoreceptor spectral sensitivities for the blue tit, which has tetrachromatic ultraviolet sensitive (UVS) color vision similar to that of parrots, and the ferret, which has dichromatic color vision similar to that of cats and dogs. All spectra are multiplied by the D65 illuminance spectrum used to model sunlight and normalized to unit area. (E) Reflectance spectra of the parrot feather red and yellow patches studied here.

In addition to measuring the individual peacock feathers described above, we also created a model peacock train using an array of 28 peacock eyespot feathers (Fig 1C) arranged to match the geometry of eyespots in actual peacock trains [41]; this was used to simulate the appearance of the train during display (when the train is erect) or during walking, perching or standing (when the train is held horizontally; see Fig 1B). In their native range in India and Pakistan, peafowl are reported to live in a variety of habitats, including open moist and dry-deciduous forest, scrub jungle, and adjacent grasslands, and their breeding season is reported to coincide with the start of the rainy season [42], after which eyespot feathers are shed by molting [43,44]. Green leaves have a generic reflection spectrum due to their dominant pigment, chlorophyll, as determined for a variety of environments [45–47] including deciduous forests and other native vegetation in India [48,49]. To simulate the native habitat of peafowl during the day, we selected background vegetation using as reference images a variety of photographs of peafowl in India, Pakistan and Sri Lanka from the Macauley Library at the

Cornell Lab of Ornithology and the Internet Bird Collection (details in S2 Appendix). The variety of background foliage selected from trees, brush and grasses from the midatlantic region of the USA (40.0093° N, 75.3057° W) (S3 Appendix) were selected to approximate the color, luminance and texture of many of the green plants found in the native environments of peafowl and many other bird species.

Vision Models

The Indian peafowl's visual system has four classes of color-sensitive (chromatic) single cone cells: violet (VS), short (SWS), medium (MWS) and long (LWS) wavelength-sensitive cones, and one type of double cone that is sensitive to brightness (luminance) [50]. In order to illustrate their spectral responses under natural illumination, Fig 2C shows the peafowl cone's spectral sensitivities $S_r(\lambda)$ for the rth photoreceptor class (including ocular media and oil droplet transmission) multiplied by the CIE D65 irradiance spectrum, $I(\lambda)$, and normalized to unit area; we used this standard illuminant because of its close match the solar irradiance spectrum for the elevation angles found for actual peacock displays [24,51]. To model the tetrachromatic UVS (ultraviolet-sensitive) vision of parrots we used blue tit (*Cyanistes caeruleus*) cone spectral sensitivities [52,53], which has peak spectral sensitivities that agree to $< 2\%$ with those of budgerigars (*Melopsittacus undulatus*), a type of parrot [54] (Fig 3D).

The visual systems of dichromatic mammalian predators have been studied for a variety of genera, and found to include S (blue- to near-UV-sensitive) [55] and L (green-sensitive) cone populations in all carnivorans studied to date, including felids [56] and canids [57]. Behavioral studies have confirmed that domestic cats [58] and dogs [59,60] have dichromatic color vision. Brightness signals in dichromatic mammals are assumed to be due to only the L cones [61]. We used ferret (*Mustela putorius*) cone spectra [53] to model dichromat vision because ferret spectral peaks agree closely with those of cats, dogs and foxes (i.e., $\leq 4.4\%$ for S and $\leq 1.4\%$ for L cones) [56,57,62] (Fig 2C).

Like many other primarily diurnal birds, peafowl are active throughout the day, with a peak in their foraging and display activities in the morning (post dawn) and late afternoon. Ecological studies have used camera trapping and radiotelemetry tracking to measure diel activity patterns of wild cats and dogs in Southeast Asia and the neotropics. The results show that many of these potential predators of peafowl and other wild birds are active and hunt during the daylight hours as well as crepuscular and nocturnal conditions, and that many are primarily diurnal [63–70]. Animal-borne video methods have shown that free-ranging feral domestic cats, a major predation threat to wild birds, often hunt during daylight hours as well [71]. Wild felids are reported to hunt primarily using vision [72]. Dholes (*Cuon alpinus*), a wild canid reported to prey on peafowl, are reported to have good vision and olfaction, and to locate prey primarily by sight rather than by scent [73]. Domestic dogs and cats, as well as coyotes (*Canis latrans*), another wild canid, have been shown to use both vision and olfaction in a variety of contexts (e.g., [60,74,75] and references therein), including to locate prey [76]. Thus, it is relevant to consider whether these predators use color signals and photopic (high luminance) vision to locate prey. Under low light conditions, chromatic signals will be weak and visual signals will be dominated by luminance contrast via rod photoreceptors, which have a spectral sensitivity similar to that of the luminance channel for these mammals [77,78]. In addition, a predator's visual acuity and ability to distinguish contrast is greatly diminished under low light conditions [79–82]. Therefore, we modeled visual perception of visual signals by these predators for photopic conditions, to give the best case scenario for detection.

Reflectance Spectroscopy

We measured reflectance spectra using a model USB2000+ spectrometer and OceanView software (Ocean Optics, Largo FL, USA) over the wavelength range 300–850 nm, using 100 ms integration time, 3 pixel boxcar averaging (corresponding to the optical resolution of 6.5 pixels = 2.06 nm FWHM), and averaging over 5 samples. All spectra were recorded in a dark room. Samples were illuminated by an Ocean Optics PX-2 Pulsed Xenon Light source triggered at 200 Hz using square wave pulses from a model 330120A function generator (Agilent Technologies, Wilmington, DE, USA); the source was turned on and allowed to warm up and stabilize for 15 minutes before data collection. Light for illumination and detection was carried in P400-1-UV-VIS optical fibers transparent to 200 nm (Ocean Optics). We used two PTFE white standards with flat 99.0% reflectance over 300–700 nm: a Spectralon USRS-99-010-EPV (Labsphere, North Sutton, NH USA) and a model SM05CP2C (Thorlabs). White standard and dark currents were measured every fifteen minutes. For each feather and each measurement geometry, raw reflectance spectral data were recorded for each feather color patch sample radiance, AR, white standard radiance, AR_r and dark current,

D. The reflectance spectrum, $R(\lambda) = \frac{AR-D}{AR_r-D}$, was smoothed over a wavelength interval of 20 nm using Savitzky-Golay smoothing in Origin; this reduced high frequency noise but did not change reproducible features of the spectra peak shapes.

Transmission spectra for the filters used in multispectral imaging were measured by recording the spectrum of light reflected from the white standard with and without the filter inserted into the light path with its face at normal incidence to the incident light. Reflectance values for color and gray standards were measured using a RPH-SMA reflectance probe stand (Thorlabs, Newton NJ USA) with the illuminating light at 45 deg to normal incidence and detected at normal incidence. The reflectance goniometer for feather measurements used (S2 Fig) was adapted from previously published designs [83,84] but with an additional angular degree of freedom to allow measurement of the bidirectional reflectance distribution function, in which the angle of observation and illumination are not confined to the specular reflection geometry [85]. Both the illumination and detection optical pathways were focused using a 74-UV lens (Ocean Optics) to a 2 mm diameter spot at about 5 cm from the output surface of the lens. The feather samples were realigned every time the angle of illumination and/or detection was adjusted to ensure both beams focused on the same region of the feather. To assess reproducibility of spectra for the same color patch on each feather, we measured each set of spectra three times for each sample after dismounting and remounting each sample.

Multispectral Imaging

Multispectral imaging allows the measurement of color and luminance signals from across an entire feather, or array of feathers, under natural environmental lighting conditions (including the effect of reflected light from green background vegetation); it also allows modeling of the effects of blurring due to visual acuity limitations [86]. The methods used here involve recording two images of the same sample photographed through different filters, including reflection standards to allow measurements of absolute reflectances. The different combinations of filters and camera color channels then are analyzed using visual models for the viewing animal of interest. Multispectral images were recorded using a GoPro Hero 4 Silver Edition camcorder (GoPro Inc, San Mateo, CA USA) modified for full spectral imaging by replacing its original lens and infrared (IR) filter with a quartz lens transparent to < 300 nm [87,88]. Because the spectral response of this camera's IMX117 Exmor-R CMOS sensor (Sony Corp., Tokyo, Japan) is sensitive throughout the visible and near-UV, these cameras have been used in multispectral imaging [89,90] (S1 Fig). Multispectral photographs were recorded at 3000 × 2250 pixel resolution and the GoPro settings medium field of view, Protune CAM-RAW mode (for no white balance compensation), flat color, low sharpness, ISO 400, exposure -2, night mode (to enable shutter speed control), auto shutter and spot meter on. Each sample was photographed twice for each geometry and illumination condition to give two multispectral images: 1) a UV image using an Andrea-UV filter (< 1% transmission for > 400 nm; UVIROptics, Eugene, Oregon USA; 2) a visible RGB (red, green, blue) image using two UV-IR cut filters to pass 400–700 nm light (Hoya Corp., Tokyo Japan). Filter transmission spectra were measured using the methods described in "Reflectance Spectroscopy" (S1 Fig). The camera's large depth-of-field eliminated the need for refocusing between visible and UV images. To maintain constant camera alignment between photographs, we mounted the camera rigidly using optical mounts (Thorlabs, Newton NJ, USA) and attached filters using quick-release Xume magnetic adapters (Panalpina Inc., Port Reading, NJ, USA); all images were taken using a remote trigger. Each feather image included a model Micro FSS08 8-step grayscale diffuse reflectance standard (Avian Technologies, New London, NH USA) mounted level with the sample plane for calibrating absolute reflectance [53]. Images of the model train included a larger 6-step grayscale and color checker chart (DGK Color Tools WDKK Waterproof, Digital Image Flow, Boston MA USA). Reflectance spectra for each grayscale in each filter and camera color channel combination were measured using the methods described in "Reflectance Spectroscopy". Each image also included an object of known size for spatial calibration.

All samples were mounted on a tripod for imaging. Three sets of multispectral images each were obtained with the model train held erect and held horizontal viewed from the side. Peacock eyespots were oriented with their rachis vertical to simulate their average orientation in the erect train during courtship displays and the model train was oriented in a variety of directions to simulate the variation in appearance of the iridescent train eyespot feathers during courtship display, standing and walking. The camera was mounted on a second tripod a distance 20.0 ± 1.0 cm from feather samples and 1.70 to 2.00 ± 0.05 m from the model train. For feather samples, the camera was oriented to record images at normal observation angle ($\theta = 0 \pm 2$ deg) with respect to the feather sample plane (Fig 4). The size of feather sample images was 55 mm x 67 mm, corresponding to 7.3

pixel/mm. Images were captured during June-July 2018 in the Haverford College Arboretum (latitude, longitude: 40.0093° N, 75.3057° W) for 24.2 ± 0.2 deg C and 55.5 ± 1.5% relative humidity. All feather samples were illuminated by direct sunlight with an azimuthal angle $\Psi = 45 \pm 3$ deg clockwise from the camera's optical axis and at solar elevation angles $\Phi = 30 \pm 3$ deg, corresponding to an angle $\alpha = 52 \pm 3$ deg between the observation and illumination directions (Fig 4). These illumination and observation angles agree with those measured for female peafowl observing courtship displays in the morning and late afternoon [24]; they also agree with the angle found to enhance reflectance contrast between the two largest color patches in the peacock's eyespot [23]. In general, these solar angles hold for the morning times when most birds are most active [91–93]. Optimal color contrasts for non-iridescent feathers have been found to correspond to the range of observation-illumination angles α used in this study [94]; this is relevant because pigment-based colors can appear in combination with structural coloration [38]. In addition, for this observation geometry, the bird's body subtends the greatest visual angle. The peacock eyespot feather samples were surrounded by additional loose green barbs to simulate their setting in the actual train, while the parrot feather samples were surrounded by saucer magnolia (*Magnolia x soulangeana*) leaves picked ≤ 1 hour before image capture. We also imaged a variety of green leaves for comparison (S3 Fig). Black velvet fabric was mounted behind the feather samples to limit backscattered light and a lens hood was used to reduce lens flare. The model peacock train was photographed against a variety of foliage backgrounds for solar elevation angle between 37 to 55 deg.

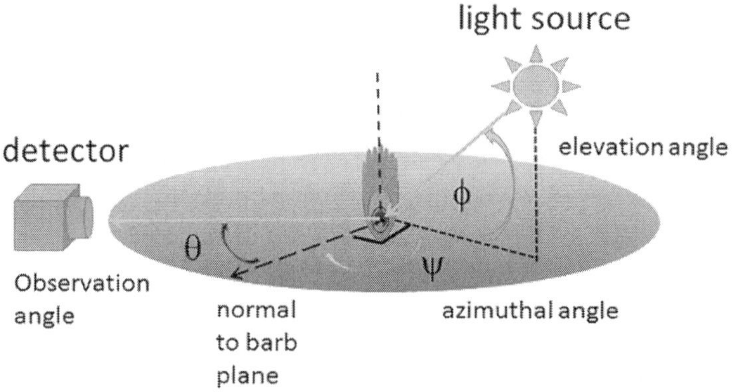

Figure 4. Multispectral imaging geometry showing the angles of observation and illumination. Multispectral images were first processed using custom scripts written in MATLAB v15a with the Machine Vision, Signal Process-

ing and Fitting toolboxes (MathWorks, Natick MA USA); all code is available on figshare at https://figshare.com/s/688fb19dad98b6273324. Images stored as jpeg files were calibrated and corrected for lens distortions using the MATLAB Camera Calibration application, and then corrected for perspective distortions using MATLAB's *fitgeotrans* and *imwarp* commands. Images captured using the UV and visible filters were checked for alignment by hand and then converted into linearized and normalized measures of reflectance, as explained under "Quantitative visual signal analysis" below.

To account for distance-dependent blurring due to each viewing species' visual acuity [95,96], multispectral images with linearized intensities were spatially filtered before analysis to model the effect of viewing distance on contrasts between feathers and background foliage, and its effect on contrasts within the patterned eyespot feathers (See details in S4 Appendix). While peahens view peacock courtship displays at nearby distances ≥ 1 to 2 m [24], we also modeled a variety of greater viewing distances (2, 4, 8 and 16 m). Color patches were defined by hand in the original images and used for each modeled distance for uniformity. Following [97], to ensure that intensity samples were independent, we sampled each image using a square grid with spacing equal to a visual acuity disk, after spatial filtering and before color and brightness analysis. To model the effect of spatial filtering on the peacock's blue head, neck and breast plumage, we used an image with green foliage background with an approximately peacock-shaped cutout of the blue plumage superimposed; spatial filtering was performed using peacock body dimensions [98] to define the composite image's effective spatial scale.

Quantitative Visual Signal Analysis

The color and brightness contrast perceived by an animal depends on the reflectance of adjacent patches as well as the sensitivities of the animal's photoreceptor cone types. Quantitative models for computing these contrasts have been developed and well-validated (see review in [99]). We computed the color contrast, ΔS_c, between color patches in the feathers and background vegetation in our multispectral images using the receptor noise limited color opponent model, which has been shown to predict behavioral thresholds for visual signals in birds, humans and insects [100]. All calculations were performed using a custom MATLAB script, which was tested by verifying that it computed the same values as the multispectral analysis software package MICA version 1.22 [53]. First, intensity values, V, from each multispectral image were corresponded to the actual reflected irradiance, R, for this camera by an S-log transformation:

$$R(V) = Ae^{-\frac{V}{T_o}} + C. \qquad (1)$$

The parameters A, T_o and C were obtained from nonlinear least squares fits in MATLAB (adjusted-$R^2 \geq 0.997$) of the measured V and R values for each pixel in each RGB channel of the image for the 8-step grayscale. The resulting fits then were used to convert measured intensity values for each pth color patch into linearized and normalized reflected intensities (range [0,1]) for each combination of filter and RGB image channel. To compute the color and brightness contrasts, these intensities were converted into the cone quantum catch values, Q_{pr}, for each of the viewer's rth cone photoreceptors:

$$Q_{pr} = \int I(\lambda)R_p(\lambda)S_r(\lambda)d\lambda / \int I(\lambda)S_r(\lambda)d\lambda, \qquad (2)$$

where $I(\lambda)$ is the illumination spectrum, $S_r(\lambda)$ is the rth cone receptor's normalized spectral sensitivity and $R_p(\lambda)$ is the pth patch's reflectance spectrum. Because birds and mammals are known to achieve color constancy under a wide variety of illumination conditions [101,102], this equation also incorporates the von Kries transformation, a mechanism for maintaining color constancy [103]. To accomplish this conversion, we used MICA to compute the parameters of a polynomial cone mapping between the UV filter blue channel and the visible filter RGB channels of the multispectral images recorded by our filter-camera system and the corresponding cone quantum catches, Q_{pr} [53,86]. This software finds the optimal mapping using our measured filter transmission and camera RGB spectral response curves with either the dichromatic ferret or tetrachromatic peafowl cone spectral sensitivities, the CIE D65 illumination spectrum and a large database of natural spectra. The net effect is to combine all measured values of linearized and normalized reflectance to compute the quantum catch, Q_{pr}, of each rth cone (r = S or L for dichromats and r = VS, SWS, MWS, or LWS for tetrachromats) for the pth sample color patch. Using a linear 2-way interaction cone mapping model, we obtained a near perfect fit for each visual system: ferret ($R^2 \geq 0.999$), peafowl ($R^2 \geq 0.996$) and blue tit (≥ 0.990 UVS cone, ≥ 0.998 all other cones).

The resulting cone quantum catch values, Q_{pr}, can be used to compute normalized color space coordinates, for the pth color patch: qp=Qpr∑rQpr. For tetrachromats, the receptor index r = VS or UVS, SWS, MWS, LWS and q_p = (v,s,m,l), while for dichromats r = S, L and q_p = (sw,lw). After normalization, this corresponds to a three-dimensional tetrachromat color space for birds and a one-dimensional colorspace for dichromats, here chosen to rely on sw. Following [104], to validate the results of our multispectral imaging code, we compared dichromat color space sw coordinates computed by both MICA and our MATLAB code (sw_M) from our multispectral images with those computed directly from reflectance spectra (sw_R) for six color chart squares. Use of the camera and UV/visible filter cone mapping model was validated for multispectral image analysis by the goodness of the linear fit, zero intercept and unit slope, between the two sets of color space measures gave $sw_M = (0.018 \pm 0.031) + (1.00 \pm 0.07) \times sw_R$ + (adjusted-$R^2 = 0.993$) (S1 Fig).

To compute color contrasts, ΔS_c, we first computed the rth cone's log-linear quantum catch (Weber-Fechner), $\log Q_{rp}$, for each pth patch. This was used to compute the difference in rth cone response for the pqth patch pair, $\Delta_{rpq} = \log Q_{rp} - \log Q_{rq}$. The color contrast then is computed from differences between opponent cone pairs weighted by receptor noise. Dichromats have only S/L receptor opponency, so for them, $\Delta S_C = |\Delta_{Lpq} - \Delta_{Spq}| / \sqrt{e_L^2 + e_S^2}$ [100]. The corresponding equation for color contrast in tetrachromats is more complicated because all six possible combinations of the four single cones pairs should be considered [105]:

$$\Delta S_C^2 = \frac{\begin{pmatrix} (e_S e_{VS})^2(\Delta_L - \Delta_M)^2 + ((e_M e_{VS})^2(\Delta_L - \Delta_S)^2) + \\ (e_S e_M)^2(\Delta_L - \Delta_{VS})^2 + (e_S e_L)^2(\Delta_M - \Delta_{VS})^2 + \\ (e_S e_M)^2(\Delta_L - \Delta_{VS})^2 + (e_L e_M)^2(\Delta_{VS} - \Delta_S)^2 \end{pmatrix}}{(e_S e_M e_L)^2 + (e_{VS} e_M e_L)^2 + (e_{VS} e_S e_L)^2 + (e_{VS} e_S e_M)^2} \qquad (3)$$

For bright illumination levels, receptor noise is assumed to be a constant determined only by the Weber fraction, w_f and the relative population density, g_r, for each rth cone class [99]: $e_r = w_f / \sqrt{g_r}$. For peafowl, we used the value for chromatic Weber fractions of $w_f = 0.06$ for L cones for domestic chickens based on color discrimination [106]. Receptor noise values for the other single cone classes were estimated using mean peafowl relative population densities $g_r = (0.477, 0.892, 1.047, 1)$ for (VS, SWS, MWS, LWS) [107], yielding $e_r = (0.087, 0.064, 0.06, 0.06)$. For parrots, we used $g_r = 0.25{:}0.33{:}1.05{:}1$ and $w_f = 0.105$ found for spectral sensitivity in budgerigars [54], corresponding to $e_r = (0.210, 0.182, 0.102, 0.105)$. Because color discrimination has not been measured for non-human mammals [81], following [108] we used $w_f = 0.22$ found for brightness discrimination in domestic dogs (range 0.22–0.27) [109]. The relative cone population fractional densities measured for domestic cats [110] give a mean $g_r(S,L) =$

(0.12,1); similar ratios have been reported for various wild felids [111] and domestic dogs [112]. This gives the estimated predator receptor noise for color discrimination as $(e_S, e_L) = (0.64, 0.22)$.

The brightness contrast, ΔS_L, between each pq[th] pair of color patches was computed from the quantum catches, Q_{Lp} for the pth color patch for the spectral response for the luminance channel (double cones for birds and L cones for dichromat predators) using $\Delta S_L = (\log Q_{Lp} - \log Q_{Lq})/w_f$, where w_f is the Weber fraction for brightness discrimination. For birds, we used $w_f = 0.18$ measured for double cones in budgerigars [113]; for comparison, lower values 0.10 have been found for pigeons [114] and higher values ≥ 0.24 for chicks of the domestic chicken [115]. For predators, we used $w_f = 0.22$ for brightness discrimination in domestic dogs as explained above; for comparison, $w_f = 0.10$ in humans [81,116] and $w_f = 0.42$–0.45 for the horse, the only other terrestrial mammal for which data exist [81,117].

Color and brightness contrasts are interpreted in units of just noticeable distances (JND), with JND = 1 corresponding to the threshold for two patches to be discriminable under ideal illumination and viewing conditions when suitable data exist for the visual system being modeled [100,106]. Behavioral studies have shown that birds detect colorful fruit at a rate that correlates with increasing color (but not brightness) contrast for values >> 1 JND [118], while in lizards, the probability of discriminating a color from its background was found to be < 20% at 1 JND and to scale approximately linearly over the range 1 ≤ JND ≤ 12 [119]. Behavioral tests in zebra finches have found that color contrast detection thresholds range from JND = 1 to 2.5 to 3.2, depending on background color [120]. Following [121], we therefore assume that the contrast detection threshold is approximately JND = 1 and we define contrasts in the range 1 < JND ≤ 3 as weakly detectable.

Simulated Avian and Mammalian Predator Images

Using MATLAB, we also generated "false color" images intended to provide human viewers with a simulation of the colors, intensity and contrasts perceived by a different visual system. The intensity in each color channel on a false color image was calculated from the square-root transformed quantum catch measured for the appropriate cone species; this intensity mapping corresponds approximately to the human perception of brightness [122]. Following [53], we represented the tetrachromatic vision of birds using two images: 1) an RGB image created from the transformed LWS, MWS and SWS cone catch data, respectively; 2) a fourth grayscale image created from the transformed VS or UVS cone catch. After [19], we represented the two color channels of dichromatic mammals using a single RGB image by setting the blue channel equal to the transformed S cone quantum catch and the yellow (red + green) channel equal to the transformed L cone quantum catch. To accompany the texture analysis, we also generated an additional luminance-only grayscale image using the transformed luminance cone quantum catch.

To compensate for the greatly reduced color and brightness contrast discrimination value for the dichromatic predators relative to humans, we reduced the brightness and color contrasts of dichromat images as follows. Dichromat false color images generated as described above were first converted into CIE L*a*b* colorspace [122]. We then adjusted the brightness contrast for both the luminance and yellow-blue images. To do so, we rescaled the L* (luminance) channel to a new value L^*_{fc} so as to maintain the same mean intensity over the entire image while reducing the contrast. This was accomplished using the formula: $L^*_{fc} = w_{Lh} / w_{Ld} \times (L^* - L^*_{m}) + L^*_{m}$, where L^*_{m} = the mean luminance in the image, and the brightness contrast Weber fractions were $w_{Lh} = 0.14$ for humans and $w_{Ld} = 0.22$ for dogs [81]. For the yellow-blue images only, we also rescaled the color contrast by multiplying the b (blue-yellow opponency) channel in the L*a*b* image by w_{Ch} / w_{Cd}, where $w_{Ch} = 0.06$, the color contrast Weber thresholds for humans [106], and $w_{Cd} = 0.22$, the value for dogs: $b_{fc} = b \times w_{Ch} / w_{Cd}$. (The red-green opponency channel, a, was zero everywhere in these false color images since the red and green channels are set equal.) The resulting images reflected the reduced contrast perceived by a dichromatic mammal with the given Weber fraction.

Pattern Analysis

To model the perception of visual texture of the peacock's train viewed against foliage, we also performed granularity pattern analysis on the model peacock train photographs, using MATLAB code adapted from [34] and MICA's granularity texture analysis package [53]. In granularity analysis, an image based on the luminance channel is filtered using an FFT bandpass filter centered at a series of spatial frequencies (granularity bands). For each bandpass-filtered image, the "pattern energy" (a measure of information at each spatial scale) is computed as the standard deviation of its pixel intensity values. The "granularity spectrum" then is defined as pattern energy vs granularity band. Granularity analysis was performed on the model peacock train images processed for the dichromatic predator luminance channel as explained above. Granularity spectra were computed for polygonal regions of interest (ROI) encompassing the entire model train and each type of surrounding vegetation (i.e., tall grass, brush or trees). To compensate for the effect of ROI shape and background, we used the following method adapted from MICA. For each ROI, we first computed a masked image in which all regions outside the ROI were replaced by a black background. Next, we created a mean masked image in which the region inside the ROI in the masked image was replaced by the mean intensity within the ROI. Identical granularity calculations were performed on both images and their difference was used to create a shape-independent granularity spectrum.

We also computed granularity spectra and summary statistics for comparing textures of the model train and its background [32,34]. The summary statistics comprised: total energy (the energy summed across all filter bands, which increases as pattern contrast increases), peak filter size (the granularity band at peak energy; larger peak filter size corresponds to smaller most prevalent feature size), and proportion energy (the maximum energy divided by the total energy, a measure of how much of the spectral energy lies at the most prevalent feature size; this decreases with increasing pattern scale diversity). Granularity spectra were plotted as "normalized energy" (pattern energy divided by total energy) vs granularity band to give a measure of how pattern information is distributed across spatial scales. Images with a uniform distribution of pattern scales have correspondingly uniform granularity spectra, while images dominated by a single feature scale should have strongly peaked spectra.

Edge detection of the luminance channel image provides an alternative measure of pattern complexity. The basic idea is that image processing algorithms can find edges in good agreement with human perception by calculating the local maxima of image intensity gradients [122]. We used the Canny edge filter in MATLAB to find edges using sigma = 3 and threshold = 0.15 to 0.20 (relative to maximum luminance image intensity set to 1) [80,123]. Model train images were first log transformed and then processed using contrast-limited adaptive histogram equalization [124] (*adapthisteq* in MATLAB) to detect texture edges in regions of widely differing illumination. The edge fraction (percentage of edge pixels in each ROI) was then used to compare the model train with various types of vegetation in the background [36]; higher edge fractions indicate a more complex pattern with more spatial features.

To provide a source of images for pattern analysis that correspond directly to foliage in the peacock's native habitat, we searched for photographs of Indian peacocks in the Macaulay Library at the Cornell Lab of Ornithology and the Internet Bird Collection [125] that showed peacocks with full trains in a variety of orientations with eyespots visible against native green vegetation. Photographs were used only if they did not appear to have been selectively contrast or color enhanced, and if images of both birds and vegetation were unblurred and of sufficient resolution. A total of 14 images satisfied these conditions. While we could not use these photographs for multispectral analysis because they lack the required calibrations, we could determine a good approximate to the luminance using the green image channel because the green channels of digital cameras have spectral sensitivities that approximate that of L cones (e.g., Fig 2D and [86]). Consequently, granularity and edge detection analyses were performed on grayscale images based on the green channel of each photograph. Because photographs were taken from a variety of distances from the peacocks, we normalized the peak filter size (most prominent spatial feature scale) for each photograph by the peak filter size for the peacock's train to allow comparison between photographs.

Statistical Analysis

Our analysis of color and brightness contrasts followed the two-step process recommended in [126]. First, to determine whether the mean color and brightness contrasts between each patch pair had a statistically significant difference given their variances, we used PERMANOVA modified for non-normal, heterogeneous data [127] implemented in the software package FATHOM [128] using 1000 bootstrap samples. Note that is it possible for this difference to be statistically significant, but to have a value too small for it to be perceptually distinguishable. We therefore determined the effect size (how perceptually distinct each color patch pair) as follows. We first drew with replacement 1000 bootstrapped sample pairs using the MATLAB command *datasample*, and computed the mean ΔS_C and ΔS_L for this bootstrap resample. These mean contrasts were averaged over all images to get the mean and s.e.m. for each color patch pair for each sample; the grand mean and s.e.m. then was computed by averaging over all replicates for each feather type. Grand means and s.e.m. for the texture summary statistics were calculated from the mean of each statistic taken over all model train data for the train, grass, brush, and tree foliage. All results are reported as grand mean [95% CI = 2 s.e.m].

Data Accessibility

All data and software required in order to replicate all of our results are archived either in the supplemental materials or at https://figshare.com/s/688fb19dad98b6273324.

RESULTS

Literature review of Predation on Peafowl

A review of the literature did not uncover evidence that peacocks with full trains experience an enhanced predation risk, but rather supports the idea that wild adult peafowl are preyed on infrequently by mammals in their native habitats and that the highest risk of predation is primarily during the first year [129]. (Note that peafowl have different plumage in their first year and that males do not develop the fully grown train and eyespots (the focus of this study) until their third or fourth year.) For example, one study found that wild peafowl are preyed on far less by leopards than expected given that they were the most abundant prey species in the region studied [130], another that dholes and tigers preyed on peafowl less than expected based on their density [131] and a third that peafowl were a "significantly avoided prey species" in a review of prey selection by tigers [132]. In addition to these results for wild peafowl in native habitats, one survey of a feral peafowl population [133] found that, among adult peacocks with full-grown trains, the males that were preyed on tended to have relatively small trains and lower mating success. Another study of feral peafowl found that adult peacocks preyed upon by domestic dogs and red foxes (*Vulpes vulpes*) had half the predation risk as adult female peafowl and the same number of eyespots and mating success and a slightly longer train ($5.2 \pm 2.4\%$, based on 3 predated males) as surviving adult males [134]. Adult peacocks are reported to have several effective anti-predator strategies, including running [135], flight [136], fighting with their sharp spurs [28], hiding in dense thickets [29,129,135], roosting in high trees chosen for their protection against predators at dusk [129,137], and using group vigilance along with alarm calling [138]. Consistent with this, three studies have shown that the peacock's train does not significantly hinder locomotor performance during flight or walking; indeed, peacocks walking on a treadmill were efficient compared to other avian bipeds for which data exist and had a reduced metabolic cost during the breeding season when their trains were fully grown [136,139,140].

Reflectance Spectroscopy

Reflectance spectra for peacock eyespot feather color patches (Fig 2D) and peacock iridescent blue plumage (Fig 2E) had spectral peaks consistent across repeated measured to 4 to 16 nm (95% CI) and exhibited similar spectral peaks and overall shape to those measured for zero elevation angle [22,23,30]. A comparison of cone spectral sensitivity data (Fig 2C) with these reflectance spectra show that the peacock SWS cone is well matched to reflectance from its iridescent peacock blue plumage. While the spectral peaks for the bronze (BZ), blue-green (BG) and outer loose green (GB) barbs agree well with its SWS, MWS and LWS cone spectral ranges, all three also coincide with the same L cone sensitivity for the predator. Both the predator and peafowl VS cone spectral sensitivities also overlap with the reflectance spectrum from the eyespot's central BB and PB dark violet patches. However, reflectance from these features is weak compared to the other color patches, all of which also reflect weakly in the UV. Fig 2E shows the reflectance spectrum of a representative green leaf, illustrating how its peak at approximately 550 nm and its overall spectra resemble that of the peacock's loose green barbs; similar spectra for green background foliage have been reported in (Cazetta, Schaefer, and Galetti 2009; Loyau et al. 2007).

Reflectance spectra for parrot feathers (Fig 3E) agreed with previously published values [141], confirming a good spectral match between yellow and red feather reflectance with MWS and LWS avian cone sensitivities. The predator L cone sensitivity spans a spectral range corresponding to longer wavelength reflection from both yellow and red pigments (Fig 3D). Both the yellow and red patches also have reflectance peaks in the UV with a better overlap with bird UVS cone sensitivity than that of VS cones.

A comparison of the peak spectral sensitivities of S and L cones for canids, domestic cats and ferrets with those for the four single cone populations of 21 bird species from 8 different orders [142] (Fig 5) illustrates that predator S cones have a peak response similar to that of bird VS, but not UVS, cones. The predator S peak values lie between most peaks of the avian UVS/VS and SWS cone populations, whereas predator M peak values lie between avian MWS and LWS peak values.

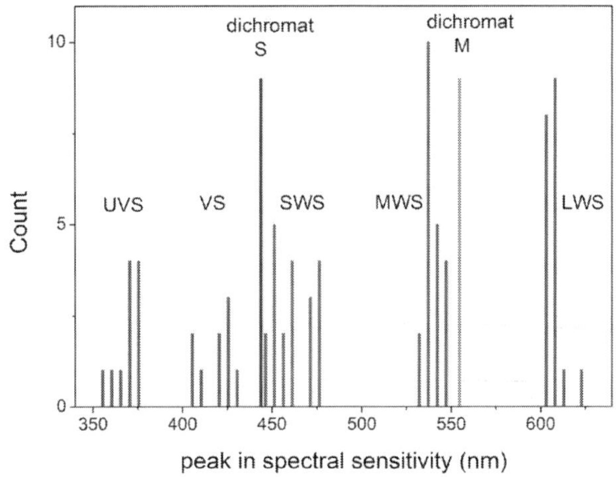

Figure 5. Comparison of peak spectral responses of predator and bird cones. Peak single cone spectral sensitivities for ferret S and L cones (Douglas & Jeffery, 2014) and for bird VS/UVS, SWS,MWS and LWS cones from Fig 5B in (Hart & Hunt, 2007) for 21 species of birds from 9 orders.

Color and brightness contrast analysis

False color images and analyses using the receptor noise model of visual discrimination are shown in Figs Figs6–9; all data and PERMANOVA pseudo-F and P values are reported in S1–S4 Datasets.

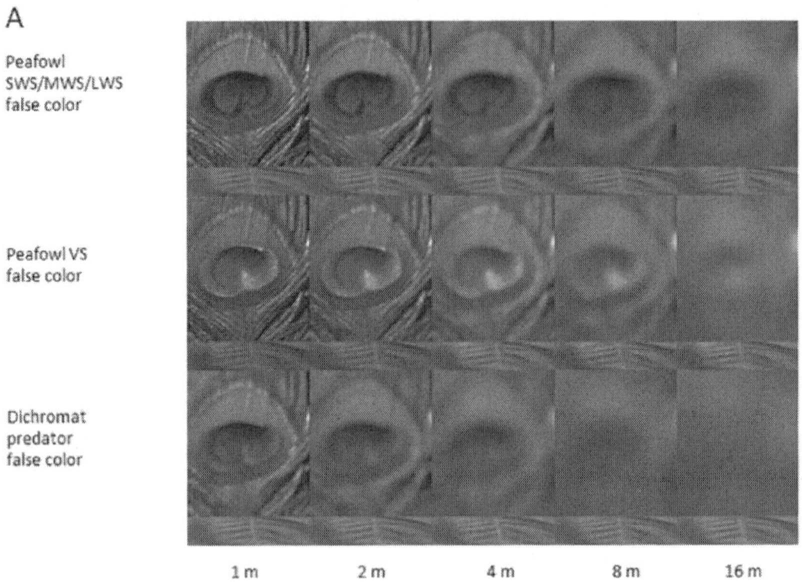

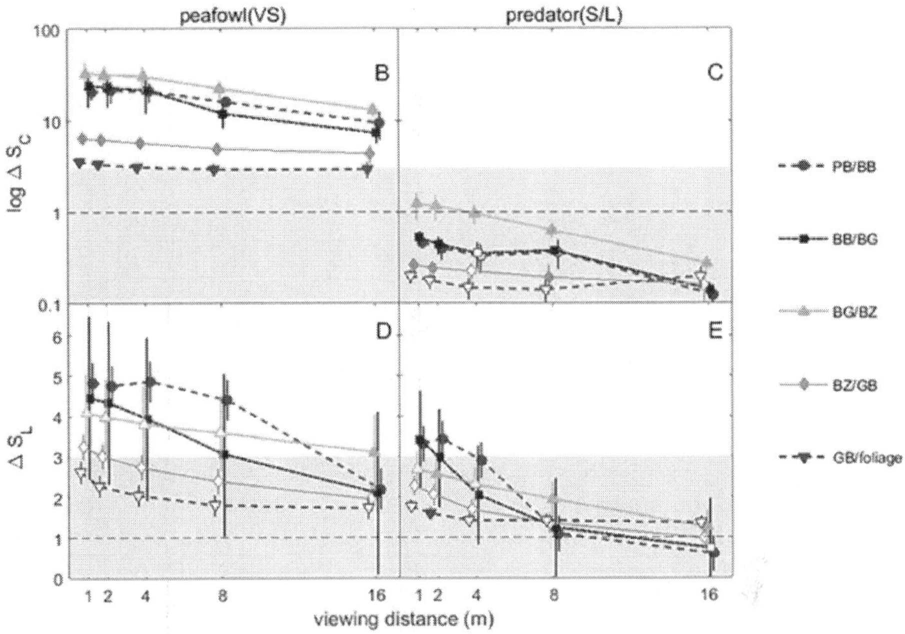

Figure 6. False color images and color and brightness contrast analysis of peacock eyespot feathers. (A) False color images modelling peafowl and dichromatic mammalian predator vision of peafowl eyespot and green leaf (inset at bottom of eyespot image) for different viewing distances. Note that the false color images should be considered as a relative guide and not an absolute indication of the detectability of contrasts because humans have better contrast thresholds by a factor of 3.7 for color and 2 for brightness compared to dichromatic mammals. (B)-(E) Estimated color (delta S_C) and brightness (delta S_L) contrasts for adjacent color patches on the peacock eyespots and green vegetation, over a range of viewing distances. All data are shown as grand means with 95% CI error bars. Contrasts corresponding to the same distance have been displaced by horizontal jitter to avoid overlap. Data above the 1 JND line are above the expected threshold for discrimination and contrasts within the grey shaded regions are at most weakly detectable. Closed symbols indicate contrasts that are statistically significant in each organism's colorspace (i.e., PERMANOVA P < 0.05); note that contrasts that are not statistically significant (closed symbols) due to their large, overlapping variances in the corresponding colorspace may still have mean values greater than the detection threshold.

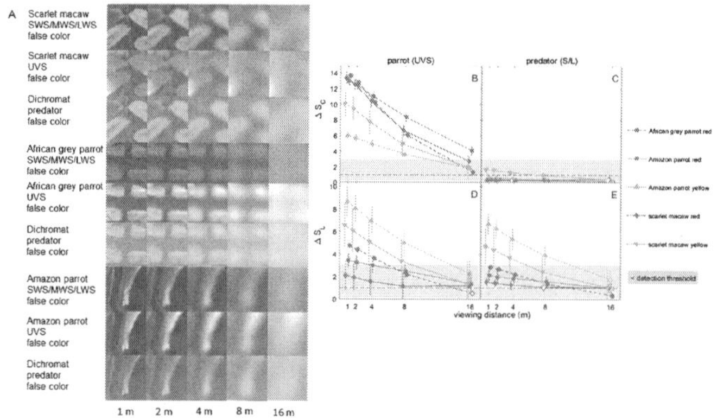

Figure 9. False color images and color and brightness contrast analysis of parrot feathers. (A) False color images in parrot ultraviolet sensitive (UVS) and dichromatic mammalian predator vision of scarlet macaw, African grey parrot and Amazon parrot red and yellow feathers vs green leaf for different viewing distances. (B)-(E) Color and luminance contrasts for parrot feather colors relative to green vegetation, over a range of viewing distances. See Fig 6 caption for further details.

False color images for various simulated viewing distances are displayed for peacock eyespot feathers in Fig 6A and visual signals are plotted vs distance in Fig 6B–6E. In peafowl vision, all pairs of adjacent color patches in the peacock's eyespot give large, statistically significant color contrasts > 3 JND for all distances. The greatest color contrasts were between the blue-green patch and surrounding rings and between the two central pupil-like patches; for some distances ≤ 8 m these same pairs of color patches also had statistically significant brightness contrasts in the 1–3 JND low detectability range. By contrast, in dichromat vision none of the eyespot patch pairs had color contrasts above 1 JND, and only the three innermost pairs of eyespot patches had brightness contrasts that were in the weakly detectable 1–3 JND range.

For peafowl vision, at all distances the model peacock train had statistically significant color and brightness contrasts that were > 3 JND for brush and trees, but not grass (Fig 7; additional false color images in S4 Fig). In dichromat predator vision, all color contrasts for the model train were < 1 JND and brightness contrasts were in the weakly detectable 1–3 JND range. Peacock blue plumage was found to be perceptually detectable by conspecifics at all distances and to lie in the weakly detectable 1–3 JND range for dichromat vision (Fig 8). The false color images of peacock feathers (Figs (Figs,6, ,7 and and,8) demonstrate how color signals relative to background vegetation are diminished when the single dichromat L cone replaces the separate SWS/MWS/LWS cones for birds, especially at larger distances.

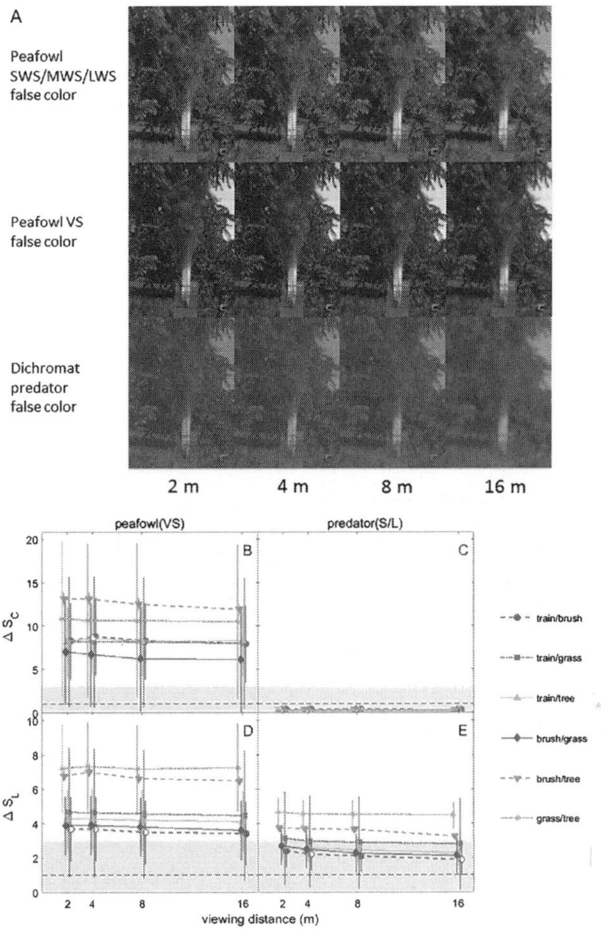

Figure 7. False color images and color and brightness contrast analysis of the peacock model train photographed against various types of vegetation backgrounds. (A) False color images in peafowl and dichromatic mammalian predator vision of peafowl model train for different viewing distances. (B)-(E) Color and luminance contrasts for the model train and features of vegetation, over a range of viewing distances. All data are shown as grand means with 95% CI error bars. See Fig 6 caption for further details.

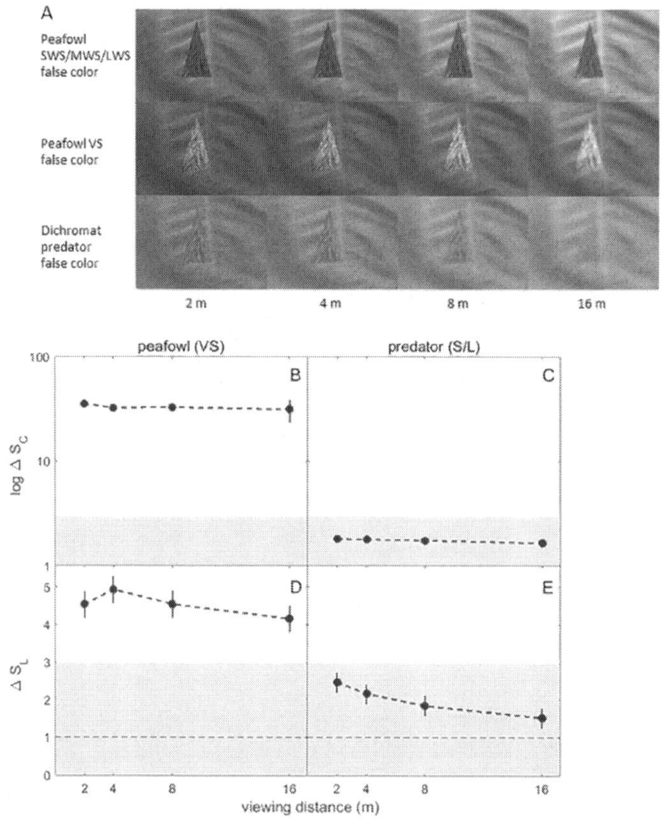

Figure 8. False color images and color and brightness contrast analysis of peacock blue neck feathers used to model the body's appearance against green foliage. (A) False color images in peafowl and dichromatic mammalian predator vision of peacock blue breast plumage vs green foliage for different viewing distances. (B)-(E) Color and luminance contrasts for the blue plumage relative to green vegetation, over a range of viewing distances. See Fig 6 caption for further details.

Red and yellow parrot feather color patches exhibited large, statistically significant contrasts in avian UVS vision in general, with color contrasts > 3 JND for ≤ 8 m and brightness contrasts > 1 JND for most samples (Fig 9). By contrast, for dichromat vision, none of the red parrot patches and the African grey parrot and scarlet macaw yellow patches had color contrasts > 1 JND, and the Amazon parrot feather yellow patches just exceeded 1 JND for ≤ 4 m. For distances ≤ 8 m, red parrot feather patches had brightness contrasts in the weakly detectable 1–3 JND range for ≤ 8 m and yellow patches had mean values in the range 2.4–6.7 JND.

We also measured the contrasts between leaves from the background leaf used (saucer magnolia) and seven other plant species with different shades of green. For dichromat mammal vision, the contrasts for these leaf pairs were $\Delta S_C = 0.55$ [0.42, 0.69] and $\Delta S_L = 2.53$ [2.01, 3.06] (mean [95% CI]). (S3 Fig). These values correspond to an additional source of uncertainty for the feather visual contrasts. These between-leaf color contrasts are too small to change the conclusions for the feather color contrasts discussed above. The values for brightness contrast suggest that the brightness contrasts for very dark or very light feathers viewed against very light or very dark green foliage, respectively, could correspond to the readily detectable range. The corresponding values for tetrachromatic bird vision ($\Delta S_C = 2.75$ [2.23, 3.28] and $\Delta S_L = 3.30$ [2.64, 3.97]) are also small compared to most of the corresponding feather-foliage JND values, consistent with birds being able to detect the high JND feather patches relative to leaves over many distances.

Pattern Analysis

Fig 10 and S4 Fig show false color images and granularity spectra for the multispectral images of the model train and different types of background vegetation (tall grass, brush and trees) for the various viewing distances modeled (S5 Dataset). Table 1 gives summary statistics for the pattern analysis of online photographs and S6 Fig shows the corresponding granularity spectra. Because the spatial frequency of objects in an image as well as an animal's visual field depend on distance, we would expect objects with similar textures observed at different distances to have similarly-shaped spectra, but possibly different frequency peaks and widths. Granularity spectra for the model train (Fig 10B) indeed had the same shape as those for background vegetation in that each had a single broad peak for granularity band values > 3; similar results were found for trains of live peacocks photographed against native habitats (S6 Fig). The peak spatial frequencies of each granularity spectrum moved to lower values as viewing distance increased, as expected from the blurring of fine scale features (Fig 10C). For all distances, the model train and background vegetation had values of proportion energy, peak frequency and total energy that agreed at the 95% CI, with the only exception that the proportion energy for distances > 4 m differed between the model train and trees (Fig 10D–10F). Similarly, the summary statistics for pattern analysis of photographs of the trains of live peacocks and native vegetation all agreed at the 95% CI level (Table 1). Collectively, these results demonstrate that the peacock's train is an excellent match for the predominant feature size distribution, overall contrast and pattern scale diversity of a variety of background vegetation. Moreover, visual examination of the edge detected images (Fig 10A) supports the finding that the calculated edge fractions for the model train and trains of live peacocks and background foliage agreed at the 95% CI level at all distances (Fig 10G, Table 1).

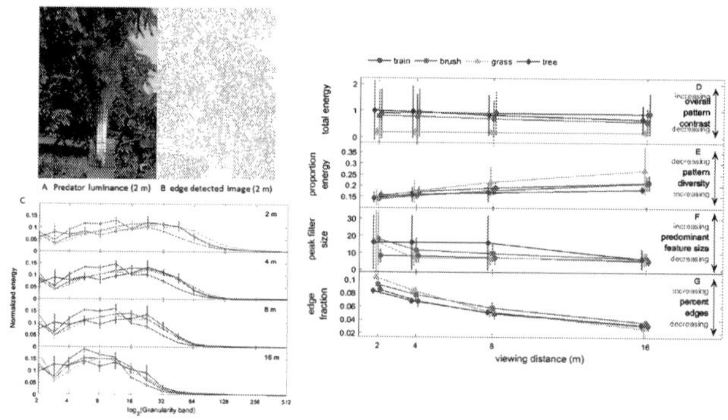

Figure 10. Texture analysis of the model peacock train photographed against various vegetation backgrounds. (A) Image based on dichromatic mammalian predator luminance channel and (B) result of edge detection on the luminance image. (C) Granularity spectrum for the model peacock train and three different regions of vegetation in the background vs viewing distance. (D) Total spectral energy summed over the granularity spectrum, which gives a measure of overall pattern contrast. (E) Proportional energy, a measure of how much the dominant feature size dominates and hence pattern diversity; (F) spatial frequency at peak spectral energy, which is inversely proportional to the predominant feature size; (G) edge fraction, the proportion of the image corresponding to edges. Data are grand means for all model train images and error bars show 95% CI.

Table 1. Texture analysis results for images from online sources (grand means [95% CI]).

	Total energy	Proportion energy	Peak filter (normalized)	edge fraction
train	0.47 [0.05, 0.88]	0.13 [0.12, 0.14]	1	0.036 [0.015, 0.056]
brush	0.65 [0.34, 0.96]	0.13 [0.12, 0.14]	2.0 [1.0, 3.0]	0.037 [0.014, 0.061]

DISCUSSION

The results of our study show that sexually-selected color signals readily detectable by conspecifics are not necessarily conspicuous to mammalian predators. Instead, for all distances considered, the color and brightness contrasts for all feather samples studied here relative to green foliage were much greater for birds than for dichromatic mammals. For all viewing distances modeled here, most feather samples had color contrasts in dichromatic predator vision that were perceptually indistinguishable from background vegetation; two exceptions were in the low detectable range: the peafowl's blue plumage (color contrast range [1.6, 1.8] JND) and the yellow scarlet macaw feather patches (color contrasts [1.6, 1.7] for ≤ 2 m). Unsurprisingly, the same feathers were highly conspicuous to conspecifics: their color contrasts were comparable to values found for avian visual modeling for fruit viewed against green foliage [118,143]. The brightness contrasts for these feathers vs background foliage in dichromatic predator vision were on the whole greater than the corresponding color signals, although only values for yellow exceeded the weakly detectable 1–3 JND range. This suggests that patterns with high brightness contrast, such as those created by white and dark melanin-pigmented plumage, might be more readily detectable by dichromat predators than color signals, and thus represent a greater detection risk [144]; such brightness-based visual signals also would presumably be more readily detectable for low light conditions. While the interpretation of supra-threshold color and brightness contrasts is still debated [145], our results show that such supranormal stimuli remain detectable by conspecifics and other birds even at large distances where carnivores cannot perceive them.

Peacock Plumage in Dichromatic Predator Vision

Focusing now on peacock eyespots, the large color contrasts for peafowl vision arise from spectral tuning between the reflectance spectra of each peacock eyespot color patch and peafowl single cone spectral sensitivities, similar to the agreement reported earlier between red and yellow pigment reflectance spectra and tetrachromatic UVS cone responses for parrot plumage and vision [141]. It is especially notable that the greatest color contrasts are due to the blue-green ring, since its iridescence has been found to correlate with peacock mating success [22,23], and its chromatic contrast was calculated to be the most salient signal in images of a displaying peacock [146].

When we computed measures for the model peacock train against a foliage background in dichromatic predator vision, the train feathers were found to have below detection threshold color contrasts and brightness contrasts in the low detectability 1–3 JND range, similar in magnitude to those for various types of green vegetation. Pattern analysis also indicated that peacock train feathers have similar textures to many types of vegetation in their native habitats. Taken together with the eyespot and textural analysis results, this indicates that dichromatic mammalian predators are likely unable to discriminate the peacock's train from green vegetation during foraging, although the eyespot's innermost features create low detectable brightness contrasts at nearby distances.

The peacock's blue plumage had large, detectable levels of color contrast at all distances for peafowl vision, though both color and brightness contrasts were in the low detectable range for dichromatic predator vision. Thus, the blue head, neck and breast contour feathers may represent a greater visual signal for distant conspecifics, as well as a greater predation risk, than the much larger train; however, all of these values are likely less conspicuous when forest shade diminishes their blue hue. Given that noniridescent blue feathers from other birds have been shown to have similar reflectance spectra to peafowl blue plumage, these results are likely generalizable to blue feathers from other species of birds [147].

In addition to color and pattern effects, salience can also be influenced by particular shapes. For example, some authors have speculated that the concentric shapes on the peacock's train plumage might be especially salient because of the resemblance to eyes [148]. Indeed, it has been shown that eyes and eye-like patterns are highly salient visual signals for birds, humans and domestic dogs [149–151]. However, our false color images show that peafowl eyespots do not always appear to have a central dark, circular pupil when viewed at typical display distances, either in peafowl or dichromatic mammalian predator visual models (Fig 11). Therefore, it is not obvious that peacock ocelli appear eye-like to nearby viewers. On the other hand, blurring at larger distances ≤ 8 m causes eyespots to appear to have pupil-like dark centers in both peafowl and dichromatic predator vision, indicating that they may indeed appear eyelike to predators when viewed at this intermediate range (Fig 11D–11F).

Figure 11. False color images of peacock eyespots for different visual models. At typical viewing distances during courtship displays, peacock eyespots have central features with brightness comparable to the surrounding rings in false color images for peafowl vision in the (A) LWS/MWS/SWS and (B) VS cone channels, as well as for dichromatic mammals (C). False color image of the model train at 2m in peafowl (D) LWS/MWS/SWS and (E) VS vision, and for (F) dichromatic mammalian vision show that the eyespots feature a darker pupil-like center.

Taken as a whole, these results do not support the hypothesis that the peacock's eyespot feathers are highly conspicuous for all viewers. The measured color and brightness contrasts and pattern textures indicate that eyespot feathers could serve as disruptive camouflage [31] under some conditions (e.g., for peacocks viewed in their native scrub jungles) by breaking up the train's outline and making it more difficult to distinguish from surrounding foliage. Indeed, [152,153] noted that even to humans the peacock's train can be well camouflaged against foliage in its native habitats. This could be useful when wild peafowl roost in trees at night and when they spend the hottest hours of the day in dense thickets of brush [129,135,154]. Based on such reports, it has been suggested [155] that peafowl eyespots originated as a form of camouflage for their native dappled light environments. Note, however, that during much of the day, peafowl perform many actions (e.g., foraging, calling and displaying) in open spaces, and during these times other cues are likely to render them conspicuous to predators. Our results suggest that their colors are not likely to add appreciably to their predation risk during these times.

Motion cues during peafowl displays and other behaviors might enhance the detectability of their visual signals, although evidence is mixed whether motion increases or decreases visual contrast thresholds [156]. On the other hand, the motions of the flexible loose green barbs in the train might also simulate that of background brush and grasses, a visual illusion studied in insects, crabs, spiders and lizards but not yet in birds [157]. To elucidate the effect of motion on visual signals during displays, future video studies could utilize cameras adapted for multispectral imaging provided with a filter that transmits light from the near-UV to 700 nm, allowing modeling of dynamic visual signals in dichromat vision.

These conclusions are consistent with the lack of evidence in the literature that peacocks with trains experience a significant, let alone an enhanced, risk of predation compared to, e.g., either peacocks without trains or peahens. The available evidence thus indicates that any handicap suffered by adult peacocks is likely to be incurred by factors other than visual signals created by their eyespot train feathers. For example, peacocks spend a large percentage of their time maintaining their plumage [158] and displaying [24,29]. Thus, the elaborate courtship displays of peacocks may correspond to handicaps due to visual signals from their blue plumage, time lost from foraging for food due to plumage maintenance and courtship displays, the male's likely inattention to predators during displays, and the metabolic demands of the male's courtship displays [159].

Red and Yellow Pigmented Plumage in Dichromatic Predator Vision

Considering now red and yellow feathers, we note that red plumage is at best weakly detectable given its sub-threshold color contrasts and low brightness contrasts when viewed by dichromatic mammals against green foliage (although yellow parrot feathers have brightness contrasts that should be more readily detectable by mammalian predators at close distances). A consideration of cone spectral sensitivities and feather reflectance spectra suggests two reasons for this difference. First, these yellow feathers had an overall higher reflectance than the corresponding red feathers, resulting in their having a higher brightness contrast relative to leaves. Second, yellow feather pigments reflect considerable light in the UV compared to red pigments, whereas UV reflectance is low for green plants. Yellow feathers thus stimulate both predator S and L cones, while green plants primarily stimulate the L cones, providing a mechanism for distinguishing yellow feathers from green foliage backgrounds.

These conclusions should hold for other birds with red and yellow plumage given that a wide variety of species of birds have similar color vision to the species considered here (Fig 5), and that feathers colored with carotenoid pigments have very similar reflectance spectra to the pigment psittacofulvin found in parrot feathers [37–39]. Thus, our findings indicate that many species of red and yellow feathered birds that appear conspicuous to other birds and humans may in fact be cryptic or poorly visible to predators because of background matching [31]. Our findings also have broader implications for interpreting how color cues, camouflage and possible eye mimicry appear to the majority of mammals. Trichromacy in primates has been suggested to have evolved for a variety of reasons [160], including detecting ripe fruit and immature leaves [161], breaking camouflage (e.g., during foraging for eggs) [35], sexual or social signaling [162], and predator detection [163]. Our results suggest that the evolution of trichromacy may also have provided catarrhine primates, howler monkeys and some marsupials with an advantage in detecting colorful birds, reptiles, amphibians and insects.

Visual Modeling: Limitations and Future Directions

The largest source of uncertainty in our analysis is the lack of behaviorally-measured Weber fractions for color contrast for terrestrial carnivorans for the conditions considered here, and the relatively few species for which peak cone spectral wavelengths have been measured for carnivoran mammals. This data also would be valuable for studies that often have had to rely on human visual modeling in analyzing egg camouflage [36] and the relationship between plumage, brightness and antipredator vigilance [164]. The most relevant measures would involve behavioral tests to determine whether these mammals can detect feathered model birds when other cues (e.g., olfactory) are controlled for. Any such studies ought to be sure to use illumination sources that include UV, as well as color cues that closely match the reflectance spectra of natural objects [108].

In spite of these limitations, it is important to note several factors that indicate feathers may have even lower visual contrasts in predator vision than computed here. First, our viewing and illumination geometries were chosen to optimize color and brightness cues, but predators will encounter prey under suboptimal conditions. Second, when feathers are viewed at low illumination levels, they also are likely to have lower color and brightness contrasts [165,166] and be more blurred due to reduced visual acuity [80]. Third, since UV reflectance helps distinguish feathers from green foliage, the reduction in UV irradiance in forest shade is likely to render feathers less detectable in forest shade than in direct sunlight [167]. Fourth, we also modeled only distance-dependent blurring due to visual acuity (retinal sampling), but a more complete treatment would use each species' behaviorally measured contrast sensitivity function (CSF) [80] to account for the optics of the eye and other factors that have been determined for our study species [168]. While we lacked the data to perform this additional analysis, the additional blurring would make dichromatic predators even less likely to be able to detect the feathers than our estimated contrasts indicate. Fifth, we chose to compare feathers with relatively dark green leaves. The small variation in color contrasts measured between leaves with varying shades of green indicates our findings are generalizable. The variation in brightness contrasts between leaves suggests that bright feathers should be even less visible against light green foliage. Thus, our results may overestimate the detectability of these feathers by dichromatic mammals.

It is also worth considering why these models predict that dichromatic predators and birds perceive plumage color so differently. Ultraviolet vision per se does not result in these differing visual signals: these dichromatic mammalian predators have similar near-UV S cone spectral sensitivity to the VS cones of birds [55,169]. Indeed, as noted above, since red and yellow parrot feathers and the central patches on peacock feathers reflect appreciable UV light this may make these feathers more detectable by dichromats. It is therefore important to include UV reflectance in modeling of visual signals in dichromatic mammalian visual systems, as opposed to relying on image processing of human visible RGB photographs (Pongrácz et al. 2017). These results also are not merely a consequence of birds having more types of cones than carnivores: given the similar spectral response of dichromatic mammal S and L cones and avian VS and MWS cones, color patches could in principle generate similar contrasts in both visual systems. Instead, these feathers have low contrast in dichromatic mammal vision due to a combination of low visual acuity, higher receptor noise levels and poorer spectral discrimination over the L cone response range.

CONCLUSION

Darwin stated that "Even the bright colors of many male birds cannot fail to make them conspicuous to their enemies of all kinds" [2]. On the contrary, our study implies that some species of birds that appear vividly colorful to humans and other birds may appear drab and inconspicuous in the eyes of mammalian predators. This conclusion is consistent with our review of the literature on predation by wild felids and canids on peafowl and other birds. The detectability of colorful plumage by predator vision depends on specifics of pigmentation, photoreceptor response, and environmental context, as suggested by sensory drive theory [170]. Thus, the predation risk incurred by colorful plumage needs to be assessed on a case-by-case basis using behavioral studies in combination with measurements of visual signals for the predator of interest.

Predators have a variety of other means of detecting prey, including visual motion perception and sensing acoustic, tactile and olfactory cues. Our results highlight the importance of understanding how dynamic behaviors during multimodal displays, foraging and other activities make birds more apparent to mammals and other predators than do their seemingly-conspicuous colors alone.

ACKNOWLEDGMENTS

We wish to thank Cassie Stoddard for providing us with her edge detection code, Charles Chubb and Roger Hanlon for their granularity code, and Robert Beyer for instrumentation design.

REFERENCES

1. Andersson MB. Sexual selection. Princeton University Press; 1994.
2. Darwin C. The descent of man and selection in relation to sex. Murray; 1888.
3. Jabr F. How beauty is making scientists rethink evolution. New York Times Magazine. 2 Jan 2019: 22–48.
4. Ranjith V, Jose B. Habitat preference of Indian peafowl (Pavo cristatus) in selected areas of Palakkad district, Kerala, India. Current Science. 2016;110: 2177–2182.
5. Zahavi A. Mate selection—A selection for a handicap. Journal of Theoretical Biology. 1975;53: 205–214. 10.1016/0022-5193(75)90111-3
6. Endler JA. Natural Selection on Color Patterns in Poecilia Reticulata. Evolution. 1980;34: 76–91. 10.1111/j.1558-5646.1980.tb04790.x
7. Møller AP, Erritzøe J, Nielsen JT. Causes of interspecific variation in susceptibility to cat predation on birds. Chinese Birds. 2010;1: 97–111.
8. Cain KE, Hall ML, Medina I, Leitao AV, Delhey K, Brouwer L, et al. Conspicuous Plumage Does Not Increase Predation Risk: A Continent-Wide Test Using Model Songbirds. The American Naturalist. 2019;193: 359–372. 10.1086/701632
9. Munson ES. How Birds use Conspicuous Plumage to Elude Predators. The Passenger Pigeon. 1990;52: 311–314.
10. Ruiz-Rodríguez M, Avilés JM, Cuervo JJ, Parejo D, Ruano F, Zamora-Muñoz C, et al. Does avian conspicuous colouration increase or reduce predation risk? Oecologia. 2013;173: 83–93. 10.1007/s00442-013-2599-6
11. Outomuro D, Söderquist L, Johansson F, Ödeen A, Nordström K. The price of looking sexy: visual ecology of a three-level predator–prey system. Functional Ecology. 2017;31: 707–718. 10.1111/1365-2435.12769
12. Loss S, Will T, Marra P. Direct mortality of birds from anthropogenic causes. Annual Review of Ecology, Evolution, and Systematics. 2015;46: 99–120.
13. Cronin TW, Johnsen S, Marshall NJ, Warrant EJ. Visual Ecology. Princeton University Press; 2014.

14. Endler JA, Thery M. Interacting Effects of Lek Placement, Display Behavior, Ambient Light, and Color Patterns in Three Neotropical Forest-Dwelling Birds. The American Naturalist. 1996;148: 421–452. 10.1086/285934
15. Miller PE, Murphy CJ. Vision in dogs. Journal-American Veterinary Medical Association. 1995;207: 1623–1634.
16. Håstad O, Victorsson J, Ödeen A. Differences in color vision make passerines less conspicuous in the eyes of their predators. PNAS. 2005;102: 6391–6394. 10.1073/pnas.0409228102
17. Théry M, Debut M, Gomez D, Casas J. Specific color sensitivities of prey and predator explain camouflage in different visual systems. Behav Ecol. 2005;16: 25–29. 10.1093/beheco/arh130
18. Endler JA. Interactions between predator and prey. Behavioural ecology. 1991; 169–202.
19. Chiao C-C, Wickiser JK, Allen JJ, Genter B, Hanlon RT. Hyperspectral imaging of cuttlefish camouflage indicates good color match in the eyes of fish predators. PNAS. 2011;108: 9148–9153. 10.1073/pnas.1019090108
20. Nokelainen O, Hubbard N, Lown AE, Wood LE, Stevens M. Through predators' eyes: phenotype–environment associations in shore crab coloration at different spatial scales. Biol J Linn Soc. 2017;122: 738–751. 10.1093/biolinnean/blx101
21. Sumner P, Mollon JD. Colors of primate pelage and skin: Objective assessment of conspicuousness. American Journal of Primatology. 2003;59: 67–91. 10.1002/ajp.10066
22. Loyau A, Gomez D, Moureau B, Théry M, Hart NS, Jalme MS, et al. Iridescent structurally based coloration of eyespots correlates with mating success in the peacock. Behav Ecol. 2007;18: 1123–1131. 10.1093/beheco/arm088
23. Dakin R, Montgomerie R. Eye for an eyespot: how iridescent plumage ocelli influence peacock mating success. Behav Ecol. 2013;24: 1048–1057. 10.1093/beheco/art045
24. Dakin R, Montgomerie R. Peacocks orient their courtship displays towards the sun. Behav Ecol Sociobiol. 2009;63: 825–834. 10.1007/s00265-009-0717-6

25. Freeman AR, Hare JF. Infrasound in mating displays: a peacock's tale. Animal Behaviour. 2015;102: 241–250. 10.1016/j.anbehav.2015.01.029
26. Dakin R, McCrossan O, Hare JF, Montgomerie R, Kane SA. Biomechanics of the peacock's display: How feather structure and resonance influence multimodal signaling. PLOS ONE. 2016;11: e0152759 10.1371/journal.pone.0152759
27. Yorzinski JL, Patricelli GL, Babcock JS, Pearson JM, Platt ML. Through their eyes: selective attention in peahens during courtship. Journal of Experimental Biology. 2013;216: 3035–3046. 10.1242/jeb.087338
28. Petrie M, Tim H, Carolyn S. Peahens prefer peacocks with elaborate trains. Animal Behaviour. 1991;41: 323–331. 10.1016/S0003-3472(05)80484-1
29. Harikrishnan S, Vasudevan K, Sivakumar K. Behavior of Indian peafowl Pavo cristatus Linn. 1758 during the mating period in a natural population. The Open Ornithology Journal. 2010;3: 13–19.
30. Yoshioka S, Kinoshita S. Effect of macroscopic structure in iridescent color of the peacock feathers. Forma-Tokyo. 2002;17: 169–181.
31. Stevens M, Merilaita S. Animal Camouflage: Mechanisms and Function. Cambridge University Press; 2011.
32. Stoddard MC, Stevens M. Pattern mimicry of host eggs by the common cuckoo, as seen through a bird's eye. Proceedings of the Royal Society of London B: Biological Sciences. 2010; rspb20092018. 10.1098/rspb.2009.2018
33. Akkaynak D, Siemann LA, Barbosa A, Mäthger LM. Changeable camouflage: how well can flounder resemble the colour and spatial scale of substrates in their natural habitats? Royal Society Open Science. 2017;4: 160824 10.1098/rsos.160824
34. Barbosa A, Mäthger LM, Buresch KC, Kelly J, Chubb C, Chiao C-C, et al. Cuttlefish camouflage: The effects of substrate contrast and size in evoking uniform, mottle or disruptive body patterns. Vision Research. 2008;48: 1242–1253. 10.1016/j.visres.2008.02.011
35. Troscianko J, Wilson-Aggarwal J, Griffiths D, Spottiswoode CN, Stevens M. Relative advantages of dichromatic and trichromatic color vision in camouflage breaking. Behav Ecol. 2017;28: 556–564. 10.1093/beheco/arw185

36. Stoddard MC, Kupán K, Eyster HN, Rojas-Abreu W, Cruz-López M, Serrano-Meneses MA, et al. Camouflage and Clutch Survival in Plovers and Terns. Scientific Reports. 2016;6: 32059 10.1038/srep32059
37. McGraw KJ, Nogare MC. Distribution of unique red feather pigments in parrots. Biology Letters. 2005;1: 38–43. 10.1098/rsbl.2004.0269
38. Shawkey MD, Hill GE. Carotenoids need structural colours to shine. Biology Letters. 2005;1: 121–124. 10.1098/rsbl.2004.0289
39. Toral GM, Figuerola J, Negro JJ. Multiple ways to become red: Pigment identification in red feathers using spectrometry. Comparative Biochemistry and Physiology Part B: Biochemistry and Molecular Biology. 2008;150: 147–152. 10.1016/j.cbpb.2008.02.006
40. Dalrymple RL, Hui FKC, Flores-Moreno H, Kemp DJ, Moles AT. Roses are red, violets are blue–so how much replication should you do? An assessment of variation in the colour of flowers and birds. Biol J Linn Soc. 2015;114: 69–81. 10.1111/bij.12402
41. Dakin R, Montgomerie R. Peahens prefer peacocks displaying more eyespots, but rarely. Animal Behaviour. 2011;82: 21–28. 10.1016/j.anbehav.2011.03.016
42. Gokula V. Display behaviour of Indian Peafowl Pavo cristatus (Aves: Galliformes) during the mating season in viralimalai, TamilNadu, India. TAPROBANICA: The Journal of Asian Biodiversity. 2015;7.
43. Sharma IK. Ecological Studies of the Plumes of the Peacock (Pavo cristatus). The Condor. 1974;76: 344–346. 10.2307/1366352
44. Beebe W. A Monograph of the Pheasants [Internet]. Witherby & Co; 1918. https://www.jstor.org/stable/4072814
45. Gates DM, Keegan HJ, Schleter JC, Weidner VR. Spectral Properties of Plants. Appl Opt, AO. 1965;4: 11–20. 10.1364/AO.4.000011
46. Jensen JR. Remote sensing of the environment: An earth resource perspective 2/e. Pearson Education India; 2009.
47. Regan B. C., Julliot C., Simmen B., Viénot F., Charles–Dominique P., Mollon J. D. Fruits, foliage and the evolution of primate colour vision. Philosophical Transactions of the Royal Society of London Series B: Biological Sciences. 2001;356: 229–283. 10.1098/rstb.2000.0773
48. Jeganathan C, Dash J, Atkinson PM. Mapping the phenology of natural vegetation in India using a remote sensing-derived chlorophyll

index. International Journal of Remote Sensing. 2010;31: 5777–5796. 10.1080/01431161.2010.512303

49. Lee DW. Canopy dynamics and light climates in a tropical moist deciduous forest in India. Journal of Tropical Ecology. 1989;5: 65–79. 10.1017/S0266467400003229
50. Hart NS. Vision in the peafowl (Aves: Pavo cristatus). Journal of Experimental Biology. 2002;205: 3925–3935.
51. Spitschan M, Aguirre GK, Brainard DH, Sweeney AM. Variation of outdoor illumination as a function of solar elevation and light pollution. Scientific Reports. 2016;6: 26756 10.1038/srep26756
52. Hart NS, Partridge JC, Cuthill IC, Bennett ATD. Visual pigments, oil droplets, ocular media and cone photoreceptor distribution in two species of passerine bird: the blue tit (Parus caeruleus L.) and the blackbird (Turdus merula L.). J Comp Physiol A. 2000;186: 375–387. 10.1007/s003590050437
53. Troscianko J, Stevens M. Image calibration and analysis toolbox–a free software suite for objectively measuring reflectance, colour and pattern. Methods in Ecology and Evolution. 2015;6: 1320–1331. 10.1111/2041-210X.12439
54. Lind O, Chavez J, Kelber A. The contribution of single and double cones to spectral sensitivity in budgerigars during changing light conditions. J Comp Physiol A. 2014;200: 197–207. 10.1007/s00359-013-0878-7
55. Douglas RH, Jeffery G. The spectral transmission of ocular media suggests ultraviolet sensitivity is widespread among mammals. Proceedings of the Royal Society of London B: Biological Sciences. 2014;281: 20132995 10.1098/rspb.2013.2995
56. Guenther E, Zrenner E. The spectral sensitivity of dark- and light-adapted cat retinal ganglion cells. J Neurosci. 1993;13: 1543–1550. 10.1523/JNEUROSCI.13-04-01543.1993
57. Jacobs GH, Deegan JF, Crognale MA, Fenwick JA. Photopigments of dogs and foxes and their implications for canid vision. Visual Neuroscience. 1993;10: 173–180. 10.1017/S0952523800003291
58. Clark DL, Clark RA. Neutral point testing of color vision in the domestic cat. Experimental Eye Research. 2016;153: 23–26. 10.1016/j.exer.2016.10.002

59. Neitz J, Geist T, Jacobs GH. Color vision in the dog. Visual Neuroscience. 1989;3: 119–125. 10.1017/S0952523800004430
60. Kasparson AA, Badridze J, Maximov VV. Colour cues proved to be more informative for dogs than brightness. Proceedings of the Royal Society of London B: Biological Sciences. 2013;280: 20131356 10.1098/rspb.2013.1356
61. Osorio D, Vorobyev M. Photoreceptor spectral sensitivities in terrestrial animals: adaptations for luminance and colour vision. Proceedings of the Royal Society of London B: Biological Sciences. 2005;272: 1745–1752. 10.1098/rspb.2005.3156
62. Calderone JB, Jacobs GH. Spectral properties and retinal distribution of ferret cones. Visual Neuroscience. 2003;20: 11–17. 10.1017/S0952523803201024
63. Azlan JM, Sharma DSK. The diversity and activity patterns of wild felids in a secondary forest in Peninsular Malaysia. Oryx. 2006;40: 36–41. 10.1017/S0030605306000147
64. Carter N, Jasny M, Gurung B, Liu J. Impacts of people and tigers on leopard spatiotemporal activity patterns in a global biodiversity hotspot. Global Ecology and Conservation. 2015;3: 149–162. 10.1016/j.gecco.2014.11.013
65. Di Bitetti MS, De Angelo CD, Di Blanco YE, Paviolo A. Niche partitioning and species coexistence in a Neotropical felid assemblage. Acta Oecologica. 2010;36: 403–412. 10.1016/j.actao.2010.04.001
66. Hernández-SaintMartín AD, Rosas-Rosas OC, Palacio-Núñez J, Tarango-Arámbula LA, Clemente-Sánchez F, Hoogesteijn AL. Activity patterns of jaguar, puma and their potential prey in San Luis Potosi, Mexico. ACTA ZOOLÓGICA MEXICANA (NS). 2013;29: 520–533.
67. Kamler JF, Johnson A, Vongkhamheng C, Bousa A. The diet, prey selection, and activity of dholes (Cuon alpinus) in northern Laos. J Mammal. 2012;93: 627–633. 10.1644/11-MAMM-A-241.1
68. Karanth KU, Sunquist ME. Behavioural correlates of predation by tiger (Panthera tigris), leopard (Panthera pardus) and dhole (Cuon alpinus) in Nagarahole, India. Journal of Zoology. 2000;250: 255–265.
69. Lynam AJ, Jenks KE, Tantipisanuh N, Chutipong W, Ngoprasert D, Gale GA, et al. Terrestrial activity patterns of wild cats from camera-trapping. Raffles Bulletin of Zoology. 2013;61.

70. Odden M, Wegge P. Spacing and activity patterns of leopards Panthera pardus in the Royal Bardia National Park, Nepal. wbio. 2005;11: 145–152. 10.2981/0909-6396(2005)11[145:SAAPOL]2.0.CO;2
71. Horn JA, Mateus□Pinilla N, Warner RE, Heske EJ. Home range, habitat use, and activity patterns of free-roaming domestic cats. The Journal of Wildlife Management. 2011;75: 1177–1185. 10.1002/jwmg.145
72. Sunquist M, Sunquist F. Wild cats of the world. University of chicago press; 2002.
73. Castelló JR. Canids of the World: Wolves, Wild Dogs, Foxes, Jackals, Coyotes, and Their Relatives. Princeton University Press; 2018.
74. Lawson R, Fogarty D, Loss S. Use of visual and olfactory sensory cues by an apex predator in deciduous forests. Canadian Journal of Zoology. 2019;
75. Lawson R, Fogarty D, Loss S. Use of visual and olfactory sensory cues by an apex predator in deciduous forests. Can J Zool. 2019; 10.1139/cjz-2018-0134
76. Hughes NK, Price CJ, Banks PB. Predators Are Attracted to the Olfactory Signals of Prey. PLOS ONE. 2010;5: e13114 10.1371/journal.pone.0013114
77. Kelber A, Yovanovich C, Olsson P. Thresholds and noise limitations of colour vision in dim light. Philosophical Transactions of the Royal Society B: Biological Sciences. 2017;372: 20160065 10.1098/rstb.2016.0065
78. Khokhlova TV. Current views on vision in mammals. Biol Bull Rev. 2013;3: 347–361. 10.1134/S207908641305006X
79. Kang I, Reem RE, Kaczmarowski AL, Malpeli JG. Contrast Sensitivity of Cats and Humans in Scotopic and Mesopic Conditions. Journal of Neurophysiology. 2009;102: 831–840. 10.1152/jn.90641.2008
80. Melin AD, Kline DW, Hiramatsu C, Caro T. Zebra Stripes through the Eyes of Their Predators, Zebras, and Humans. PLOS ONE. 2016;11: e0145679 10.1371/journal.pone.0145679
81. Olsson P, Lind O, Kelber A. Chromatic and achromatic vision: parameter choice and limitations for reliable model predictions. Behav Ecol. 2017;29: 273–282. 10.1093/beheco/arx133
82. Pasternak T, Merigan WH. The luminance dependence of spatial vision in the cat. Vision Research. 1981;21: 1333–1339. 10.1016/0042-6989(81)90240-6

83. Meadows MG, Morehouse NI, Rutowski RL, Douglas JM, McGraw KJ. Quantifying iridescent coloration in animals: a method for improving repeatability. Behav Ecol Sociobiol. 2011;65: 1317–1327. 10.1007/s00265-010-1135-5

84. Van Wijk S, Bélisle M, Garant D, Pelletier F. A reliable technique to quantify the individual variability of iridescent coloration in birds. Journal of Avian Biology. 2016;47: 227–234. 10.1111/jav.00750

85. Vukusic P, Stavenga DG. Physical methods for investigating structural colours in biological systems. Journal of The Royal Society Interface. 2009;6: S133–S148. 10.1098/rsif.2008.0386.focus

86. Stevens M, Párraga CA, Cuthill IC, Partridge JC, Troscianko TS. Using digital photography to study animal coloration. Biol J Linn Soc. 2007;90: 211–237. 10.1111/j.1095-8312.2007.00725.x

87. Prutchi D. Exploring Ultraviolet Photography. Amherst Media; 2016.

88. Igoe D, Parisi A, Carter B. Characterization of a Smartphone Camera's Response to Ultraviolet A Radiation. Photochemistry and Photobiology. 2013;89: 215–218. 10.1111/j.1751-1097.2012.01216.x

89. Yun HS, Park SH, Kim H-J, Lee WD, Lee KD, Hong SY, et al. Use of Unmanned Aerial Vehicle for Multi-temporal Monitoring of Soybean Vegetation Fraction. Journal of Biosystems Engineering. 2016;41: 126–137. 10.5307/JBE.2016.41.2.126

90. Vogt MC, Vogt ME. RESEARCH ARTICLE: Near-Remote Sensing of Water Turbidity Using Small Unmanned Aircraft Systems. Environmental Practice. 2016;18: 18–31. 10.1017/S1466046615000459

91. Amlaner CJ, Ball NJ. A Synthesis of Sleep in Wild Birds. Behaviour. 1983;87: 85–119.

92. Lang SDJ, Mann RP, Farine DR. Temporal activity patterns of predators and prey across broad geographic scales. Behav Ecol. 10.1093/beheco/ary133

93. Robbins CS. Effect of time of day on bird activity. Studies in Avian Biology. 1981;6: 275–286.

94. Barreira AS, García NC, Lougheed SC, Tubaro PL. Viewing geometry affects sexual dichromatism and conspicuousness of noniridescent plumage of Swallow Tanagers (Tersina viridis). The Auk. 2016;133: 530–543. 10.1642/AUK-15-170.1

95. Barnett JB, Michalis C, Scott-Samuel NE, Cuthill IC. Distance-dependent defensive coloration in the poison frog Dendrobates tinctorius, Dendrobatidae. PNAS. 2018;115: 6416–6421. 10.1073/pnas.1800826115
96. Caves EM, Johnsen S. AcuityView: An r package for portraying the effects of visual acuity on scenes observed by an animal. Methods in Ecology and Evolution. 2018;9: 793–797. 10.1111/2041-210X.12911
97. Endler JA. A framework for analysing colour pattern geometry: adjacent colours. Biol J Linn Soc. 2012;107: 233–253. 10.1111/j.1095-8312.2012.01937.x
98. Talha MMH, Mia MM, Momu JM. Morphometric, productive and reproductive traits of Indian peafowl (Pavo cristatus) in Bangladesh. International Journal of Development Research. 2018;8: 19039–19043.
99. Renoult JP, Kelber A, Schaefer HM. Colour spaces in ecology and evolutionary biology. Biological Reviews. 2015;92: 292–315. 10.1111/brv.12230
100. Vorobyev M, Osorio D. Receptor noise as a determinant of colour thresholds. Proceedings of the Royal Society of London B: Biological Sciences. 1998;265: 351–358. 10.1098/rspb.1998.0302
101. Olsson P, Wilby D, Kelber A. Quantitative studies of animal colour constancy: using the chicken as model. Proc R Soc B. 2016;283: 20160411 10.1098/rspb.2016.0411
102. Kelber A, Osorio D. From spectral information to animal colour vision: experiments and concepts. Proceedings of the Royal Society of London B: Biological Sciences. 2010;277: 1617–1625. 10.1098/rspb.2009.2118
103. Stoddard MC, Prum RO. Evolution of Avian Plumage Color in a Tetrahedral Color Space: A Phylogenetic Analysis of New World Buntings. The American Naturalist. 2008;171: 755–776. 10.1086/587526
104. Troscianko J, Wilson-Aggarwal J, Stevens M, Spottiswoode CN. Camouflage predicts survival in ground-nesting birds. Scientific Reports. 2016;6: 19966 10.1038/srep19966
105. Kelber A. Colour in the eye of the beholder: receptor sensitivities and neural circuits underlying colour opponency and colour perception. Current Opinion in Neurobiology. 2016;41: 106–112.

10.1016/j.conb.2016.09.007

106. Olsson P, Lind O, Kelber A. Bird colour vision: behavioural thresholds reveal receptor noise. Journal of Experimental Biology. 2015;218: 184–193. 10.1242/jeb.111187
107. Hart NS. Variations in cone photoreceptor abundance and the visual ecology of birds. J Comp Physiol A. 2001;187: 685–697. 10.1007/s00359-001-0240-3
108. Stoddard MC, Miller AE, Eyster HN, Akkaynak D. I see your false colours: how artificial stimuli appear to different animal viewers. Interface Focus. 2019;9: 20180053 10.1098/rsfs.2018.0053
109. Pretterer G, Bubna-Littitz H, Windischbauer G, Gabler C, Griebel U. Brightness discrimination in the dog. Journal of Vision. 2004;4: 10–10. 10.1167/4.3.10
110. Linberg KA, Lewis GP, Shaaw C, Rex TS, Fisher SK. Distribution of S- and M-cones in normal and experimentally detached cat retina. Journal of Comparative Neurology. 2001;430: 343–356. 10.1002/1096-9861(20010212)430:3<343::AID-CNE1035>3.0.CO;2-U
111. Ahnelt PK, Schubert C, Kübber-Heiss A, Schiviz A, Anger E. Independent variation of retinal S and M cone photoreceptor topographies: A survey of four families of mammals. Visual Neuroscience. 2006;23: 429–435. 10.1017/S095252380623342X
112. Mowat FM, Petersen-Jones SM, Williamson H, Williams DL, Luthert PJ, Ali RR, et al. Topographical characterization of cone photoreceptors and the area centralis of the canine retina. Mol Vis. 2008;14: 2518–2527.
113. Lind O, Karlsson S, Kelber A. Brightness Discrimination in Budgerigars (Melopsittacus undulatus). PLOS ONE. 2013;8: e54650 10.1371/journal.pone.0054650
114. Hodos W, Bessette BB, Macko KA, Weiss SRB. Normative data for pigeon vision. Vision Research. 1985;25: 1525–1527. 10.1016/0042-6989(85)90231-7
115. Jones CD, Osorio D. Discrimination of oriented visual textures by poultry chicks. Vision Research. 2004;44: 83–89. 10.1016/j.visres.2003.08.014

116. Maertens M, Wichmann FA. When luminance increment thresholds depend on apparent lightness. Journal of Vision. 2013;13: 21–21. 10.1167/13.6.21

117. Geisbauer G, Griebel U, Schmid A, Timney B. Brightness discrimination and neutral point testing in the horse. Can J Zool. 2004;82: 660–670. 10.1139/z04-026

118. Cazetta E, Schaefer HM, Galetti M. Why are fruits colorful? The relative importance of achromatic and chromatic contrasts for detection by birds. Evol Ecol. 2009;23: 233–244. 10.1007/s10682-007-9217-1

119. Fleishman LJ, Perez CW, Yeo AI, Cummings KJ, Dick S, Almonte E. Perceptual distance between colored stimuli in the lizard Anolis sagrei: comparing visual system models to empirical results. Behav Ecol Sociobiol. 2016;70: 541–555. 10.1007/s00265-016-2072-8

120. Lind O. Colour vision and background adaptation in a passerine bird, the zebra finch (Taeniopygia guttata). Royal Society Open Science. 2016;3: 160383 10.1098/rsos.160383

121. Siddiqi A, Cronin TW, Loew ER, Vorobyev M, Summers K. Interspecific and intraspecific views of color signals in the strawberry poison frog Dendrobates pumilio. Journal of Experimental Biology. 2004;207: 2471–2485. 10.1242/jeb.01047

122. Russ JC. The Image Processing Handbook [Internet]. CRC Press; 2016. 10.1201/b18983

123. Lovell PG, Ruxton GD, Langridge KV, Spencer KA. Egg-Laying Substrate Selection for Optimal Camouflage by Quail. Current Biology. 2013;23: 260–264. 10.1016/j.cub.2012.12.031

124. Zuiderveld K. Contrast Limited Adaptive Histogram Equalization Graphics Gems. 1994; 474–485.

125. Lynx Promocions, S.L. Photos of Indian Peafowl (Pavo Cristatus). In: The Internet Bird Collection [Internet]. [cited 25 Feb 2019]. https://www.hbw.com/ibc/species/53521/photos?title=indian%20peafowl&uid=&sort_by=value&sort_order=DESC&items_per_page=20&page=1

126. Maia R, White TE. Comparing colors using visual models. Behav Ecol. 2018;29: 649–659. 10.1093/beheco/ary017

127. Anderson MJ, Walsh DCI, Clarke KR, Gorley RN, Guerra-Castro E. Some solutions to the multivariate Behrens–Fisher problem for dissimilarity-based analyses. Australian & New Zealand Journal of

Statistics. 2017;59: 57–79. 10.1111/anzs.12176

128. Jones DL. FATHOM: A Matlab toolbox for ecological and oceanographic data analysis [Internet]. St. Petersburg, FL USA: College of Marine Science, University of South Florida; 2017. https://www.marine.usf.edu/research/matlab-resources/

129. de Silva PK, Santiapillai C, Dissanayake S. Some aspects of the population ecology of the blue peafowl, Pavo cristatus, in Ruhuna National Park, Sri Lanka. Journal of South Asian Natural History. 1996;2: 113–126.

130. Mondal K, Gupta S, Qureshi Q, Sankar K. Prey selection and food habits of leopard (Panthera pardus fusca) in Sariska Tiger Reserve, Rajasthan, India. mammalia. 2011;75: 201–205. 10.1515/mamm.2011.011

131. Johnsingh AJT. Large mammalian prey-predators in Bandipur. Journal of the Bombay Natural History Society. 1983;80: 1–57.

132. Hayward MW, Jędrzejewski W, Jêdrzejewska B. Prey preferences of the tiger Panthera tigris. Journal of Zoology. 2012;286: 221–231. 10.1111/j.1469-7998.2011.00871.x

133. Petrie M. Peacocks with low mating success are more likely to suffer predation. Animal Behaviour. 1992;44: 585–586. 10.1016/0003-3472(92)90072-H

134. Takahashi M, Arita H, Hiraiwa-Hasegawa M, Hasegawa T. Peahens do not prefer peacocks with more elaborate trains. Animal Behaviour. 2008;75: 1209–1219. 10.1016/j.anbehav.2007.10.004

135. Ali S, Ripley SD. Handbook of the Birds of India and Pakistan: Together with Those of Bangladesh, Nepal, Bhutan and Sri Lanka Volume 2: Megapodes to Crab Plover. Oxford University Press; 1981.

136. Askew GN. The elaborate plumage in peacocks is not such a drag. Journal of Experimental Biology. 2014;217: 3237–3241. 10.1242/jeb.107474

137. Trivedi P, Johnsingh AJT. Roost selection by Indian Peafowl (Pavo cristatus) in Gir Forest, India. Journal of the Bombay Natural Histroy Society. 1996;93: 25–29.

138. Yasmin S, Yahya HSA. Group size and vigilance in Indian peafowl. Journal of the Bombay Natural History Society. 2000;97: 425–428.

139. Thavarajah NK, Tickle PG, Nudds RL, Codd JR. The peacock train does not handicap cursorial locomotor performance. Scientific Reports.

2016;6: 36512 10.1038/srep36512

140. Wilkinson H, Thavarajah N, Codd J. The metabolic cost of walking on an incline in the Peacock (Pavo cristatus). PeerJ. 2015;3: e987 10.7717/peerj.987
141. Tinbergen J, Wilts BD, Stavenga DG. Spectral tuning of Amazon parrot feather coloration by psittacofulvin pigments and spongy structures. Journal of Experimental Biology. 2013; jeb.091561. 10.1242/jeb.091561
142. Hart N, Hunt D. Avian visual pigments: characteristics, spectral tuning, and evolution. The American Naturalist. 2007;169: S7–26. 10.1086/510141
143. Fadzly N, Burns KC, Zuharah WF. Evaluating Frugivore-fruit Interactions Using Avian Eye Modelling. Trop Life Sci Res. 2013;24: 31–50.
144. Montgomerie R, Lyon B, Holder K. Dirty ptarmigan: behavioral modification of conspicuous male plumage. Behav Ecol. 2001;12: 429–438. 10.1093/beheco/12.4.429
145. Stuart-Fox D. Opening the "black box" of modeling animal color vision: a comment on Olsson et al. Behav Ecol. 2018;29: 284–284. 10.1093/beheco/arx154
146. Pike TW. Quantifying camouflage and conspicuousness using visual salience. Methods in Ecology and Evolution. 2018;9: 1883–1895. 10.1111/2041-210X.13019
147. Osorio D, Ham AD. Spectral reflectance and directional properties of structural coloration in bird plumage. Journal of Experimental Biology. 2002;205: 2017–2027.
148. Ridley M. How the peacock got his tail. New Scientist. 1981;91: 398–401.
149. Somppi S, Törnqvist H, Kujala MV, Hänninen L, Krause CM, Vainio O. Dogs Evaluate Threatening Facial Expressions by Their Biological Validity–Evidence from Gazing Patterns. PLOS ONE. 2016;11: e0143047 10.1371/journal.pone.0143047
150. Stevens M, Ruxton GD. Do animal eyespots really mimic eyes? Current Zoology. 2014;60: 26–36.
151. Yorzinski JL, Platt ML, Adams GK. Eye-spots in Lepidoptera attract attention in humans. R Soc Open Sci. 2015;2 10.1098/rsos.150155

152. Hillgarth N. Social organization of wild peafowl in India. World Pheasant Assoc J. 1984;9: 47–56.
153. Johnsingh AJT, Murali S. The ecology and behaviour of the Indian peafowl (Pavo cristatus) Linn. of Injar. J Bombay Nat Hist Soc. 1978;75: 1069–1079.
154. Dookia S. Ecology and Behaviour of Indian Peafowl (Pavo cristatus) in Keoladeo National Park, Bharatpur, Rajasthan, India. International Journal of Fauna and Biological Studies. 2015;2: 97–103.
155. Ridley MW, Lelliott AD, Rands MRW. The courtship display of feral peafowl. Journal of World Pheasant Association. 1984;9: 57–68.
156. Hodos W, Potocki A, Ghim MM, Gaffney M. Temporal modulation of spatial contrast vision in pigeons (Columba livia). Vision Research. 2003;43: 761–767. 10.1016/S0042-6989(02)00417-0
157. Peters R, Hemmi J, Zeil J. Image motion environments: background noise for movement-based animal signals. J Comp Physiol A. 2008;194: 441–456. 10.1007/s00359-008-0317-3
158. Walther BA. Do peacocks devote maintenance time to their ornamental plumage? Time budgets of male blue peafowl Pavo cristatus. Lundiana. 2003;4: 149–154.
159. Vehrencamp SL, Bradbury JW, Gibson RM. The energetic cost of display in male sage grouse. Animal Behaviour. 1989;38: 885–896. 10.1016/S0003-3472(89)80120-4
160. Carvalho LS, Pessoa DMA, Mountford JK, Davies WIL, Hunt DM. The Genetic and Evolutionary Drives behind Primate Color Vision. Front Ecol Evol. 2017;5 10.3389/fevo.2017.00034
161. Melin AD, Kline DW, Hickey CM, Fedigan LM. Food search through the eyes of a monkey: A functional substitution approach for assessing the ecology of primate color vision. Vision Research. 2013;86: 87–96. 10.1016/j.visres.2013.04.013
162. Hiramatsu C, Melin AD, Allen WL, Dubuc C, Higham JP. Experimental evidence that primate trichromacy is well suited for detecting primate social colour signals. Proc R Soc B. 2017;284: 20162458 10.1098/rspb.2016.2458
163. Pessoa DMA, Maia R, De A Ajuz RC, Moraes PZPMRD, Spyrides MHC, Pessoa VF. The adaptive value of primate color vision for predator detection. American Journal of Primatology. 2014;76: 721–729. 10.1002/ajp.22264

164. Pascual J, Senar JC, Domènech J. Plumage brightness, vigilance, escape potential, and predation risk in male and female Eurasian Siskins (Spinus spinus). The Auk. 2014;131: 61–72.
165. Freitag FB, Pessoa DMA. Effect of luminosity on color discrimination of dichromatic marmosets (Callithrix jacchus). J Opt Soc Am A, JOSAA. 2012;29: A216–A222. 10.1364/JOSAA.29.00A216
166. Sicsú P, Manica LT, Maia R, Macedo RH. Here comes the sun: multimodal displays are associated with sunlight incidence. Behav Ecol Sociobiol. 2013;67: 1633–1642. 10.1007/s00265-013-1574-x
167. Endler JA. The Color of Light in Forests and Its Implications. Ecological Monographs. 1993;63: 1–27. 10.2307/2937121
168. Jarvis JR, Wathes CM. On the calculation of optical performance factors from vertebrate spatial contrast sensitivity. Vision Research. 2007;47: 2259–2271. 10.1016/j.visres.2007.04.015
169. Stevens M, Cuthill IC. Hidden Messages: Are Ultraviolet Signals a Special Channel in Avian Communication? BioScience. 2007;57: 501–507. 10.1641/B570607
170. Cummings ME, Endler JA. 25 Years of sensory drive: the evidence and its watery bias. Curr Zool. 2018;64: 471–484. 10.1093/cz/zoy043

6

Are Stripes Beneficial? Dazzle Camouflage Influences Perceived Speed and Hit Rates

Bettina von Helversen[1,2], Lael J. Schooler[1] and Uwe Czienskowski[1]

[1] Max Planck Institute for Human Development, Center for Adaptive Behavior and Cognition, Berlin, Germany,

[2] Department of Psychology, University of Basel, Basel, Switzerland, University of Sussex, United Kingdom,

ABSTRACT

In the animal kingdom, camouflage refers to patterns that help potential prey avoid detection. Mostly camouflage is thought of as helping prey blend in with their background. In contrast, disruptive or dazzle patterns protect moving targets and have been suggested as an evolutionary force in shaping the dorsal patterns of animals. Dazzle patterns, such as stripes and zigzags, are thought to reduce the probability with which moving prey will be captured by impairing predators' perception of speed. We investigated how different patterns of stripes (longitudinal—i.e., parallel to movement direction–and vertical–i.e., perpendicular to movement

Citation: Von Helversen B, Schooler LJ, Czienskowski U (2013) Are Stripes Beneficial? Dazzle Camouflage Influences Perceived Speed and Hit Rates. PLoS ONE 8(4): e61173. https://doi.org/10.1371/journal.pone.0061173

direction) affect the probability with which humans can hit moving objects and if differences in hitting probability are caused by a misperception of speed. A first experiment showed that longitudinally striped objects were hit more often than unicolored objects. However, vertically striped objects did not differ from unicolored objects. A second study examining the link between perceived speed and hitting probability showed that longitudinally and vertically striped objects were both perceived as moving faster and were hit more often than unicolored objects. In sum, our results provide evidence that striped patterns disrupt the perception of speed, which in turn influences how often objects are hit. However, the magnitude and the direction of the effects depend on additional factors such as speed and the task setup.

INTRODUCTION

The concept of camouflage refers to the ability of patterns to impair the chances that a target will be caught. Camouflage usually applies to stationary objects. In contrast, the concept of dazzle has been introduced to describe how disruptive patterns can protect moving targets in the same way. Dazzle has mostly been studied in the animal kingdom; for instance, researchers have investigated the impact of dorsal patterns on escape chances in snakes [1]–[5]. But dazzle can also influence human perception. In both world wars, navies painted their warships with high-contrast patterns in the hope of confusing the enemy [6]. There is no direct evidence that the dazzle designs indeed protected the ships, but recent research has demonstrated that dazzle patterns can influence the ability of humans to hit moving objects [7]. The mechanisms, however, with which dazzle patterns induce higher escape rates are not yet well understood. In general, it is thought that patterns impair the ability to accurately perceive speed, which in turn decreases the ability to capture objects [1]–[3], [7], [8]. The idea that objects with dazzle patterns disrupt the perception of speed is supported by a study showing that objects with disruptive patterns, such as zigzags or checks, were perceived as moving more slowly than unicolored objects [9].

Although the assumption that a misperception of speed should affect the ability to hit a moving target is intuitively appealing, some psychological research suggests that perceived speed plays only a minor role when hitting moving targets [10], [11]. Studies have revealed that perception and action are often based on different kinds of information [12]–[14]. Furthermore, a computerized experiment that manipulated perceived speed without affecting the object's position on the screen [10] found that perceived speed did not affect the trajectory of the hand movement to catch the object. Thus, even though dazzle patterns have been found to influence the ability to hit a moving object and to influence perceived speed, the effect of pattern on hitting probability may not be directly linked to the effect of dazzle patterns on perceived speed.

Our main goals were (1) to investigate if dazzle patterns influence (a) the ability to hit moving targets and (b) the perceived speed of the objects and (2) to test if perceived speed is related to hitting probability. A secondary goal was to investigate how specific dazzle patterns, that is, longitudinal and vertical stripes, influence the probability of hitting a moving target and perceived speed. Researchers largely agree that stripes can influence the perception of speed, but it is less clear if the orientation of the stripes affects whether objects are perceived as moving faster or slower (see [1], [5], [8], [15]). Stripes can be oriented either parallel or perpendicular to movement direction. In the following, we refer to stripes parallel to movement direction as longitudinal stripes, and stripes perpendicular to movement direction as vertical stripes. It has been debated if longitudinal stripes increase or decrease perceived speed and if longitudinally striped objects are perceived as faster than vertically striped objects. On the one hand, Gabor patches aligned with the direction of movement were perceived as moving faster than patches set at an angle to the direction of movement [16], suggesting that longitudinal patterns may be perceived as faster than vertical stripes (see also [15]). On the other hand, it has been argued that longitudinal stripes may have an advantage over vertical stripes in straight, high-speed flight because they make a snake appear to move more slowly than it actually does [1]–[3]. In addition, vertical stripes–although they have good cryptic properties if the snake stays still–are thought to provide reference points [1] that enhance movement detection and speed estimation [5]. In support of this theory, studies [1], [3] have shown a genetic correlation between flight behavior and dorsal patterns in garter snakes. In contrast, others did not find a difference in how easily vertically and longitudinally striped objects were hit [7], [8] or in perceived speed [9].

To explore the influence of dazzle patterns, we conducted two experiments. In Experiment 1 we investigated how the orientation of stripes (longitudinal vs. vertical) influences the probability with which moving targets are hit. In Experiment 2 we focused on the effect of pattern on perceived speed and the link between perceived speed and hit rate.

EXPERIMENT 1

In Experiment 1 we investigated the ability of humans to hit unicolored and striped objects in a computer game. In the game, objects moved from left to right across the screen in a straight line, starting at different heights. Objects could be hit by moving a cursor over the object with a joystick and pressing the fire button while the object was under the cursor. The cursor was positioned at ¾ of the width of the screen and could only be moved vertically (see Figure 1). Thus the task was to move the cursor to the vertical location of the object on the screen and to press the fire button when the object reached the target zone. We used three different surface patterns for the objects: unicolored, longitudinally striped, and vertically striped. If dazzle patterns have an advantage, the striped objects should be hit less often than the unicolored objects. Furthermore, if stripes differ in their effect on perceived speed and hit rates, and longitudinal stripes have an advantage over vertical stripes in straight flight, longitudinally striped objects should be hit less often than vertically striped objects.

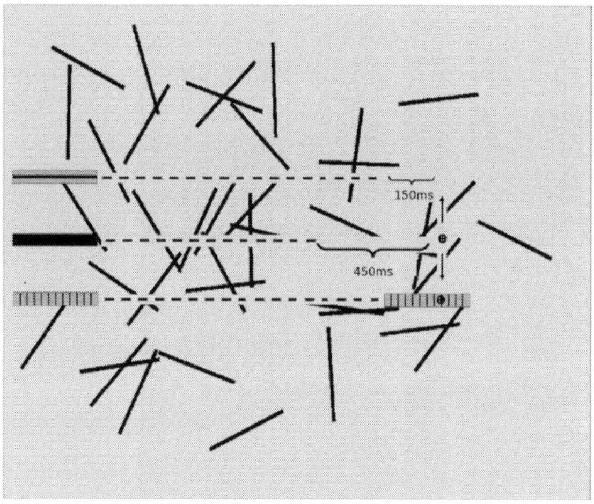

Figure 1. Experimental layout

The longitudinally striped, unicolored, and vertically striped objects are shown on the left. During the experiment, objects appeared on the left side of the screen at varying heights. The cursor (black circle) marks the position of the cursor at the beginning of a trial. The arrows above and below indicate the direction of the cursor movement. The dotted lines indicate possible paths of the objects and where they can be intercepted with the cursor. The brackets indicate the duration of the time interval during which objects would disappear.

Besides pattern, we varied the speed with which objects moved and a time interval (disappearance duration) during which objects were not visible before they had to be hit. We varied objects' speed so we could test if the effect of pattern is robust across different speeds. We introduced the disappearance duration because in natural conditions, a snake, for instance, may vanish in the grass and out of the field of view of a predator. In addition, having the objects disappear makes it more difficult for participants to rely on an object's current position [17], which should encourage them to rely on their perception of the object's speed.

METHODS

Ethics Statement

The study was conducted in accordance with the Declaration of Helsinki and approved by the ethics committee of the Max Planck Institute for Human Development in Berlin. All participants were blind to the hypotheses of the experiments and participated voluntarily. Written informed consent was received from all participants.

Participants

Fifty-eight students from Berlin universities participated in Experiment 1. Thirty were male; mean age was 24.7 years (*SD*=2.76). The experiment took about 30 min and participants received €4 for participating. Depending on performance, they earned an additional €3 on average.

Design, Procedure, and Material

In Experiment 1, the participants' task was to hit objects moving from left to right on a computer screen with a cursor controlled by a joystick. We varied the pattern of the objects, pitting longitudinally striped and vertically striped against unicolored black objects. To hit an object, participants had to move the cursor over the object and press the fire button. The cursor was positioned at ¾ of the width of the screen and could only be moved vertically (see Figure 1). Because objects disappeared shortly before they reached the point where they could be hit (see [11] for a similar design), participants had to press the fire button when they thought the object would be under the cursor without seeing the object. Objects reappeared after participants pressed the trigger. If participants fired while the cursor was over the object it counted as a hit, and otherwise a miss was recorded.

The experiment was conducted in the laboratory of the Max Planck Institute for Human Development. The room was artificially lit (standard fluorescent laboratory lighting) and lighting was kept approximately constant for all participants. The experiment was conducted on two computers, both with Microsoft Sidewinder joysticks and with 17-inch CRTs with refresh rates of 85 Hz and a resolution of 1,024×768 pixels and default brightness and contrast settings. Participants were seated in front of the screen, with their heads approximately 60 cm from the screen. The experimental software was implemented using the C# language under Microsoft.NET 2.0 with Managed DirectX. The screen background was stylized "grass," light grey (color code in the red, green, blue [RGB] color model: 220, 220, 220) with thirty 40 mm×2 mm (3.82×0.19 degrees of visual angle) black lines (RGB color code: 0, 0, 0). The angle, orientation, and location of the lines were randomly determined. We created 20 different backgrounds. For each participant we randomly selected one background for the game. Objects were rectangles of 39 mm length and 7 mm width (3.72×0.67 degrees of visual angle). Eighty percent of the surface of the two patterned objects was medium grey (RGB color code: 180, 180, 180) and the remaining 20% was black (RGB color code: 0,0,0). The longitudinally striped object had one black stripe running across the middle (1.4 mm×39 mm, 0.13×3.72 degrees of visual angle). The vertically striped object had 11 black stripes (0.7 mm×7 mm, 0.07×0.67 degrees of visual angle) equidistantly distributed over the object (see Figure 1). The unicolored object was black (RGB color code: 0,0,0).

In addition to object pattern we varied the objects' speed and the duration of the time during which objects disappeared from sight. Objects were either slow (12 cm/s, 11.42 degrees of visual angle per second [deg/s] given the distance from the screen) or fast (16 cm/s, 15.19 deg/s). We selected these speeds based on a pilot. In the pilot we used an adaptive testing procedure to determine the speed at which participants hit unicolored objects with a probability of 0.75 [18]. The pilot was similar to the task used in Experiment 1, but objects did not disappear from the screen. We then selected speeds in the medium and lower portion of the speed distribution that were comparable to speeds used in the literature [11].

In the experimental task the objects disappeared before they could be hit. The *disappearance duration* specified the time objects needed to travel from the point they vanished to the point where they could be hit. We varied the disappearance duration in two conditions: long: 450 ms vs. short: 150 ms. This resulted in a Speed (slow vs. fast) × Disappearance Duration (short vs. long) × Pattern (longitudinally striped, vertically striped, unicolored) within-subject design. Pattern, speed and disappearance duration were varied randomly from trial to trial.

The experiment consisted of a training phase of 50 trials and a test phase of 360 trials, with 30 trials in each condition. During training participants practiced with unicolored black objects to familiarize themselves with the task. In the training phase, to obtain an average hit rate of 50%, object speed was set by an adaptive step algorithm [18], i.e. if the object was hit, the object in the next trial moved faster, but if an object was missed, the object in the next trial moved slower.

Participants received visual and auditory feedback. If the object was hit, a tone sounded, the object turned red, and the trial ended. If it was missed, it continued on its path until it vanished off the right side of the screen. Participants were given one shot per trial. After each trial the cursor returned to the starting position and could not be moved until the next trial was started with the space bar. To reduce error variance, 80% of the objects appeared in the two middle quarters of the screen and 20% fell into the two extreme quarters.

Results and Discussion

Experiment 1 varied the pattern of the objects, the objects' speed, and the time interval between when the objects disappeared and when they could be hit (disappearance duration). We analyzed whether pattern influenced the frequency with which the objects were hit with a repeated measurement analysis of variance (ANOVA) with percentage of hits as the dependent variable and pattern, speed, and disappearance duration as within-subject factors. To minimize variance, we focused on the 288 experimental trials that were located in the middle quarters of the screen (24 trials per condition). We then calculated contrasts comparing the longitudinally striped and the vertically striped pattern with unicolored objects (see Figure 2).

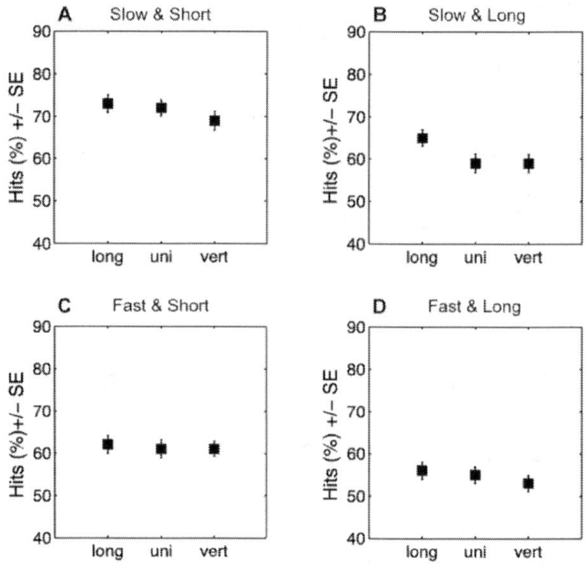

Figure 2. Mean percentages of hit rate by pattern, speed, and disappearance duration. (A) Slow speed and a short disappearance duration. (B) Slow speed and long disappearance duration. (C) Fast speed and short disappearance duration. (D) Fast speed and long disappearance duration. Vert = vertically striped object; long = longitudinally striped object, uni = unicolored object. Error bars denote the standard error.

Overall, we found that participants hit objects on average in 62% of the trials. Objects that disappeared for 450 ms were hit less often than objects that disappeared for 150 ms (ANOVA, $F_{1,57}$=93.3, P<0.001), and fast objects were hit less often than slow ones (ANOVA, $F_{1,57}$=4.78, P=0.03). The effect of speed was larger with a short disappearance duration than with a long one, as indicated by a significant interaction between speed and disappearance duration (ANOVA, $F_{1,57}$=4.78, P=0.03).

More importantly, we found a significant main effect of pattern (ANOVA, $F_{2,114}$=6.18, P=0.003). Contrasts showed that longitudinally striped objects were hit more often than unicolored objects (Contrast, $F_{1,57}$=5.63, P=0.02), but vertically striped objects were not statistically significantly different from unicolored ones (Contrast, $F_{1,57}$=1.64, P=0.21). None of the interactions–between speed and pattern (ANOVA, $F_{2,114}$=2.42, P=0.09), between disappearance duration and pattern (ANOVA, $F_{2,114}$=1.56, P=0.21), or between pattern, speed, and disappearance duration (ANOVA,

$F_{2,114}$=2.18, P=0.12) –reached significance. As illustrated in Figure 2, separate follow-up tests for speed and disappearance duration, however, found a significant effect of pattern with a long disappearance duration and slow speed (ANOVA, $F_{2,114}$=9.56, P<0.001; see Figure 2B) and with a short disappearance duration and slow speed (ANOVA, $F_{2,114}$=3.20, P=0.04, see Figure 2A). The effects of pattern were nonsignificant when objects moved at high speed (Figure 2C: short disappearance duration and high speed, ANOVA, $F_{2,114}$=0.05, P=0.95; Figure 2D: long disappearance duration and high speed, ANOVA, $F_{2,114}$=1.30, P=0.28).

In sum, we found that objects were hit less frequently when they moved faster and when they disappeared for longer. These results are not surprising, as both factors increase the difficulty of the task. More importantly, the experiment showed a significant effect of pattern on hit rates. In contrast to the dazzle hypothesis, however, longitudinally striped objects were hit more often than vertically striped or unicolored ones. These results resonate with research reporting that dazzle patterns influence the probability with which moving objects can be hit [1], [7] but suggest that the effect need not be beneficial. We will discuss this point in the General Discussion.

Overall, the effect of pattern on hitting probability seemed reliable, as we did not find a significant interaction between pattern and disappearance duration or pattern and speed. Follow-up tests, however, did not find a difference between objects' patterns for fast-moving objects, suggesting that the effect of pattern may be limited to lower speeds. The effect of pattern could diminish when objects move fast, because at high speeds the perception of speed may be less reliable in general, masking effects caused by pattern. Furthermore, at high speeds vertical stripes can blend together, giving the impression of a unicolored object and eliminating effects of vertical stripes. This blurring is known as the flicker fusion phenomenon [19]–[22].

To replicate the effects found in Experiment 1, we conducted a further experiment (Experiment 1.1) with the same experimental setup but with a reduced short disappearance duration. The results showed a similar pattern to that of Experiment 1. Details for Experiment 1.1 can be found in the supplementary online material, Experiment S1.

EXPERIMENT 2

In Experiment 1 we investigated the effect of pattern on the probability with which objects can be hit, but we did not measure perceived speed. To clarify the relationship between perceived speed and hitting probability we conducted a second experiment where we measured perceived speed and hit rates in the same experiment. Additionally, we simplified the hitting task such that all objects appeared at the same vertical position on the screen, eliminating the need to adjust the cursor to the correct vertical position. Having to move the cursor could be a potential source of error that could mask the effects of perceived speed, particularly because pattern may influence other aspects of stimulus perception [23], [24].

Methods

Ethics Statement

The study was conducted in accordance with the Declaration of Helsinki and approved by the ethics committee of the Max Planck Institute for Human Development in Berlin. All participants were blind to the hypotheses of the experiments and participated voluntarily. Written informed consent was received from all participants.

Participants

Thirty students from Berlin universities participated in Experiment 2. Seventeen were female, with a mean age of 25.23 years (SD=3.23). Participants received on average €7 for their participation. We excluded one participant from the analysis because she performed more than three standard deviations below the mean, but the pattern of results does not change if she is included.

Design and Procedure

The study consisted of two parts: a perceptual task with the goal of comparing the speed of two objects and a motor task with the goal of hitting the objects. The order of the tasks was randomized with half of the participants beginning with the speed comparison task and half with the hitting task. The same materials as in Experiment 1 were used and the experiment was conducted in the same laboratory and under similar lighting conditions, kept approximately constant between participants. However, we used the same background for all participants and the experiment was run on a computer with a 17-inch LCD screen with a resolution of 1,280×1,024 pixel. Brightness and contrast were set at default values. This resulted in a displayed size of the objects of 31 mm length×5 mm width (2.96×0.48 degrees of visual angle).

We measured luminance with a luminance meter (Gossen Mavo-Monitor, Germany). The luminance values were (in cd m^{-2}): Black=0.4, medium grey (objects' stripes)=75, light grey (background)=125. Unfortunately we cannot provide luminance measures for the first experiment, because the monitors used in this experiment had been replaced before we acquired a luminance meter.

Speed Comparison

In the speed comparison task participants saw two objects moving one after the other across the screen. After both objects had vanished, participants indicated if the second object was slower, equally fast, or faster than the first object. Each comparison consisted of a target object and a comparison object. The comparison object was a unicolored black object that moved with a speed of 12 cm/s (11.42 deg/s). The target object was unicolored or striped vertically or longitudinally. Whether the target or the comparison object was seen first was randomly determined from trial to trial. The speed of the target object was adjusted using an adaptive staircase algorithm [18]. If the participant indicated that the target object had been faster than the comparison object, the target object's speed was reduced in the next trial. If the participant indicated that the target object had been slower than the comparison object, the target object's speed was increased in the next trial. If the participant indicated the objects moved at equal speed, the target object's speed was decreased or increased depending on the direction of the last speed change. For example, if the speed had been reduced in the trial before and the participant now indicated that the speed was equal, the target object's speed was further reduced in the subsequent trials until the participant indicated that it was now slower than the comparison object. The speed comparison in each staircase continued until six reversals in speed adjustment (i.e., the speed of the target object was increased in one trial and decreased in the next trial) took place. In the first rounds the speed was adjusted by 1 cm/s (0.95 deg/s). After three reversals it was adjusted by 0.5 cm/s (0.48 deg/s). For each target object we had four staircases with different starting speeds: 8, 10, 14, and 16 cm/s (7.62, 9.53, 13.31, and 15.19 deg/s). Pattern and staircase were varied randomly from trial to trial.

Hitting Task

The hitting task was similar to the setup in Experiment 1 (see Figure 1). After a short practice period, participants played 180 trials of a hitting task.

Objects appeared on the left side of the screen and moved at a constant speed of 12 cm/s (11.42 deg/s) in a direct line across the screen. The participants' task was to fire when objects passed under the cursor position, but similar to Experiment 1, the objects disappeared shortly before they reached the cursor location. In contrast to Experiment 1, the objects always appeared at the same vertical location. Thus in this version of the hitting task, it was not necessary to adjust the cursor position and no joystick was required. Instead participants were instructed to fire by pressing the return key when they thought the object was located under the cursor. We varied the pattern of the object and the disappearance duration, resulting in a within-subject design with two factors. Disappearance duration varied with two levels, 350 ms and 150 ms. We reduced the long disappearance duration in comparison to Experiment 1 to investigate if the effect was reliable at different disappearance durations and to ensure that the objects were visible long enough to allow participants to form an accurate perception of speed. There were three types of patterns: the objects were unicolored, vertically striped, or longitudinally striped (see Figure 1). In each condition participants performed 30 trials. The pattern and disappearance duration were varied randomly from trial to trial. To reduce variance in where participants tried to hit the objects, we instructed participants to aim for the middle of the object and paid them depending on performance. They received 5 points for hitting the center (the area less than 2.5 mm/0.24 degrees of visual angles from the center), 2 points for hitting close to the center (the areas more than 2. 5 mm/0.24 degrees of visual angles and less than 9 mm/0.86 degrees of visual angles from the center), and 1 point for hitting the object at the front or the back (the area close to the front or back and more than 9 mm/0.86 degrees of visual angles from the center).

Results and Discussion

To obtain an estimate of perceived speed, we took the average speed in the trials where participants had responded that the target object moved with the same speed as the unicolored comparison object for each pattern separately. We analyzed the data with a repeated measurement ANOVA with pattern as within-subject factor and perceived speed as dependent variable. We additionally calculated contrasts comparing striped objects with unicolored objects. As illustrated in Figure 3A, the results showed that for both vertically striped (Contrast, $F_{1,28}=6.86$, $P=0.02$) and longitudinally striped (Contrast, $F_{1,28}=9.96$, $P=0.005$) objects the average speed at which they were perceived as moving equally fast as the comparison object was slower than for the unicolored objects. This suggests that at the same speed, the striped objects were perceived as moving faster than unicolored objects.

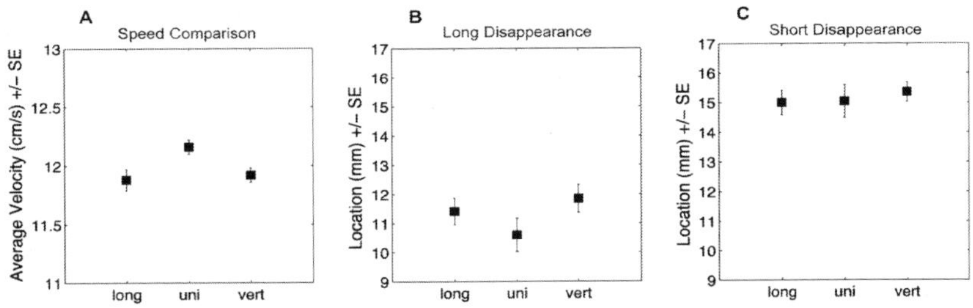

Figure 3. (A) Average perceived speed of unicolored and striped objects. (B) Average location of the object in the long disappearance condition. (C) Average location of the object in the short disappearance condition. Vert = vertically striped object; long = longitudinally striped object, uni = unicolored object. Error bars denote a standard error.

As a measure of hitting success, we measured the frequency with which participants hit the objects (for means and standard deviations see Table 1). We analyzed the data with a repeated measurement ANOVA with pattern and disappearance duration as within-subject factors and hit rate as dependent variable. Objects that disappeared for 350 ms were hit less often than objects disappearing for 150 ms (ANOVA, $F_{1,28}$=25.91, P<0.001). There was no main effect of pattern on hit rates (ANOVA, $F_{2,56}$=1.56, P=0.22), but we found a significant interaction between disappearance duration and pattern (ANOVA, $F_{2,56}$=4.72, P=0.01). Follow-up analyses for both disappearance durations separately revealed that when objects disappeared for 350 ms, unicolored objects were hit significantly less than longitudinally striped objects (Contrast, $F_{1,28}$=4.44, P=0.04) and vertically striped objects (Contrast, $F_{1,28}$=6.39, P=0.02). There was no effect with a short disappearance duration (ANOVA, $F_{2,56}$=1.39, P=0.26).

Table 1. Means and Standard Deviations of Hit Rate by Disappearance Duration and Pattern.

	Disappearance duration					
	150 ms			350 ms		
	Pattern			Pattern		
	Long	Uni	Vert	Long	Uni	Vert

Hit rate	0.97 (0.05)	0.98 (0.03)	0.98 (0.04)	0.96 (0.04)	0.94 (0.05)	0.96 (0.04)

Note. N=29. Hit rate gives the percentage of hits. Long: longitudinally striped; uni: unicolored; vert: vertically striped.

Additionally we measured where the object was, when participants fired (i.e. pressed the return button), with a location of zero corresponding to the front of the object being at the cursors location (i.e. the object would be hit close to the front) and a location of 31 mm corresponding to the back of the object being at the cursors location (i.e. the object was hit close to the back, see Figure 3). We analyzed objects' location with a repeated measurement ANOVA with location as dependent variable and pattern and disappearance duration as independent variables. The analysis of the data showed that all objects had moved along less (i.e. were hit closer to the front) with a long disappearance duration than with a short disappearance duration (ANOVA, $F_{1,28}=197.6$, $P<0.001$), but there was no main effect of pattern (ANOVA, Greenhouse–Geisser corrected, $F_{1.52,42.64}=2.20$, $P=0.14$). Separate ANOVAs for a short and a long disappearance duration indicated that pattern influenced objects' location if the disappearance duration was long (ANOVA, $F_{2,56}=3.79$, $P=0.03$). As illustrated in Figure 3 (middle panel), unicolored objects were hit closer to the front than vertically striped objects (Contrast, $F_{1,28}=6.70$, $P=0.02$), but not closer to the front than longitudinally striped objects (Contrast, $F_{1,28}=2.43$, $P=0.13$). We did not find an effect of pattern with a short disappearance duration (ANOVA, $F_{2,56}=0.33$, $P=0.72$; see Figure 3 right panel).

Overall, the results suggest that pattern had an influence on the objects perceived speed, the objects' location when participants fired, and on the frequency with which they were hit. To test if the effect of perceived speed was related to the location of the hit, we correlated the difference in perceived speed with the location where the striped objects were hit. We found that the faster vertically striped objects were perceived to be moving compared to unicolored objects, the more to the back they were hit in the long disappearance duration condition (N=29, Pearson correlation, r=0.44, P=0.02). The location where longitudinally striped objects were hit did not significantly correlate with perceived speed (N=29, Pearson correlation, r=0.16, P=0.42).

In sum, Experiment 2 showed that pattern had an influence on perceived speed, the probability with which objects were hit, and the objects' location when participants fired. Participants perceived striped objects that moved slower than unicolored objects as equally fast and hit striped objects more to the back compared to unicolored objects. Furthermore, we found a correlation between the degree with which people misperceived the speed of vertically striped objects and the location of the hit, providing further support that stripes influence the perception of speed, which in turn influences how the objects are hit. This supports the idea that dazzle patterns can affect the ability to hit moving targets and that this effect is caused by a misperception of speed, resonating with earlier research [7]–[9]. Somewhat surprisingly, striped objects were hit more to the back than unicolored objects, although they were perceived as moving faster. If objects move slower than perceived one would expect more hits to the front of the object. The correlation between differences in perceived speed and the objects' location, however, suggests that the effect was related to speed perception. One possible explanation is that dazzle patterns decrease confidence in speed perception, which in turn decreased reaction times.

Similar to in Experiment 1, where the effect of pattern was stronger when objects disappeared for 450 ms than when they disappeared for 150 ms, the effect of pattern appeared only if the objects vanished for 350 ms in the "grass." This suggests that pattern may influence hit rate only when objects are out of sight before they can be hit. The effect of pattern could be stronger with longer disappearance durations, because the disappearance of the objects forced participants to rely on perceived speed to estimate when the object would reach the target zone. When objects do not disappear or disappear for only a short time, people may instead use the objects' position to decide when to fire, which in turn could reduce the influence of pattern on hit rates. This explanation resonates with research suggesting that people strongly rely on position when hitting moving targets [10]–[11].

GENERAL DISCUSSION

The Influence of Dazzle Patterns on Perceived Speed and Hit Rate

In two experiments we demonstrated that dazzle patterns influence the perception of speed, which in turn influences the rate at which moving targets are hit. In both experiments, longitudinally striped objects were hit more often than unicolored objects; and in Experiment 2, but not in Experiment 1, vertically striped objects were hit more often than unicolored objects. In addition, Experiment 2 offered direct evidence that striped objects were perceived as moving faster than unicolored objects. These results are consistent with the observation that longitudinally striped snakes make the perception of speed more difficult [1] and with prior research reporting effects of dazzle patterns on hit rates [7]–[8] and perceived speed [9]. In contrast to [1] and [9] our results, however, suggest that striped patterns can be perceived as moving faster than unicolored objects. This suggests that the relation between perceived speed and dazzle pattern may be more complex and could depend on factors such as the type of comparison object, which was white in the study by [9], but black in our study.

The effect of pattern, however, differed somewhat between the two experiments. In Experiment 1, we did not find an effect of vertical stripes, but we did find an effect in Experiment 2. These differences could be due to chance, but also differences in the task setup of the experiments could have contributed. In Experiment 1 participants not only had to press the fire button at the right time (i.e., when they believed the object reached the horizontal position of the cursor), but they also had to adjust the cursor to the correct vertical position. In Experiment 2 the vertical position of the cursor was fixed. If vertical stripes impeded the perception of the object's vertical position [24], this could have reduced the hit rate for vertically striped objects in the first experiment.

A possible limitation of our experiment is that the luminance and contrast of the unicolored object and the striped objects differed. The unicolored black object had a lower luminance and higher contrast than the striped objects (which both were 80% grey and only 20% black). Visual contrast can influence perceived speed [25]–[27] and thus differences in contrast could have contributed to the differences in perceived speed and hit rates. Speaking against this explanation is that low contrast patterns are usually found to be moving slower than high contrast pattern and a similar effect for high- and low-contrast patterns on hit rates was found [7]. Furthermore, the effect of pattern on perceived speed was found to be more pronounced with high-contrast patterns than with low-contrast patterns [9].

Benefits of Dazzle Patterns

In our experiments we found that dazzle patterns influenced the perception of speed, and had an impact on the probability of hitting the objects. However, in both experiments striped objects were hit more often than or equally often as unicolored objects, casting doubt on the proposed general benefit of dazzle patterns. There are at least two possible explanations for why in our task striped objects were hit more often than unicolored objects, whereas other research has found that dazzle patterns reduce how frequently objects are hit. For one, the way objects are captured could influence the results. For instance, it has been argued [28] that a relation between dorsal pattern and flight behavior in lizards was an adaptation to predation from the land versus the air. In our experiments participants were required to hit objects by waiting until they reached the target zone. However, many predators may hunt their prey by chasing them, which could in turn influence how dazzle pattern affect the ability to hit moving prey.

Second, the effect of pattern on hit rates could depend on the type of movement. Research on snakes suggests that one type of camouflage works well when a snake follows a curved flight path and another when the snake follows a straight path. As even the straight path of a fleeing snake is more serpentine than the strictly linear movement path of the objects in our task, this interaction could figure into our results.

In sum, future tests of the effects of patterns on chances of escape should take into account the natural situation in which these markings are found (see also [29]). It may not be enough to determine whether a particular pattern causes a misperception of an object's speed. Rather, it is essential to consider how this misperception interacts with the habitat and how the object moves or is pursued.

CONCLUSION

In our experiments we found reliable but small differences between striped objects and a unicolored object, which we could trace to a misperception of speed caused by dazzle patterns. Striped objects were hit about 2% more often than unicolored objects. Although this may sound like a negligible effect, from an evolutionary perspective an increase in survival rate of 3% can result in a strong selection pressure [30]. Furthermore, the differences between striped and unicolored objects observed in our experiments are comparable to the effect sizes found in previous research [9], where it was argued that a misperception of this magnitude could be sufficient to grant protection from a roving predator.

ACKNOWLEDGMENTS

We thank Julia Schooler and Anita Todd for editing a draft of this manuscript.

REFERENCES

1. Brodie ED (1992) Correlation selection for color pattern and antipredator behavior in the garter snake *Thamnophis ordinoides* . Evolution 46: 1284–1298.
2. Brodie ED (1993) Consistency of individual differences in anti-predator behaviour and colour pattern in the garter snake, *Thamnophis ordinoides* . Anim Behav 45: 851–861.
3. Brodie ED (1989) Genetic correlations between morphology and antipredator behaviour in natural populations of the garter snake *Thamnophis ordinoides* . Nature 342: 542–543.
4. Creer DA (2005) Correlations between ontogenetic change in color pattern and antipredator behavior in the racer, *Coluber constrictor* . Ethology 111: 287–300.
5. Jackson JF, Ingram W, Campbell HW (1976) The dorsal pigmentation patterns of snakes as an antipredator strategy: A multivariate approach. Am Nat 110: 1029–1053.
6. Williams D (2001) Naval camouflage 1914–1945. Barnsley, UK: Pen and Sword Books. 256 p.
7. Stevens M, Searle WTL, Seymour JE, Marshall KLA, Ruxton GD (2011) Motion dazzle and camouflage as distinct anti-predator defenses. BMC Biol 9: 81.
8. Stevens M, Yule DH, Ruxton GD (2008) Dazzle coloration and prey movement. Proc Biol Sci 275: 2639–2643.
9. Scott-Samuel NE, Baddeley R, Palmer CE, Cuthill IC (2011) Dazzle camouflage affects speed perception. PLoS One 6: 1–5.
10. Smeets JBJ, Brenner E (1995) Perception and action are based on the same visual information: Distinction between position and velocity. J Exp Psychol Hum Percept Perform 21: 19–31.
11. Brouwer A-M, Brenner E, Smeets JBJ (2002) Hitting moving objects: Is target speed used in guiding the hand? Exp Brain Res 143: 189–211.
12. Brouwer AM, Franz VH, Thornton IM (2004) Representational momentum in perception and grasping: Translating versus transforming objects. J Vis 4: 575–584.
13. Creem SH, Proffitt DR (2001) Grasping objects by their handles: A necessary interaction between cognition and action. J Exp Psychol Hum Percept Perform 27: 218–228.

14. Proffitt DR, Bhalla M, Gossweiler R, Midgett J (1995) Perceiving geographical slant. Psychon Bull Rev 2: 409–428.
15. Brown JF (1931) The visual perception of velocity. Psychol Forsch 1: 14–34.
16. Georges S, Seriès P, Frégnac Y, Lorenceau J (2002) Orientation dependent modulation of apparent speed: Psychophysical evidence. Vision Res 42: 2757–2772.
17. Brouwer A-M, Brenner E, Smeets JBJ (2000) Hitting moving objects. The dependency of hand velocity on the speed of the target. Exp Brain Res 133: 242–248.
18. Leek MR (2001) Adaptive procedures in psychophysical research. Percept Psychophys 63: 1279–1292.
19. Endler J (1978) A predator's view of animal color patterns. Evol Biol 11: 319–364.
20. Srygley RB (1999) Incorporating motion into investigations of mimicry. Evol Ecol 13: 691–708.
21. Lindell LE, Forsman A (1996) Sexual dichromatism in snakes: Support for the flicker-fusion hypothesis. Can J Zool 74: 2254–2256.
22. Pough FH (1976) Multiple cryptic effects of crossbanded and ringed patterns of snakes. Copeia 1976: 834–836.
23. Thompson P (2008) Does my butt look big in this? Horizontal stripes, perceived body size and the Oppel-Kundt illusion. J Vis 8: 822–822.
24. Taya S, Miura K (2007) Shrinkage in the apparent size of cylindrical objects. Perception 36: 3–16.
25. Thompson P, Stone LS (1997) Contrast affects flicker and speed perception differently. Vis Res 37: 1255–1260.
26. Stone LS, Thompson P (1992) Human speed perception is contrast dependent. Vis Res 32: 1535–1549.
27. Anstis S (2003) Moving objects appear to slow down at low contrasts. Neural Netw 16: 933–938.
28. Carretero MA, Vasconcelos R, Fonseca M, Kaliontzopoulou A, Brito JC, et al. (2006) Escape tactics of two syntopic forms of the *Lacerta perspicillata* complex with different colour patterns. Can J Zool 84: 1594–1603.
29. Blakemore MR, Snowden RJ (2000) Textured backgrounds alter perceived speed. Vis Res 40: 629–638.

30. Haldane JBS (1927) A mathematical theory of natural and artificial selection, part V: Selection and mutation. Math Proc Camb Philos Soc 23: 838–844.

7

How the Knowledge of Interactions between Meningococcus and the Human Immune System Has Been Used to Prepare Effective Neisseria meningitidis Vaccines

R. Gasparini[1], D. Panatto[1], N. L. Bragazzi[1], P. L. Lai[1], A. Bechini[2], M. Levi[2], P. Durando[1] and D. Amicizia[1]

[1]Department of Health Sciences, University of Genoa, Via Pastore 1, 16132 Genoa, Italy
[2]Department of Health Sciences, University of Florence, Viale G.B. Morgagni 48, 50134 Florence, Italy

ABSTRACT

In the last decades, tremendous advancement in dissecting the mechanisms of pathogenicity of *Neisseria meningitidis* at a molecular level has been achieved, exploiting converging approaches of different disciplines, ranging from pathology to microbiology, immunology, and omics sciences (such as genomics and proteomics). Here, we review the molecular biology of the infectious agent and, in particular, its interactions with the immune system, focusing on both the innate and the adaptive responses. Meningococci

Citation: How the Knowledge of Interactions between Meningococcus and the Human Immune System Has Been Used to Prepare Effective Neisseria meningitidis Vaccines. R. Gasparini, D. Panatto, N. L. Bragazzi, P. L. Lai, A. Bechini, M. Levi, P. Durando, and D. Amicizia.

exploit different mechanisms and complex machineries in order to subvert the immune system and to avoid being killed. Capsular polysaccharide and lipooligosaccharide glycan composition, in particular, play a major role in circumventing immune response. The understanding of these mechanisms has opened new horizons in the field of vaccinology. Nowadays different licensed meningococcal vaccines are available and used: conjugate meningococcal C vaccines, tetravalent conjugate vaccines, an affordable conjugate vaccine against the *N. menigitidis* serogroup A, and universal vaccines based on multiple antigens each one with a different and peculiar function against meningococcal group B strains.

INTRODUCTION

The immune system protects humans from attack by microorganisms such as bacteria, viruses, protozoa, fungi, parasites, and organisms such as helminths. The skin is the first barrier and its protective action is enhanced by bodily secretions, such as sweat and sebum, which exert a broad antimicrobial activity [1, 2]. The mucous membranes are protected by external and internal secretions, such as tears, saliva, and mucus, which contain molecules that can neutralize bacteria. Tissues such as the skin and mucous membranes are populated by immune cells, which can act against the microorganisms that circumvent the first physical and biochemical barriers.

The immune system is very complex and its defensive response is subdivided into innate and adaptive responses [3]. The innate response triggers an immediate, nonspecific, general action and is activated by typical signs of infection. The adaptive response is able to develop a highly specific, extremely accurate action, which is stored in the so-called immune memory.

This paper provides an overview of the interaction between the immune system and Gram-negative bacteria with particular reference to *Neisseria meningitidis* in the perspective of developing new vaccines against this pathogen.

RAM-NEGATIVE BACTERIA AND IMMUNITY

Outer Membrane Components

Over thousands of years, bacteria have developed several mechanisms whereby they can circumvent the immune system. Specifically, Gram-negative bacteria possess a complex of envelopes, which allow the selective passage of nutrients into the cell and the excretion of metabolic waste outside. Structurally, Gram-negative bacteria possess an outer membrane (OM), which, together with the peptidoglycan and inner membrane (IM), delimits the periplasm and cytoplasm compartments. Many molecules of glycolipids, especially lipopolysaccharide (LPS), emerge from the outer leaflet of the OM, while, from the inner layer of the OM, lipoproteins reach the peptidoglycan, with which they engage. Moreover, proteins such as porins cross the OM; these are very important for the active, passive, and selective permeability of small molecules, ions, and water [4]. Most porins have a trimeric structure and an oval shape. The bacterial porins perform many functions; indeed, they help the microorganism to adhere to the cells of the host tissue and to evade the defence mechanisms of the human body, thereby favouring invasion of the host. They are also able to elicit both innate and adaptive immunity. Porins can inhibit phagocytic activity [5] and activate the complement system by means of both classic and alternative pathways [6]. For instance, Neisserial porins can activate the transport of NF-κB into the nucleus of B and dendritic cells (DCs) [7]. The DNA/NF-κB complex then recalls other proteins, such as coactivators and RNA polymerase, which transcribe the DNA into mRNA; finally, this mRNA is exported to the cytosol and translated into proteins. This leads to a change in the function of the cell; for example, the cell may begin to produce proinflammatory cytokines.

Porins are clearly involved in the induction of proinflammatory activity, although it is not known which toll-like receptors recognize them. By contrast, it is known that LPS stimulates toll-like receptors 2 and 4 [8]. Three distinct regions characterize LPS, namely, lipid A, which fixes the molecule to the outer leaflet of the OM, the core polysaccharide which binds to lipid A by means of a disaccharide phosphate bridge, and antigen O, which is the most distal portion. The general structure of LPS is fully conserved, while the core oligosaccharide is highly variable.

Toll-like receptors are a family of conserved signal transducers able to induce an innate immune response. To date, at least 11 mammalian TLRs have been identified. Their stimulation by bacterial components activates the innate immune response. TLR2 recognizes peptidoglycan, lipopeptides, and bacterial proteins. However, it is interesting that LPS can overstimulate the innate immune response, thereby eliciting inflammation. As a result, the normal defences may not function correctly. Furthermore, it should be borne in mind that TLR5 recognizes flagellin, which is the main component of bacterial flagella [9]. For example, mutations of the TLR4 gene contribute to development of severe meningococcal infections [10]. In addition, through the recognition of *N. meningitidis* DNA, TLR9 exerts strong protection against the microorganism [11].

Innate and Adaptive Immune Responses

The innate immune system is able to detect other conserved microbial components, called pathogen-associated molecular patterns (PAMPs), such as nucleic acid structures, lipoteichoic acid, and peptidoglycan [12]. The pattern recognition receptors (PRRs) of immune cells include, in addition to TLRs, the NOD-like receptors (NLRs) and the RIG-1-like receptors (RLRs), which are able to recognize microbial components in the cytosol [13]. TLRs, NLRs, and RLRs are able to activate mitogen-activated protein kinase (MAPK) and the transcription of NF-κB factor. A different set of NLRs helps to activate caspase-1 and the consequent assembly of inflammasomes [14]. The granulocytes and macrophages are the first cells that participate in the activation of the innate immune response. Shortly afterwards, the DCs and natural killer cells are activated. Specifically, neutrophils produce antimicrobial proteins, such as LL37, alpha and beta defensins, enzymes [15], interferons (IFN) alpha, beta, and gamma, C-reactive protein, and chemokines, contribute to activating the complement cascade. Macrophages produce reactive oxygen species (ROS) (e.g., H_2O_2) and reactive nitrogen species (RNS). Subsequently, DCs, which can also be activated by TLR2 and TLR4, activate natural killer (NKs) cells [16] and induce maturation of CD4+ T cells [17–22].

Many bacterial components are able to stimulate the adaptive human immune response. Porins can activate the translation of NF-κB in the nucleus of B and DCs, while class I Pilin E induces highly specific antibodies (Abs) and class II induces cross-reacting Abs. Furthermore, complement cascade activation, as well as particularly C3b activation, opsonizes antigens, thereby enabling APCs to activate the adaptive response.

NEISSERIA MENINGITIDIS AND IMMUNITY

Meningococcal Genome

Meningococci have developed several "immunoescape" strategies [23], the molecular bases of which can be understood by taking into account the nature of the Neisserial genome. Progress in the field of molecular biology and the introduction of high-throughput technologies (HTTs) have tremendously advanced our understanding of the complexity of the Neisserial machinery. By using sophisticated approaches such as whole-genome sequencing (WGS) and microarrays, functional genomics investigations have uncovered the mechanisms that facilitate or hinder *N. meningitidis* growth, colonization, and invasion and have helped to explain its extraordinary intrastrain variation and adaptation to the environment. Other techniques, such as genome-wide association studies (GWAS), have shed light on the pathogen-host interaction and the host's susceptibility to the microbe. Genomics and postgenomics have not only increased our knowledge of the biology and pathogenesis of *N. meningitidis* but proved to be extremely useful in discovering candidate antigens and in developing effective new vaccines [24].

Being a naturally competent pathogen, *N. meningitidis* has a highly dynamic, plastic, and flexible genome with a size range of only more than 2,000 kilobases [25]. This genome differs from other microbial genomes in that it lacks some of the typical two-component systems and sigma factors [26]. Despite being relatively small and compact, it has elaborated a variety of mechanisms that contribute to explaining its high adaptability both to host and to environment. Meningococcus is usually polyploid, containing up to 2–5 genomes—polyploidy being a sign of virulence—while *N. lactamica* is monoploid [27]. Neisserial pathogenicity is intrinsically polygenic [28] and is given by a variety of different pathogenic islands (PAIs or genomic islands, GEIs), including gonococcal genetic islands (GGIs) [29] and a recently discovered meningococcal disease-associated (MDA) island [30].

The nature of the Neisserial chromosome and the presence of extrachromosomal material contribute to explaining an important immunoescape strategy, known as structural or antigenic variation, which consists of camouflage of the Neisserial repertoire expressed. Basically, it can involve horizontal or lateral gene transfer (HGT/LGT) (mainly via transformation and, to a lesser extent, via conjugation and phage transduction) and allelic exchange/rearrangement of genes or gene portions taken up from the environment (Table 1). In addition, as its genome contains multiple copies of certain genes, for example, opacity factor proteins and pilins [31], homologous intragenic recombination also results in frequent surface structural variation.

Table 1. An overview of the most important immunoescape strategies exploited by *Neisseria meningitidis*.

Immunoescape mechanism	Details	References
Structural/antigenic variation	It consists in the modified expression of domains, which are antigenically different within a clonal population, by which the pathogen is able to escape the host immunity selection and circumvent the immune surveillance It usually involves LOS/LPS, opacity, and pilin proteins LOS/LPS and opacity factor structural/antigenic variation depends essentially on phase variation Pili antigenic variations depend on RecA-mediated recombination	[31, 75, 76]
Autolysis	It is mediated by OMPLA	[77]
Blebbing and microvesicles formation	The blebs originate as evaginations of the outer layer	[78]
Capsule switching	Due to microevolution, there is shift from serogroup B to serogroup C, from serogroup C to W-135, from serogroup Y to W-135, and from serogroup Y to B; nanostructured materials such as MWNTs and mesoporous silica increase transformational capacity	[30, 79–87]
Capsule modification	For example, modification of lipid A of meningococcal LOS/LPS with phosphoethanolamine protects *Neisseria* from neutrophils-mediated killing Another example is given by the O-acetylation of capsular antigens (LpxL2 gene mutants are indeed more virulent) LpxL1 gene mutants activate TLR4 less efficiently	[88]
Genome plasticity	HGT/LGT (via conjugation, transduction, and transformation) and homologous intragenic recombination	[25, 27, 30, 89]
Host modification	*Neisseria* exploits a bacterial sialyltransferase scavenging available host CMP-NANA for modifying LOS/LPS	[70]
Molecular mimicry	CP of serogroup B strain is a homopolymer of α2-8-linked sialic acid and is similar to NCAM-1 L-NNT in the lipopolysaccharide of virulent strains is similar to an antigen on red blood cells DMP19 acts as a DNA-mimic protein	[67, 69, 71–74, 90, 91]
Metabolic pathways	Examples are iron, lactate, glutamate uptake, utilization, and avoidance of neutrophil oxidation burst, ROS, and RNS	[92, 93]
Molecular decoy	FprB has an antigenic subdomain for binding antibodies, which is not essential for the functioning of the autotransporter; it also blebs with OMPs and LPS/LOS distract the immune system, directing the response away from the microbe	[94]

Immunotype switch	LPS immunotype switches from L3 to L8/L1 by lgtA, lgtC phase variation LOS immunotype can contribute to immunoescape	[95, 96]
Phages and prophages	The pathogen hosts a number of prophages, from the Mu-related family to the phage l-related group and the family of filamentous M13-like phages	[25, 30, 89]
Phase variation	High-frequency reversible changes can occur in the length of SSRs (of capsule, LOS, opacity factor, porin, adhesin, invasin, autotransporter, haemoglobin receptor, DNA mismatch repair, and pilin genes, termed as contingency genes and organized in modules called phasevarions) Other repeat sequences can be REP2, CRs, CREEs, and NIMEs Transposon-like elements can play a role Phase variation mediates resistance to antibiotics Phase variation mediates carriage persistence	[50, 52, 59]
Pilin conversion and modification	Pilin is posttranslationally modified by addition of a glycan, two phosphorylcholines, and a glyceramido acetamido trideoxyhexose residue	[97, 98]
Plasmid	Examples of plasmids that can contribute to *Neisseria* variability are pJS-A, pJS-B	[33]
Recruitment of human components of immune system	*Neisseria* escapes complement-mediated killing recruiting and sequestering fH to its surface	[91]
Temperature-regulated defence	RNA thermosensors finely tune the expression of CP components, LOS, and fHBP, thus protecting against human immune killing	[99]

CMP-NANA: cytidine 5′-monophospho-N-acetylneuraminic acid; CP: capsule; CRs: Correia repeats; CREE: Correia repeat-enclosed element; DNA: deoxyribonucleic acid; fH: complement factor H; fHBP: fH binding protein; HGT: horizontal gene transfer; lgt: prolipoprotein diacylglyceryl transferase; L-NNT: lacto-N-neotetraose; LOS: lipooligosaccharide; LPS: lipopolysaccharide; LGT: lateral gene transfer; MWNTs: multiwalled nanotubes; NCAM-1: neural cell adhesion molecule 1; NIME: Neisserial intergenic mosaic element; OMPs: outer membrane proteins; OMPLA: outer membrane phospholipase A; RecA: recombinase A; REP2: repetitive extragenic palindromic sequence; RNA: ribonucleic acid; RNS: reactive nitrogen species; ROS: reactive oxygen species; SSRs: simple sequence repeats; TLR: toll-like receptor.

Moreover, the pathogen hosts a number of prophages, from the Mu-related family to the phage l-related group and the family of filamentous M13-like phages [27, 32]. The most widely studied sites for phage integration are known as duplicated repeat sequence 3 (dRS3) [26], which belong to the family of Neisserial intergenic mosaic elements (NIMEs). Plasmids, such as pJS-A and pJS-B, also play an important role [33].

Another surface modulation occurs via phase variation, a process involving the modulation of gene expression via a mechanism of on/off switching (transition from an expressed state of the gene to an unexpressed one or vice versa). Besides this kind of "functional" phase variation, *Neisseria* can also undergo a "structural" switch, namely, a transition between two forms of a gene product. The genes, which are involved in this strategy, are termed "contingency genes" [26] and can be coupled and interlinked in structures called phasevarions (phase-variable regulons) [34], which have a regulatory function. Phase variation includes a variety of sophisticated mechanisms [35], such as slipped strand mispairing (SSM) [36], microsatellite instability [37], and reversible insertion of minimal mobile elements (MMEs) [38]. Therefore, these mechanisms can involve single nucleotides (homopolymeric repeats) or complex nucleotidic structures (short tandem repeats), occurring either upstream of a gene in the promoter or within an open reading frame (ORF)/coding sequence (CDS). Changes upstream of a gene result in modulation of its transcriptional efficiency and therefore of its final protein concentration. This is, for example, the case of Opc, porin A (*por*A), and *fet*A genes. Alterations within a gene, which insert *de novo* stop codons, alter the full translation of the gene. An example of this mechanism is provided by the opa genes and the genes coding for adhesins, such as *nad*A [39]. Phase variation of opa genes has been extensively characterized: they occur in four distinct copies and code for similar, but not identical, proteins. Phase variation can thus involve one copy or another, independently of each other, and can result in eleven variants. In this case, phase variation is therefore equivalent to antigenic variation. Besides opacity factor proteins, phase variation can involve up to hundreds of genes [40]: from the genes coding for pilins [41] or for proteins involved in genome maintenance and DNA repair [42, 43] to genes encoding proteins involved in the cell cycle control and regulation [44], autotransporters [45, 46], or enzymes like the pilin phosphorylcholine transferase *ppt*A [47] or the glycosylase *mut*Y [48], among others [49, 50]; the reader is referred to Table 2, which provides a more detailed overview of the phase-variable genes. Moreover, new mechanisms leading to phase variations have been discovered [51].

Table 2. An overview of the most important genes and gene products of *Neisseria meningitidis* involved in immunoescape mechanisms.

N. meningitidis molecule	Immunological role	Reference
*ani*A	A nitrite reductase: it protects *Neisseria* from nitrosative stress during both colonization and invasion	[90, 100–102]
App	It is phase-variable	[103]
ausI/MspA	An autotransporter and a serine protease; it is phase-variable	[45, 46]
Biofilm (and molecules involved in the biofilm synthesis, such as *aut*A or *hrp*A, or optimizing pathogen survival in biofilm, such as the alpha-peptide of IgA1 protease, *adh*C, *est*D)	Biofilm protects from macrophages; *adh*C is involved in S-nitrosoglutathione metabolism and in glutathione-dependent detoxification system; EstD is involved also in *Neisseria* colonization	[104–108]
Blebs (with OMPs and LPS/LOS) and SOMVs	They protect from neutrophils-mediated killing and NETs; they divert the immune response away from the pathogen	[78]
Capsule and molecules involved in the capsule synthesis such as *kps*C, *kps*S	It activates TLR2 pathway, it increases serum resistance, and it inhibits the classical pathway of complement	[109–111]
Cas9 and the CRISPR-Cas system	CRISPR-Cas9-mediated repression of bacterial lipoprotein expression facilitates evasion of TLR2 by the pathogen; it is involved in gene expression and regulation	[112, 113]
*cbp*A	It mediates zinc piracy and protects from nutritional immunity	[93]
Cps	As a gene, it is involved in the capsule biosynthesis; as RNA, it acts as a thermosensor; Cps gene amplification protects the pathogen	[99, 114]
CrgA	It is involved in the regulation of pili and capsule expression; it plays a major role in the infectious cycle of *Neisseria*	[114–116]
Css	As a gene, it is involved in the capsule biosynthesis; as RNA, it acts as a thermosensor	[99]
*ctr*A, *ctr*D	As genes, they are involved in the capsule export; as RNAs, they act as thermosensors; IS1301 in the IGR between *sia* and *ctr* operons mediates resistance to Abs	[99, 117, 118]

*cyc*P	It is involved in denitrification metabolism and protects *Neisseria* from nitrosative stress	[90, 119, 120]
dam	It is involved in phase variation and modulation	[42]
*dca*C	It is phase-variable	[40]
*din*B	A DNA polymerase IV belonging to the SOS regulon: it is involved in phase variation and modulation	[42]
DNA mismatch repair genes (*f pg*, *mut*L, *mut*S, *mut*Y, *rec*A, *r ec*N, *uvr*D)	They are phase-variables; they protect against oxidative stress	[42, 48, 51, 121]
drg	It is involved in phase variation and modulation	[42]
*far*A, *far*B, *far*R	They remove antimicrobial peptides, proteases, lysozyme, and acids from the bacterial cytosol and protect the pathogen	[122, 123]
*fbp*A, *fbp*B	They are involved in phase variation and modulation	[51]
Feta	It is involved in phase variation and modulation	[124–126]
fHbp (formerly known as GNA1870)	It is involved in phase variation and modulation; it protects *Neisseria* from complement-mediated killing, binding fH	[90, 127]
*frp*A, *frp*B, *frp*C	They are phase-variable; they can act as a molecular decoy	[124, 125, 128]
funZ	It is a site of bacteriophage insertion; it is phase-variable	[49]
fur	It is involved in phase variation and modulation; it tunes the gene expression of virulence genes	[102, 129]
ggt	It regulates pathogen growth	[130]
Ght	It is involved in the capsule biosynthesis and in the resistance mechanisms of the pathogen	[131, 132]
*glt*T	It favours meningococcal internalization into human endothelial and epithelial cells; it regulates pathogen growth	[133, 134]
H.8	AAEAP motifs are target for generation of blocking Abs	[135–138]
Haemoglobin-linked iron receptors (*hpuA*, *hpuB*, *hmbR*)	They are involved in phase variation and modulation	[43, 139–141]

Hfq	A RNA chaperone: it is involved in stress response and virulence and is a pleiotropic regulator of protein expression	[142]
*hsd*S	It is phase-variable	[49]
IgA protease	It cleaves secretory IgA, hinders Ab binding and function, and may play role in biofilm formation; it cleaves lysosomal LAMP1 in epithelial cells; moreover, it is phase-variable	[122, 142, 143]
*kat*A	It confers resistance to RNS, including peroxynitrite (PN), protects against ROS, and detoxifies H_2O_2	[90, 102, 122, 144]
Laz	A lipid-modified azurin: it protects against hydrogen peroxide and copper toxicity; it promotes *Neisseria* growth and survival	[135, 138, 145]
lbpA, *lbpB*	They are involved in iron acquisition and metabolism; they are phase-variable; moreover, the release of LbpB enables *Neisseria* to escape from complement-mediated killing	[90, 122, 146]
*lct*P	Its inactivation results in C3-mediated cell lysis	[102, 147, 148]
*lgt*A, *lgt*B, *lgt*C, *lgt*D, *lgt*E, *lgt*G	They are involved in LOS biosynthesis and are phase-variable; for example, *lgt*A or *lgt*C phase variation mediates LPS immunotype switch from L3 to L8/L1	[60]
LOS/LPS	It protects from macrophages; strains of the same species produce different LOS glycoforms	[122]
*lpt*A	It adds a phosphoethanolamine group to lipid A and confers resistance to defensins and cathelicidins	[90, 149]
Lst	LOS sialylation (by the enzyme Lst) prevents complement deposition and phagocytosis by neutrophils	[122, 150]
*mes*J	It is phase-variable	[49]
Msf	It binds to vitronectin; it increases serum resistance	[151]
Mip	It tunes gene expression	[102, 152, 153]
*mis*R, *mis*S	They are phase-variable; they are involved in capsule regulation and modification	[114, 154]
*mlt*A (formerly known as GNA33)	It tunes gene expression	[155]
*mnt*A, *mnt*B, *mnt*C	They protect against oxidative stress	[122, 156]

*mod*A, *mod*B	They are phase-variable	[34]
*msr*A, *msr*B	They are involved in the methionine sulfoxide reduction and they repair oxidized proteins	[122, 157]
*mtr*C, *mtr*D, *mtr*E	They protect against cationic antimicrobial peptides and toxic hydrophobic molecules	[122, 158, 159]
nadA and its regulator *nad*R	It binds to Hsp90, recruits ARF6 and Rab11, and activates human monocytes and macrophages, triggering IFN-gamma and R-848 dependent pathways; it interacts with beta1 integrins; it is phase-variable	[39, 160–165]
*nal*P	An autotransporter protease: it cleaves C3, facilitates degradation of C3b, and enhances Neisserial survival in human serum; it stabilizes the biofilm; moreover, it is involved in the processing of other proteases, such as the proteases which release LbpB, whose release enables *Neisseria* to escape from complement-mediated killing; NalP processes also App and IgA1 protease; it has an important role in the virulence of the pathogen	[24, 102, 166]
Nhba (formerly known as GNA2132)	It tunes gene expression	[167]
nhhA	It activates TLR4-dependent and independent pathways; it triggers apoptosis in macrophages; it increases serum resistance and protects from phagocytosis and complement attack; it is essential for colonization	[168, 169]
*nif*S	It is phase-variable	[49]
*nir*K	It protects *Neisseria* from nitrosative stress during colonization and invasion	[170, 171]
*nor*B	It favours the pathogen growth, enabling utilization and consumption of NO during microaerobic respiration, enhances pathogen survival, protects *Neisseria* from nitrosative stress during colonization and invasion, decreases and downregulates the production of NO-regulated cytokines, such as TNF-alpha, IL-12, IL-10, CCL5 (RANTES), and CXCL8 (IL-8), and prevents host cell S-nitrosothiol formation	[100, 119, 120, 170, 172]

*nsp*A	It binds to factor H and inhibits AP	[122, 173–175]
*nsr*R	It is involved in denitrification metabolism and protects *Neisseria* against nitrosative stress	[176, 177]
*oat*W, *oat*Y	They tune gene expression	[178]
Opa	It interacts with CEACAM, promoting endothelial cell attachment and upregulating endoglin (CD105) and cooperation with 1 integrins; it elicits innate host defences and actively suppresses adaptive immune responses that would eliminate the pathogen	[179–184]
Opc	It binds to vitronectin, it inhibits AP, and it increases serum resistance; it elicits innate host defences and actively suppresses adaptive immune responses that would eliminate the pathogen	[179–182, 184]
*oxy*R	It regulates catalase expression and is involved in the protection from oxidative stress	[185, 186]
P36	It is involved in Neisserial adhesion.	[187]
*pac*A, *pac*B	They are involved in the composition and regulation of peptidoglycan membrane	[188]
*pgl*A, *pgl*B, *pgl*G, *pgl*H	They are phase-variable	[60–62]
Pili	They alter the expression levels of human genes known to regulate apoptosis, cell proliferation, inflammatory response, adhesion, and genes for signaling pathway proteins such as TGF-beta/Smad, Wnt/beta-catenin, and Notch/Jagged	[189]
*pil*C1	It interacts with mucosal surface and mediates the crossing of the BBB	[41, 169]
P*ilE*, *pil*S	They are involved in nonreciprocal recombination-based antigenic variation	[76]
P*ilE*, *pil*V	They bind to CD147 for vascular colonization; they mediate also *Neisseria* internalization	[190, 191]
*pil*P, *pil*Q	They are involved in pilus biogenesis and outer membrane stabilization	[51, 192]

porA	It binds to fH, C3b, C4b, and C4bp (more strongly under hypotonic conditions); it increases serum resistance; it is involved in phase variation	[122, 139, 173]
*por*B	It inhibits factor H-dependent AP; it interacts with TLR1 and TLR2 and activates I B, MAPK/MAPKK, and PTK pathways, leading to CD86 up-regulation, to IL-6, IL-12, and TNF secretion in B cells and DCs, and to IgB secretion in B cells	[122, 173]
*ppt*A	It is phase-variable	[47]
Ppx	It is an exopolyphosphatase whose mutation protects *Neisseria* from complement-mediated killing; it interacts with the AP of the complement activation	[64]
*rmp*M	It is involved in phase variation and modulation	[193, 194]
Sialic acid synthase (*neu*B, *sia*A, *sia*B, *sia*C, *syn*C)	They are phase-variable	[102]
*sod*B, *sod*C	They protect from phagocytosis by human monocytes/macrophages	[102]
tbpA, *tbpB* (also known as *tbp*1, *tbp*2)	They are involved in nutritional immunity	[121]
TdfF	It is involved in intracellular iron acquisition and is found only in genomes of pathogen strains	[28]
Temperature sensors (such as RNA thermosensors located in the 5′ UTRs of genes necessary for capsule biosynthesis, the expression of fHbp, and sialylation of LOS/LPS)	Activated by coinfecting pathogens, they recruit mechanisms of resistance and immunity escape	[99]
*ton*B	It is involved in nutritional immunity, supplying energy to the pathogen	[93]
Uncharacterized proteins (NGO1686, NMB0741, NMB1436, NMB1437, NMB1438, and NMB1828)	They protect from nonoxidative factors, but their mechanisms are still not understood; NMB1436, NMB1437, and NMB1438 are putatively involved in iron metabolism	[122, 195]
Uncharacterized factor (NMA1233)	It is involved in phase variation and modulation	[26, 51]
*xse*B	It is involved in phase variation	[26]
znuD	It protects from neutrophils and nutritional immunity	[92]

Ab: antibody; AP: Alternative Pathway; ARF6: ADP-ribosylation factor 6; App: adhesion and penetration protein; BBB: blood-brain barrier; cbp: calprotectin binding protein; CEACAMs: carcinoembryonic antigen-related cell adhesion molecules; CRISPR: clustered regularly interspaced short palindromic repeats; ctr: capsule transport apparatus; dam: DNA adenine methyltransferase; drg: dam replacing gene; fur: ferric uptake regulator; ggt: gamma-glutamyl aminopeptidase; hsp: heat-shock protein; IgA: immunoglobulin A; lbp: lactoferrin binding protein; lct: lactate permease; LOS: lipooligosaccharide; Mip: macrophage infectivity potentiator; mltA: membrane-bound lytic transglycosylase A; IGR: intergenic region; Msf: meningococcal surface fibril; Msr: methionine sulfoxide reductase; NadA: *Neisseria* adhesion A; NhhA: *Neisseria* hia homologue A; oat: O-acetyltransferase; OMV: outer membrane vesicle; opa: opacity-associated protein a; opc: opacity-associated protein c; pac: peptidoglycan O-acyltransferase; pil: pilin; por: porin; RNA: ribonucleic acid; RNS: reactive nitrogen species; Sod: superoxide dismutase; SOMVs: spontaneously released OMVs; Tbp: transferrin-binding protein; TLR: toll-like receptor; UTRs: untranslated regions; uvr: ultraviolet resistant.

The mechanism implying MMEs involves different kinds of genetic elements, such as the Correia repeats (CRs) and the Correia repeat-enclosed elements (CREEs), known also as the *Neisseria* miniature insertion sequences (NEMIS) [52, 53], which constitute about 2% of the Neisserial genome [54]. Other genetic elements are the insertion sequence (IS) elements, such as IS1016-like, IS1106, IS1301 [55, 56], and IS1655 [57].

It is worth noting that the number of genes involved in phase variation is enormously greater than for any other pathogen studies so far [58]. Some genes are "phasotypes"; that is to say, they play a role in carriage and are downregulated, favouring host persistence [59].

As already mentioned, in some cases antigenic/structural variation and phase variation, albeit conceptually two distinct mechanisms, cooperate in increasing the genetic complexity of the Neisserial genome. Antigenic variation of LPS, for example, can derive from phase variation of one or more enzymes involved in the synthesis of the oligosaccharide chain by SSM, or by modification of LPS, for example, by glycosylation [60–62], sialylation [63, 64], or acetylation [65, 66], which, moreover, confer resistance to neutrophil-mediated killing.

Thus, both antigenic and phase variations concur in enabling *Neisseria* to evade the immune system [26, 27].

Meningococcal Capsule

LPS and themeningococcal capsule (CP) are the two major virulence factors of *N. meningitidis*. Specifically, the capsule displays a large variability of surface antigens, on the basis of which 13 different *N. meningitidis* serogroups have been identified. The CP contributes in an important way to the camouflage of the microorganism, which thus can better circumvent the immune system's defences. The clearest expression of this phenomenon is given by the molecular mimicry [67]. This can be seen in the nature of the polysaccharide CP of serogroup B meningococcus, a homopolymer of α2-8-linked sialic acid, which is identical to a neural cell adhesion molecule, NCAM-1 [68]. Moreover, lacto-N-neotetraose (L-NNT) in the lipopolysaccharide of virulent strains is similar to an antigen expressed on red blood cells [69–73]. Further mechanisms of molecular mimicry have been recently discovered and described [74].

During the first phase of infection, meningococcus has to avoid the surface defences of the nasopharynx, such as the peptides secreted at the mucosal surface [90] and IgA secretory Abs [90, 196]. To this end, the meningococcus can aggregate into clusters and produce abundant OM vesicles (OMVs), thus managing to hide its surface antigens and to deflect the action of the surface defences from the bacterial cell [196]. In addition, the CP protects *Neisseria* from cationic antimicrobial peptides (CAMPs), including cathelicidin [196]. Conversely, the presence of capsular polysaccharide restrains the invasion and colonization of the nasopharyngeal barrier by hiding the adhesins and invasins of the meningococcus [143, 158]. On the other hand, the presence of the capsule may allow the microorganism to pass unharmed through the intracellular environment by exploiting the system of cell microtubules, at least in the case of serogroup B *N. meningitidis* [197]. Moreover, the CP is essential for meningococcal growth and survival in the bloodstream and cerebrospinal fluid, increasing serum resistance. During the different stages of infection, the capsule may hinder or promote the survival of *N. meningitidis* in the human host; indeed, the microorganism can modulate the production of capsule components, which depends on 4 genes, three of which—*sia*A, *sia*B, and *sia*C—are situated in the *cps* locus. The *sia*D gene induces the production of polysialyltransferase, which allows the polymerization of sialic acid. For instance, in the early stages of infection, the production and assembly of sialic acid are downregulated [198]. Another example of polysialyltransferase system is given by *oat*WY [178].

In addition to the above-mentioned actions, the most important virulence activity of the CP is probably the significant impairment of both Neisserial adherence to DCs and the phagocytic killing of bacteria; indeed, the CP inhibits both the opsonic and the nonopsonic phagocytosis of *N. meningitidis* [199]. It prevents the formations of effective Abs against *N. meningitidis*.

CP downregulates both classical and alternative complement pathways and prevents the proper insertion of the membrane attack complex (MAC) [200, 201]. LPS also contributes to inhibiting MAC deposition [201, 202].

Moreover, CP switching contributes to escaping detection and killing. This is a complex phenomenon due to microevolution and usually involves Neisserial strains expressing sialic acid (e.g., the shift from serogroup B to C, from serogroup C to W-135, from serogroup Y to W-135, and from serogroup Y to B) [79–82, 203]. The molecular basis is provided by the allelic replacement of the sialic acid CP polymerase.

Surprisingly, nanostructured materials such as multiwalled carbon nanotubes (MWNTs) and mesoporous silica have been found to increase *Neisseria*'s transformational capacity [83, 84].

Major and Minor Adhesion Mechanisms of *N. meningitidis*

N. meningitidis possesses a multifaceted system of adhesins, such as pili and other systems of adhesion (i.e., opacity-associated proteins OpA and OpC). Adhesion is probably a coordinated process mediated first by pili, which are composed of several proteins; the most important of these is Pilin E (PilE) [204], but Pilin C (PilC) [205] and the secretin Pilin Q (PilQ) [206] have also been described. PilE is the main constituent of the *Neisseria* type IV pilus. In 1984, Diaz et al. identified proteins I and II as the main components of the type IV pili and noted that Abs against protein I were highly specific [207]. Subsequently, Pilin E was classified as belonging to class I and class II. Class I Pilin E is highly variable, while class II Pilin E is highly conserved [208, 209]. For this reason, class II Pilin E has been suggested as a candidate antigen for a vaccine against meningococci [147]. The regulation of pilin genes is quite complex [97, 115, 116, 190].

Other components are as follows: *pil*V [210], *pil*P, *pil*D (a prepilin-processing leader peptidase), *pil*F and *pil*T (traffic NTPases), *pil*G (involved in the pilus assembly), *pil*M (pilus biogenesis protein), and *pil*W (involved in the pilus stabilization) [148], among others.

Although the interactions between type IV pili and cellular receptors are not completely known, they may interact with a membrane cofactor protein, named CD46 receptor, and with alpha 1 and alpha 2 integrins [211]. However, it is known that pili contribute to the aggregation of Neisserial cells [212]. This fact, associated with the ability of pili to act as a signalling protein by interacting directly with the β2-adrenergic receptor, contributes to the depletion of junction proteins, thus helping meningococcus to pass through the epithelial and endothelial cells and, subsequently, to cross the blood-brain barrier (BBB) [75, 213].

Although pili are essential to the first phase of Neisserial adhesion to the cells of the airways, other adhesion molecules, such as LPS and porin A, intervene to strengthen the microbial bond. In particular, OpcA and OpcC appear to be very important; indeed, OpcA binds carcinoembryonic antigen cell adhesion molecules (CEACAMs), heparan sulphate proteoglycan (HSPG) and integrins [179–183, 214]. Opc proteins can combine with HSPGs and, through vitronectin and fibronectin, with their integrin targets. Furthermore, Opa proteins are able to elicit innate human defences that favour the survival of *N. meningitidis*, while actively suppressing adaptive immune responses that would eliminate the pathogen [184]. The variability of the expression of different Opa proteins could play a major role in prolonging the state of infection by circumventing the humoral host immune response [215].

The adhesion and penetration protein (*app*) [103], which is a member of the autotransported family of secreted proteins, owes its name to its ability to adhere to human cells, thereby favouring the entry of Neisseriae. To circumvent the immune system, meningococci possess formidable machineries that allow them to secrete proteins in different manners; in particular, Neisseriae mainly use the autotransporter pathway (also known as type V secretion system) [216]. The first-described protein belonging to the autotransporter superfamily was an IgA protease from *N. gonorrhoeae* [217]. MspA (meningococcal serine protease A) is another putative autotransporter protein. Not all strains of *Neisseria gonorrhoeae/meningitidis* possess the gene for MspA/AusI (also known as NMB1998). However, Turner et al. [218] demonstrated that convalescent sera from subjects affected by serogroup B invasive disease recognized the MspA antigen. NhhA (*Neisseria* hia/hsf homologue A, also known as Hsf) and Neisserial adhesin A (NadA) also belong to the autotransporters. Nhha contributes to bacterial adhesion by binding heparin sulphate and laminin. In addition, through the activation of caspase, NhhA increases the rate of macrophage apoptosis [168, 219].

NadA, which is expressed by 50% of virulent strains [160], but only by 5% of the strains isolated from carriers, is of particular interest because it is one of the components of the 4CMenB (Bexsero) vaccine and binds beta1 integrins [220].

N. meningitidis: Avoidance Mechanisms against Reactive Oxygen Species (ROS), Reactive Nitrogen Species (RNS), and Antimicrobial Peptides (AMPs)

When in contact with the mucosa of the nasopharynx, *N. meningitidis* can implement several strategies of adhesion, but it must overcome many barriers of innate immunity. We have already mentioned how the capsule allows bacteria to mitigate the activity of DCs. However, it must elude other substances, such as the reactive oxygen species and reactive nitrogen species produced by macrophages and the antimicrobial peptides produced by activated neutrophils. As already mentioned, the capsule protects *N. meningitidis* from LL-37 cathelicidin, but LPS also contributes to the resistance of the bacterium against this cathelicidin [156].

Furthermore, the toxic action of ROS is neutralized by the secretion of enzymatic proteins, such as catalase and superoxide dismutase [144, 149, 221]. The gene that codes for catalase is katA and is regulated by OxyR [185, 186, 222], while SodB and SodC code for superoxide dismutase [223]. Laz, a lipid-modified azurin, protects the pathogen against H_2O_2 and cupper toxicity [135–138, 145, 176].

In addition, *N. meningitidis* possesses genes, which encode enzymes able to exert a denitrification action, such as aniA, CycP, nirK, nsrR, and norB. They favour the growth of the pathogen, enabling utilization and consumption of NO during microaerobic respiration, enhance pathogen survival, and protect *Neisseria* from nitrosative stress during colonization and invasion by preventing host cell S-nitrosothiol formation. Moreover, they reduce and downregulate the production of NO-dependent cytokines, such as TNF-alpha, IL-12, IL-10, CCL5 (RANTES), and CXCL8 (IL-8) [100, 101, 119, 120, 130, 157, 170, 172, 177, 224].

Contemporarily, already at the level of the mucosa, the microorganism must resist the complement system.

Another interesting mechanism is the strategy whereby *N. meningitidis* escapes the attempts of the host to sequester nutrients essential for growth and survival of the pathogen. This process has been termed "nutritional immunity" [131, 139]. The microbe is endowed with OM receptors (such as HmbR, HpuA or HpuB, TbpA or TbpB, and TdfF) for acquiring iron and other important metals [93, 124, 125, 128, 129, 140, 195, 225]. ZnuD is a high-affinity zinc uptake receptor, which plays an important role in enabling the pathogen to evade neutrophil-mediated killing [226, 227]. CbpA, a receptor for calprotectin, a protein released by neutrophils during inflammatory processes, is upregulated when *N. meningitidis* suffers from zinc limitation [226, 227]. Further examples of metabolic enzymes involved in nutritional immunity are glutamate transporters or molecules taking part in the carbon cycle [132–134, 228–230].

How Meningococcus Circumvents the Complement System

Three pathways can activate the complement functions, namely, the classic pathway, the alternative pathway, and the lectin pathway. All three of these pathways contribute to the transformation of C3 to C3b [231].

The alternative pathway acts by comparing self- with non-self-antigens and is activated by anything that differs from the markers of host cells. Specifically, factor H recognizes host-associated molecular patterns (HAMPs). Properdin, first identified in 1959, is another protein that can directly activate the alternative pathway of the complement system [232]. It has been demonstrated that properdin deficiency favours recurrent episodes of *N. meningitidis* infection [233].

Meningococci produce three different variants (1, 2, and 3) of a protein that binds factor H. This protein, named fHbp (factor H binding protein) or GNA1870, blocks activation of the alternative pathway of the complement system. Indeed, by surrounding themselves with fHbp, *N. meningitidis* cells capture and inactivate factor H. Thus, it is easy for the microorganism to survive and reproduce, especially in the bloodstream and cerebrospinal fluid. Hence, it is important that the 4CMenB (Bexsero) vaccine contains variant 1 of this protein, which is often expressed by virulent meningococcal strains [228]. Another vaccine (Trumenba), recently licensed in the USA, contains recombinant variants 1 and 2 of fHbp from *N. meningitidis* serogroup B, A and B subfamilies (A05 and B01, resp.) [234, 235]. The proteins are produced by exploiting an advanced genetic engineering technique, using *E. coli* as a vector.

The Adaptive Immune Response against *Neisseria meningitidis*

Microorganisms such as *N. meningitidis* are able to change many exposed surface proteins, while the polysaccharides, which constitute the capsule, are T-independent (TI-2) antigens and can activate B cells directly, without the intervention of the MHC. However, TI-2 antigens do not induce an efficient secondary response and do not induce the production of avid immunoglobulins. Rather, they induce the production of short-lived Abs belonging to the IgM class [236].

In addition, it is important to consider that, in infants and children, the development of the immune system is a dynamic process, which begins *in utero* and continues for months and even years after birth. This explains why many components of the immune system are inefficient or partially efficient in infancy and early childhood [237–239]. For this reason, most cases of meningitis and sepsis from *N. meningitidis* occur under the age of 4 years and particularly in the first year of life.

The critical role of bactericidal Abs against the exposed surface antigens of *N. meningitidis* has been demonstrated by several studies. Indeed, Goldschneider et al. [240] showed that only individuals without bactericidal Abs contracted the clinical disease. In addition, the successful therapeutic use of immune sera, which markedly reduced lethality when first implemented by Flexner [241], has shown the central role of these Abs in protecting against invasive disease. The opsonic activity of Abs is also very important in the protection against and the recovery from meningococcal disease, as is demonstrated by the role of neutrophils, macrophages, and DCs in combating *N. meningitidis*. It is also well known that the cerebrospinal fluid of patients contains large numbers of neutrophils full of microorganisms. These clusterings of neutrophils are known as neutrophil extracellular traps (NETs) and massively release cathepsin G. *N. meningitidis* circumvents these traps by blebbing spontaneously released OMVs (SOMVs). Other strategies that the pathogen exploits are modification of lipid A of LPS with phosphoethanolamine protected and upregulation of ZnuD [92].

The adaptive immune response has been studied in carriers and during both the invasive period leading to clinical meningitis and the convalescence period. T cell response is "two-faced"; while proinflammatory T cells may indeed blunt the invasive power of the pathogen, the induction of the Treg response, which is able to limit virulence, carries the price of the reduced effectiveness of the protective response, especially in children [242]. During infection, increased meningococcal antibody titres can be detected from the

4th day, peaking at the end of the third week or the beginning of the fourth week and showing a correlation with the severity of the disease and the age of the patient. In the acute period of the disease, the number of T cells generally drops, while that of B cells increases; by the end of the second week, IgG levels decline and IgM levels rise [243]. In particular, abnormalities in T cell response can be detected, such as an elevated percentage of CD25+ and HLA-DR+ T cells, an increase in CD4+ CD45R+ (suppressor-inducer) cells, with subsequent expression of activation antigens, and a decrease in CD4+ CDw29+ (helper-inducer) cells [244]. During convalescence, an age-associated Th response can be observed: specifically, a Th1 response (low IL-10/IFN-γ ratio) and a highly proliferative Th2 response (higher IL-10/IFN-γ ratio) can be detected in younger and older patients, respectively [245]. Generally, a significant CD4+ T central memory response, with serum bactericidal antibodies, a marker of protective immunity, can be found [246]. However, the above-mentioned age-related mucosal T effector/memory cell response may also be present without bactericidal antibodies [247].

Other Immunoescape Strategies

Temperature fluctuation plays an important, although underscored, role in microbial pathogenesis, colonization, invasion, and host evasion. In contrast to mammals that maintain constant body temperature, pathogens' and other animals' temperature oscillates on a daily basis. Loh and collaborators [99] have identified the molecular bases of this temperature-dependent strategy. They have studied three RNA thermosensors located in the 5′ untranslated regions (UTRs) of genes involved in the CP biosynthesis, the expression of fHbp, and sialylation of LOS/LPS. Increased temperature (e.g., during inflammation by coinfecting pathogens, such as influenza virus) "alarms" the meningococcus and triggers its defence mechanisms against human immune killing. This could be a key determinant for the transformation of a symbiont pathogen into a virulent one. However, the precise nature of this mechanism remains elusive.

Clustered regularly interspaced short palindromic repeats- (CRISPR-) Cas9 is another molecular tool that *Neisseria* can use in order to divert immune surveillance. It is an intrinsically ambivalent device, since, on one hand, being involved in gene expression and regulation, it restricts the possibility of editing the *Neisseria* genome via HGT/LGT or the insertion of exogenous nucleic material, and therefore it limits the microbial variability. On the other hand, CRISPR-Cas9-mediated repression of bacterial lipoprotein expression facilitates evasion of TLR2 by the pathogen [112].

Another mechanism is the molecular decoy, with which the microbe deceives the human immune system. For example, *fpr*B has an antigenic subdomain for binding antibodies, which is not essential for the functioning the autotransporter [94].

The example of *fpr*B is useful to show how *Neisseria* can use concurrently the previously described immunoescape strategies: *fpr*B is subject to a high degree of antigenic variation, is a phase-variable gene, is involved in nutritional immunity, and moreover exploits a molecular decoy. Neisserial carbohydrates, mimicking host carbohydrates, circumvent immune system and at the same time exploit their mimicry to recruit fH [91].

MENINGOCOCCAL VACCINES

Polysaccharide Vaccines

As at least six meningococci are pathogenic in humans, the development of meningococcal vaccines has been a challenge. The first step towards solving the problem was to find a common denominator among antigens that showed high variability. In 1969, Gotschlich et al. [248] correctly put the first imperfect piece of the puzzle in place by demonstrating the possibility to extract the capsule polysaccharide. However, although high molecular weight groups A and C meningococcal polysaccharides proved immunogenic in adults [249–251], important limitations of this vaccine emerged in subsequent years. Indeed, while both group A and group C vaccines proved effective in USA and Italian recruits [252, 253] after one administration, they displayed only short-term efficacy in older children and adults. Moreover, vaccines against serogroup C did not prevent disease in infants, and the efficacy of group A vaccines in children under 1 year of age was unclear. The immune response occurred 10 days after vaccination. In schoolchildren and adults, one dose of these vaccines seemed to provide protection for at least 3 years [254]. These findings are explained by the characteristics of the antigens contained in the vaccines. Indeed, polysaccharides with repeating epitopes induce an immune response with the following characteristics [255, 256]:(i) The response occurs between the ages of 3 and 18 months but is variable; in children less than 2 years of age, the response is usually poor; the affinity maturation of the Abs does not occur.(ii)The immunologic memory is not stimulated.(iii)Almost all (>90%) Abs produced belong to the IgM class and are produced in the spleen [257, 258]. Furthermore, several studies have suggested that vaccination with large amounts of polysaccharides induces immune tolerance towards these antigens [259].

Polysaccharide Conjugate Vaccines

Three immunogenic carrier proteins are generally used inpolysaccharide conjugate vaccines, namely, diphtheria or tetanus toxoid, CRM197 (a nontoxic variant of diphtheria toxin obtained by molecular biology), and a complex of outer membrane protein (OMP) mixture from *N. meningitidis*. The toxoids were selected as carriers, firstly because of their immunological potency and secondly because if the recipient had previously been immunized with the toxoid, a booster effect was predictable. Moreover, under particular conditions, suppressive effects can also occur. The conjugate vaccine against *N. meningitidis* serogroup C had great success in several countries (UK, Netherlands, Spain, Italy, etc.) [260]. Today, tetravalent conjugate vaccines ACWY are also available in both the USA and Europe [260–262].

Vaccines against *N. meningitidis* Serogroup B

The most difficult problem was to prepare a vaccine against *N. meningitidis* serogroup B strains. Indeed, the maximum expression of camouflage is found in the capsule of these strains. Specifically, the polysaccharide of serogroup B meningococcal strains is a homopolymer of sialic acid residues and has structural similarities to brain glycoproteins. For this reason, it was impossible to prepare a polysaccharide antimeningococcus B vaccine. To overcome this obstacle, OMVs-vaccines were developed and used during epidemics caused by *N. meningitidis* serogroup B strains. However, given the high variability of the proteins, such as porins, present in OMVs, these vaccines were effective only against very specific epidemic hypervirulent strains [263, 264]. In order to develop a vaccine against meningococcus B, other strategies were therefore implemented.

The most promising results were obtained through reverse vaccinology. This involves identifying the antigens for the vaccine not in the classic way—that is, from the components of the bacterium—but instead from the genes that express the proteins with the best characteristics to be good antigens for the vaccine. To obtain a universal vaccine against serogroup B meningococcal strains, complex bioinformatics software was also applied. Following the complete sequencing of the meningococcus B genome [265], researchers at Novartis Vaccines and Diagnostics identified 600 ORFs, which expressed proteins that are exposed on the surface of the bacterial cell. Subsequently, 350 proteins were successfully expressed in *E. coli*, purified, and used to immunize mice. Later, 28 novel protein antigens able to elicit Abs with bactericidal activity were identified. Finally, three of these 28 proteins were selected, namely, NHBA (GNA2132, fused with GNA1030 protein), fHbp (fused with GNA2091 protein), and *nad*A. NHBA (*Neisseria* heparin binding antigen), which is present in virtually all strains, binds heparin, which may increase the serum resistance of bacteria. fHbp (factor H binding protein) binds factor H, which enables the bacterium to survive in theblood [266, 267], thereby blocking the alternative pathway of the complement system. *nad*A (Neisserial adhesin A) promotes adherence to and invasion of human epithelial cells [161, 162]. In addition, the vaccine developed by Novartis Vaccines and Diagnostics contains a fourth component, namely, the vesicle of the OM from the New Zealand strains, which contain porin 1.4. Theoretically, this vaccine should elicit bactericidal Abs against the following:(i)NHBA, thus increasing the bactericidal activity of the serum.(ii)fHbp, thus exposing *N. meningitidis* to the alternative pathway of the complement system.(iii)NadA, thus hindering the adherence of *N. meningitidis* to epithelial cells.(iv)Porin A 1.4 and other components of the mixture of proteins contained in OMVs; indeed, OMVs can enhance the immune response by functioning as a conjugate complex of proteins or, rather, a complex of adjuvants.Several clinical trials have demonstrated the immunogenicity and safety of this vaccine, also named 4CMenB, in infants, children, adolescents, and adults [268, 269]. Consequently the vaccine has been approved by the EMA and named Bexsero [270]. It has also been approved by several national drug agencies (such as FDA), including the Agenzia Italiana del Farmaco (AIFA, Italian National Agency for Drugs) [271].

Owing to the wide variability of Neisserial antigens, a particular laboratory test (MATS) has been developed to estimate the potential effectiveness of the vaccine. Studies conducted worldwide have shown the potential effectiveness of Bexsero, which has been estimated at 87% in Italy [85].

Trumenba was approved for individuals aged 10 to 25 years in the USA in October 2014 [85]. The potential of the vaccine antigen was tested in the laboratory [272] and on a murine model [273, 274]. The immunogenicity, safety, and tolerability of this vaccine were investigated in a randomized controlled trial in infants aged 18–36 months, who were subdivided into three cohorts (receiving 20-, 60-, and 200-μg rLP2086 dose, resp.) and matched with two control groups: one vaccinated against hepatitis A virus (HAV) and the other with a saline placebo. After the vaccination cycle, seroconversions against Neisserial strains expressing LP2086 variants homologous to the vaccine antigens were found in 61.1–88.9% of toddlers and against strains expressing heterologous LP2086 variants in 11.1–44.4%. Adverse reaction rates were negligible and the vaccine proved to be safe and well tolerated [275]. However, another randomized phase 1/2 clinical study found high fever rates in toddlers receiving one 20- or 60-μg rLP2086 dose (64% and 90%, resp.) [276].

In a randomized study performed in the USA and Europe in a sample of adolescents (11–18 years of age), Trumenba proved to be highly immunogenic. The proportion of vaccinees with human serum bactericidal activity (h-SBA) titres with a ≥4-fold rise against hypervirulent Neisserial strains with different variants of fHbp was in the range of 75–100%. In another randomized clinical study, carried out in Australia, the safety and immunogenicity of the vaccine were assessed in 60 healthy adults (18–40 years of age) who received 120 μg doses at 0, 1, and 6 months. The percentage of seroprotected vaccinees was 94.3% against the homologous strains and 70–94.7% against the heterologous strains. The vaccine was well tolerated [277, 278].

As fHbp is also expressed by Neisserial serogroups other than B, the anti-fHbp Abs elicited by rLP2086 might exert a bactericidal effect on meningococci, such as those against *N. meningitidis* serogroup C, as proved by Harris and collaborators [279] and by Konar and colleagues [280]. Moreover, there is some evidence that Trumenba could, at least in part, have an effect on carriage and reduce the risk of acquiring some hypervirulent strains [281].

New Vaccines

The currently available *Neisseria* vaccines, described in the previous paragraphs, are reported and summarized in Table 3.

Table 3. An overview of the currently available *Neisseria meningitidis* vaccines.

Vaccine	Manufacturer	Sero-groups	Licensed age group	Administra-tion schedule	Components details
AC Vax	GlaxoSmith-Kline, UK	A, C	2y+	Single dose	50 μg each of menin-gococcal polysac-charides

ACWY Vax	GlaxoSmith-Kline, UK	A/C/Y/W-135	2 y+; can be given also at 2 mo+, even though less protective against C, Y, and W-135	Single dose	50 μg each of meningococcal polysaccharides
Bexsero (4CMenB)	Novartis Vaccines and Diagnostics	B	2 mo–17 y	Complex dose schedule depending on age: 3 doses + booster for 2–5 mo; 2 doses + booster at 6–23 mo; 2 doses at 2+ y	50 μg of each recombinant NHBA, NadA, fHbp fusion proteins, OMVs from strain NZ98/254 containing the PorA P1.4 (25 μg), adsorbed on 0.5 mg Al^{3+}
HexaMen and HexaMix	National Institute for Public Health and the Environment, Bilthoven, Netherlands	B	—	2, 3, and 4 mo, a booster dose at 12–18 mo	OMV from two recombinant engineered strains, each of which expressed three different PorA subtypes (P1.5-2, 10; P1.12-1, 13; P1.7-2, 4; P1.19, 15-1; P1.7, 16; and P1.5-1, 2-2)
Menactra (MenACWY-DT)	Sanofi Pasteur	A/C/Y/W-135	9 mo–55 y	Single dose	4 μg each of meningococcal polysaccharides conjugated to 48 μg of a diphtheria toxoid protein carrier
MenAfriVac (MenA-TT)	Serum Institute of India	A	1–29 y	Single dose	10 μg of meningococcal group A polysaccharides conjugated to 10 to 33 μg tetanus toxoid
MenBvac	National Institute for Public Health, Norway, and Novartis	B	—	3 doses (interval 5–12 w)	OMVs from the strain 44/76 adsorbed on Al^{3+}
MencevaxA	GlaxoSmith-Kline and RIT, Belgium	A	2 y+	Single dose	50 μg of meningococcal group A polysaccharides No conjugation
MencevaxAC	GlaxoSmith-Kline	A, C	2 y+	Single dose	50 μg each of meningococcal group polysaccharides No conjugation

MencevaxACY	GlaxoSmith-Kline	A, C, Y	2 y+	Single dose	50 μg each of meningococcal group polysaccharides No conjugation
MencevaxA-CYW	GlaxoSmith-Kline	A/C/ Y/W-135	2 y+	Single dose	50 μg each of meningococcal group polysaccharides No conjugation
Mengivac A + C (MenPS)	Sanofi Pasteur	A, C	—	—	50 μg of meningococcal group C polysaccharides
MenHibrix (HibMenCY-TT)	GlaxoSmith-Kline	C, Y	6 w–18 mo	2, 4, 6, and 12 to 15 mo	Meningococcal groups C and Y polysaccharides conjugated to tetanus toxoid
Meningitec (MenC-CRM)	Wyeth Vaccines, Canada, UK, and Australia	C	2 mo+	3 doses at 2–12 mo, 1 dose at 12 mo+	10 μg of meningococcal group C polysaccharides conjugated to 15 μg CRM_{197}
Meninvact	Sanofi Pasteur	C	2 mo+	2 doses at 2–12 mo, 1 dose at 12 mo+	Meningococcal group C polysaccharides conjugated to CRM_{197}
Menitorix (Hib-MenC-TT)	GlaxoSmith-Kline	C	6 w–12 mo	Booster at 1-2 y	Meningococcal group C polysaccharides conjugated to tetanus toxoid
Menjugate (MenC-CRM)	Novartis Vaccines and Diagnostics	C	2 mo+	3 doses at 2–12 mo; 1 dose at 12 mo+	10 μg of meningococcal group C polysaccharides conjugated to 12.5 to 25 μg CRM_{197}
Menomune	Sanofi Pasteur	A, C	2 y+	Single dose	50 μg each of meningococcal group polysaccharides No conjugation
Menomune	Sanofi Pasteur	A/C/ Y/W-135	2 y+	Single dose	50 μg each of meningococcal group polysaccharides No conjugation
Menovac	Finlay Institute	A/C/ Y/W-135	2–55 y	Single dose	Meningococcal group polysaccharides

Menveo (MenACWY-CRM197)	Novartis Vaccines and Diagnostics	A/C/Y/W-135	2–55 y	Single dose	10 μg of meningococcal group A polysaccharides and 5 μg of meningococcal groups C, Y, and W-135 polysaccharides conjugated to CRM_{197}
MeNZB	Institute for Public Health, New Zealand, Chiron, Novartis	B	—	—	OMVs from strain P1.7b, 4
NeisVac-C (MenC-TT)	Baxter BioScience	C	2 mo–65 y	2 doses at 2–12 mo (with an interval of at least 2 mo), 1 dose at 12 mo+	10 μg of meningococcal group C polysaccharides conjugated to tetanus toxoid
Nimenrix	GlaxoSmith-Kline	A/C/Y/W-135	1 y+	Single dose	5 μg each of meningococcal group polysaccharides conjugated to 44 μg tetanus toxoid
NmVac4	JN-International Medical Corporation	A/C/Y/W-135	2–55 y	Single dose	50 μg each of meningococcal group polysaccharides
Trumenba	Pfizer	B	10–25 y	3 doses (0–2–6 mo)	120 μg of recombinant fHbp adsorbed on 0.25 mg Al^{3+}
Zamevax	Imunoloski Zavod, Croatia	A, C	—	—	No conjugation

CRM_{197}: cross-reacting material 197; fHbp: factor H binding protein; mo: month; NadA: *Neisseria* adhesion A; NHBA: Neisseria heparin binding antigen, also named GNA2132; OMV: outer membrane vesicle; PorA: porin A; w: week; y: year; Al^{3+}: Aluminum.

The elucidation of immunoescape strategies and genomics have enabled scholars to discover new potential vaccine candidates, like NMB0928 [282] or NMB1468 [283], FrpB/fetA [125], LbpA and LbpB [284], adhesin complex protein (ACP) [285], NspA [286, 287], MIP [152], ZnuD [93], PilE [76] and PilQ [288], IgA protease [289], T cell stimulating protein A (*tsp*A) [289], or the CP polymerase of *Neisseria* serogroup X [290], among others [291].

Reverse vaccinology has proved to be a promising approach, enabling researchers to develop the effective vaccine Bexsero. New highly integrated approaches, which combine genomics with postgenomics, are leading to next-generation vaccines. A combination of reverse and forward vaccinology techniques, such as immunoproteome investigation via combined cell surface immunoprecipitation and mass spectrometry (MS) [153], and new bioinformatics strategies, such as the protectome approach [292], are promising in identifying highly conserved motifs in known bacterial protective antigens and using them for the design of effective universal vaccines [293, 294].

CONCLUSIONS

The development of effective vaccines against meningococcal disease has been a long and hard struggle. Early efforts yielded only partial results, with the creation of polysaccharide vaccines [295]. Subsequent research, however, led to the production of the conjugate vaccines [296]. Today, we have the conjugate meningococcal C vaccines [297], an affordable conjugate vaccine against *N. meningitidis* serogroup A (MenAfriVac) [298, 299], the tetravalent conjugate vaccines [270], and, finally, two "universal" vaccines against meningococcal group B strains [300, 301]. The critical rate of coverage required in order to eliminate the disease is probably not among the highest [302]. Indeed, the conjugate vaccine for serogroup C has resulted in dramatic reductions of cases of the disease [303] and created herd immunity that seems to have had a significant effect on the carrier status of adolescents and young adults. Thus, the prospect of dominating this very serious disease lies decidedly in the medium term.

However, it must be borne in mind that we are immersed in a constellation of *Neisseriae*, whose only survival niche is man. Although Neisseriae such as *N. lactamica*, *N. sicca*, *N. elongata*, *N. cinerea*, and *N. flavescens* are usually able to establish silent infection in normal humans, it is not inconceivable that, given the microorganism's great capacity for genetic variation, nonpathogenic *Neisseria* might become hazardous to humans [304].

The challenge is still open.

Abbreviations

Abs:	Antibodies
ACP:	Adhesin complex protein
AIFA:	Agenzia Italiana del Farmaco (Italian National Agency for Drugs)
AMPs:	Antimicrobial peptides
AP:	Alternative Pathway
app:	Adhesion and penetration protein
ARF6:	ADP-ribosylation factor 6
BBB:	Blood-brain barrier
C3:	Human complement 3 component
C3b:	Human complement 3b component
C4:	Human complement 4
C4bp:	Human complement 4b binding protein
CAMPs:	Cationic antimicrobial peptides
cas:	CRISPR-associated
cbp:	Calprotectin binding protein
CCL:	Chemokine (C-C motif) ligand
CD:	Cluster of Differentiation (CD4, CD25, CD45R, CD86, CD105, CD147, CDw29)
CDS:	Coding sequence
CP:	Capsule
CEACAMs:	Carcinoembryonic antigen related cell adhesion molecules
CMP-NANA:	Cytidine monophosphoacetyl *N*-Acetylneuraminic acid
CRs:	Correia repeats
CREEs:	Correia repeat-enclosed elements
CRISPR:	Clustered regularly interspaced short palindromic repeats
CRM197:	Cross-reacting material 197
CRP:	C-reactive protein
CXCL:	Chemokine (C-X-C motif) ligand
DCs:	Dendritic cells
dam:	DNA adenine methyltransferase
DNA:	Deoxyribonucleic acid
drg:	Dam replacing gene
dRS3:	Duplicated repeat sequence 3
E. coli:	*Escherichia coli*
EMA:	European Medicines Agency

FDA:	Food and Drug Administration
fH:	Factor H
fHbp:	Factor H binding protein
fur:	Ferric uptake regulator
GEI:	Genomic island
GGI:	Gonococcal genetic island
ggt:	Gamma-glutamyl aminopeptidase
GNA:	Genome-derived neisserial antigen (GNA33, GNA1030, GNA1870, GNA2091, GNA2132)
GWAS:	Genome-wide association studies
HAMPs:	Host-associated molecular patterns
HAV:	Hepatitis A virus
HGT:	Horizontal Gene Transfer
HLA:	Human leukocyte antigens
hpu:	Haemoglobin-haptoglobin utilization
h-SBA:	Human serum bactericidal activity
HSP:	Heat-Shock Protein
HSPG:	Heparan sulphate proteoglycan
HTTs:	High-throughput technologies
IFN:	Interferon
IgA:	Immunoglobulin A
IgG:	Immunoglobulin G
IgM:	Immunoglobulin M
IGR:	Intergenic region
I B:	Inhibitors of NF- B
IL:	Interleukin (IL6, IL8, IL10, IL12)
IM:	Inner membrane
IS:	Insertion sequence (IS1016, IS1106-like, IS1301, IS1655)
*kat*A:	Catalase A
LAMP1:	Lysosomal-associated membrane protein 1
lbp:	Lactoferrin-binding protein
lct:	Lactate permease
LGT:	Lateral Gene Transfer
lgt:	Prolipoprotein diacylglyceryl transferase (*lgt*A, *lgt*B, *lgt*C, *lg* *t*D, *lgt*E, *lgt*G)
L-NNT:	Lacto-N-neotetraose
LOS:	Lipooligosaccharides
LPS:	Lipopolysaccharides

MAC:	Membrane attack complex
MAPK:	Mitogen-activated protein kinase
MAPKK:	Mitogen-activated protein kinase Kinase
MATS:	Meningococcal Antigen Typing System
MDA:	Meningococcal disease-associated island
MHC:	Major histocompatibility complex
MIP:	Macrophage infectivity potentiator
MMEs:	Minimal mobile elements
*mlt*A:	Membrane-boundlytic transglycosylase A
mo:	Month
mRNA:	messenger RNA
msf:	Meningococcal surface fibril
msr:	Methionine sulfoxide reductase
MS:	Mass spectrometry
*msp*A:	Meningococcal serine protease A
mtr:	Multiple transferable resistance
MWNTs:	Multiwalled carbon nanotubes
N.:	Neisseria (*N. lactamica*, *N. sicca*, *N. elongata*, *N. cinerea*, and *N. flavescens*)
*nad*A:	Neisserial adhesin A
NCAM-1:	Neural cell adhesion molecule 1
NEMIS:	*Neisseria* miniature insertion sequences
NETs:	Neutrophil extracellular traps
NF-κB:	Nuclear factor kappa-light-chain-enhancer of activated B cells
N. gonorrheae:	*Neisseria gonorrheae*
NHBA:	Neisserial heparin binding antigen
nhhA:	*Neisseria* hia/hsf homologue
NIMEs:	Neisserial intergenic mosaic elements
NKs:	Natural killer cells
NLRs:	NOD-like receptors
N. meningitidis:	*Neisseria meningitidis*
NO:	Nitric oxide
NOD:	Nucleotide-binding Oligomerization Domain
NTPase:	Nucleoside triphosphatases
oat:	O-acetyltransferase
OM:	Outer membrane

OMPs:	Outer membrane proteins
OMVs:	Outer membrane vesicles
OMPLA:	Outermembrane phospholipase A
Opa:	Opacity-associated protein a
Opc:	Opacity-associated protein c
ORF:	Open reading frame
pac:	Peptidoglycan O-acyltransferase
PAI:	Pathogenic island
PAMPs:	Pathogen-associated molecular patterns
pil:	Pilin (*pil*C: OM/cell surface pilin; *pil*D: prepilin-processing leader peptidase; *pil*F: traffic NTPase; *pil*G: pilus assembly protein; *pil*M: biogenesis protein; *pil*P: pilot protein; *pil*Q: secretin; *pil*T: traffic NTPase; *pil*W: pilus stabilization protein)
PN:	Peroxynitrite
por:	Porin
*ppt*A:	Pilin phosphorylcholine transferase A
PRRs:	Pattern recognition receptors
PTK:	Protein tyrosine kinase
RANTES:	Regulated on activation, normal T cell expressed and secreted
RecA:	Recombinase A
REP2:	Repetitive extragenic palindromic sequence
RIG-I:	Retinoic acid-inducible gene 1
rLP2086:	Recombinant lipoprotein 2086
RLRs:	RIG-1-like receptors
RNA:	Ribonucleic acid
RNS:	Reactive nitrogen species
ROS:	Reactive oxygen species
sia:	Polysialic acid capsule biosynthesis protein (*sia*A, *sia*B, *sia*C, *sia*D)
SMAD:	Small Mothers Against Decapentaplegic
sod:	Superoxide dismutase (sodB, sodC)
SOMVs:	Spontaneously released OMVs
SSM:	Slipped strand mispairing
SSR:	Simple Sequence Repeat
tbp:	Transferrin-binding protein
TGF:	Transforming Growth Factor
Th:	T-cell helper
TI:	T-cell Independent

TLR:	Toll-like receptor (TLR2, TLR4, TLR5, TLR9)
TNF:	Tumor Necrosis Factor
*tsp*A:	T cell stimulating protein A
UK:	United Kingdom
USA:	United States of America
UTR:	Untranslated Region
w:	Week
WGS:	Whole-genome sequencing
WNT:	Wingless-related integration site
y:	Year
*znu*D:	Zinc uptake component D.

Conflict of Interests

R. Gasparini, D. Panatto, N. L. Bragazzi, P. L. Lai, A. Bechini, M. Levi, P. Durando, and D. Amicizia declare that they have no conflict of interests regarding the publication of this paper.

Acknowledgments

Although efforts to make a systematic overview have been made as much as possible, quoting the relevant scholarly literature, due to the interdisciplinary nature of the topic and its vastity, the authors apologize if they have missed some important contributions. The authors thank Dr. Bernard Patrick for revising the paper.

REFERENCES

1. M. Murakami, T. Ohtake, R. A. Dorschner, B. Schittek, C. Garbe, and R. L. Gallo, "Cathelicidin anti-microbial peptide expression in sweat, an innate defense system for the skin," *Journal of Investigative Dermatology*, vol. 119, no. 5, pp. 1090–1095, 2002.
2. D. R. Drake, K. A. Brogden, D. V. Dawson, and P. W. Wertz, "Thematic review series: skin lipids. Antimicrobial lipids at the skin surface," *Journal of Lipid Research*, vol. 49, no. 1, pp. 4–11, 2008.
3. R. Medzhitov and C. Janeway Jr., "Innate immunity," *The New England Journal of Medicine*, vol. 343, no. 5, pp. 338–344, 2000.
4. S. Galdiero, A. Falanga, M. Cantisani et al., "Microbe-Host interactions: structure and role of Gram-negative bacterial Porins," *Current Protein and Peptide Science*, vol. 13, no. 8, pp. 843–854, 2012.
5. A. Di Donato, G. F. Draetta, G. Illiano, M. A. Tufano, L. Sommese, and F. Galdiero, "Do porins inhibit the macrophage phagocyting activity by stimulating the adenylate cyclase?" *Journal of Cyclic Nucleotide and Protein Phosphorylation Research*, vol. 11, no. 2, pp. 87–97, 1986.
6. F. Galdiero, M. A. Tufano, L. Sommese, A. Folgore, and F. Tedesco, "Activation of complement system by porins extracted from *Salmonella typhimurium*," *Infection and Immunity*, vol. 46, no. 2, pp. 559–563, 1984.
7. P. Massari, S. Ram, H. Macleod, and L. M. Wetzler, "The role of porins in neisserial pathogenesis and immunity," *Trends in Microbiology*, vol. 11, no. 2, pp. 87–93, 2003.
8. K. Takeda and S. Akira, "Toll-like receptors in innate immunity," *International Immunology*, vol. 17, no. 1, pp. 1–14, 2005.
9. F. Hayashi, K. D. Smith, A. Ozinsky et al., "The innate immune response to bacterial flagellin is mediated by Toll-like receptor 5," *Nature*, vol. 410, no. 6832, pp. 1099–1103, 2001.
10. D. M. Underhill, A. Aderem, F. Hayashi et al., "The innate immune response to bacterial flagellin is mediated by Toll-like receptor 5," *Nature*, vol. 410, no. 6832, pp. 1099–1103, 2001.
11. J. Faber, C. U. Meyer, C. Gemmer et al., "Human toll-like receptor 4 mutations are associated with susceptibility to invasive meningococcal disease in infancy," *The Pediatric Infectious Disease Journal*, vol. 25, no. 1, pp. 80–81, 2006.
12. M. S. Sanders, G. T. J. van Well, S. Ouburg, P. S. J. Lundberg, A.

M. Van Furth, and S. A. Morré, "Single nucleotide polymorphisms in TLR9 are highly associated with susceptibility to bacterial meningitis in children," *Clinical Infectious Diseases*, vol. 52, no. 4, pp. 475–480, 2011.

13. S. Akira, S. Uematsu, and O. Takeuchi, "Pathogen recognition and innate immunity," *Cell*, vol. 124, no. 4, pp. 783–801, 2006.
14. T.-D. Kanneganti, M. Lamkanfi, and G. Núñez, "Intracellular NOD-like receptors in host defense and disease," *Immunity*, vol. 27, no. 4, pp. 549–559, 2007.
15. A. Denes, G. Lopez-Castejon, and D. Brough, "Caspase-1: is IL-1 just the tip of the ICEberg?" *Cell Death and Disease*, vol. 3, no. 7, article e338, 2012.
16. A. W. Segal, "How neutrophils kill microbes," *Annual Review of Immunology*, vol. 23, pp. 197–223, 2005.
17. J. C. Sun, S. Ugolini, and E. Vivier, "Immunological memory within the innate immune system," *The EMBO Journal*, vol. 33, no. 12, pp. 1295–1303, 33.
18. A. Kolb-Mäurer, A. Unkmeir, U. Kämmerer et al., "Interaction of *Neisseria meningitidis* with human dendritic cells," *Infection and Immunity*, vol. 69, no. 11, pp. 6912–6922, 2001.
19. S. M. Zughaier, "Neisseria meningitidis capsular polysaccharides induce inflammatory responses via TLR2 and TLR4-MD-2," *Journal of Leukocyte Biology*, vol. 89, no. 3, pp. 469–480, 2011.
20. C. L. Sokol, G. M. Barton, A. G. Farr, and R. Medzhitov, "A mechanism for the initiation of allergen-induced T helper type 2 responses," *Nature Immunology*, vol. 9, no. 3, pp. 310–318, 2008.
21. M. Merad, P. Sathe, J. Helft, J. Miller, and A. Mortha, "The dendritic cell lineage: ontogeny and function of dendritic cells and their subsets in the steady state and the inflamed setting," *Annual Review of Immunology*, vol. 31, pp. 563–604, 2013.
22. C. A. Janeway, P. Travers, M. Walport et al., *Immunobiology*, Garland Science, New York, NY, USA, 2001.
23. M. Pizza and R. Rappuoli, "*Neisseria meningitidis*: pathogenesis and immunity," *Current Opinion in Microbiology*, vol. 23, pp. 68–72, 2015.
24. E. Del Tordello and D. Serruto, "Functional genomics studies of the human pathogen *Neisseria meningitidis*," *Briefings in Functional Genomics*, vol. 12, no. 4, pp. 328–340, 2013.

25. C. Schoen, H. Tettelin, J. Parkhill, and M. Frosch, "Genome flexibility in *Neisseria meningitidis*," *Vaccine*, vol. 27, supplement 2, pp. B103–B111, 2009.

26. P. M. Power and E. R. Moxon, "Phase variation and adaptive strategies of *N. meningitidis*: insights into the biology of a commensal and pathogen," in *Handbook of Meningococcal Disease: Infection Biology, Vaccination, Clinical Management*, M. Frosch and M. C. J. Maiden, Eds., Wiley-Blackwell, 2006.

27. E. Rotman and H. S. Seifert, "The genetics of *Neisseria* species," *Annual Review of Genetics*, vol. 48, no. 1, pp. 405–431, 2014.

28. P. R. Marri, M. Paniscus, N. J. Weyand et al., "Genome sequencing reveals widespread virulence gene exchange among human *Neisseria species*," *PLoS ONE*, vol. 5, no. 7, Article ID e11835, 2010.

29. L. A. S. Snyder, S. A. Jarvis, and N. J. Saunders, "Complete and variant forms of the 'gonococcal genetic island' in *Neisseria meningitidis*," *Microbiology*, vol. 151, no. 12, pp. 4005–4013, 2005.

30. B. Joseph, R. F. Schwarz, B. Linke et al., "Virulence evolution of the human pathogen *Neisseria meningitidis* by recombination in the core and accessory genome," *PLoS ONE*, vol. 6, no. 4, Article ID e18441, 2011.

31. L. A. Cahoon and H. S. Seifert, "Focusing homologous recombination: pilin antigenic variation in the pathogenic *Neisseria*," *Molecular Microbiology*, vol. 81, no. 5, pp. 1136–1143, 2011.

32. M. Kawai, I. Uchiyama, and I. Kobayashi, "Genome comparison *In Silico* in *Neisseria* suggests integration of filamentous bacteriophages by their own transposase," *DNA Research*, vol. 12, no. 6, pp. 389–401, 2005.

33. M. W. J. van Passel, A. van der Ende, and A. Bart, "Plasmid diversity in neisseriae," *Infection and Immunity*, vol. 74, no. 8, pp. 4892–4899, 2006.

34. Y. N. Srikhanta, K. L. Fox, and M. P. Jennings, "The phasevarion: phase variation of type III DNA methyltransferases controls coordinated switching in multiple genes," *Nature Reviews Microbiology*, vol. 8, no. 3, pp. 196–206, 2010.

35. M. V. R. Weber, H. Claus, M. C. J. Maiden, M. Frosch, and U. Vogel, "Genetic mechanisms for loss of encapsulation in polysialyltransferase-gene-positive meningococci isolated from

healthy carriers," *International Journal of Medical Microbiology*, vol. 296, no. 7, pp. 475–484, 2006.

36. A. van der Ende, C. T. P. Hopman, and J. Dankert, "Multiple mechanisms of phase variation of PorA in *Neisseria meningitidis*," *Infection and Immunity*, vol. 68, no. 12, pp. 6685–6690, 2000.
37. P. Martin, K. Makepeace, S. A. Hill, D. W. Hood, and E. R. Moxon, "Microsatellite in instability regulates transcription factor binding and gene expression," *Proceedings of the National Academy of Sciences of the United States of America*, vol. 102, no. 10, pp. 3800–3804, 2005.
38. R. Tobes and E. Pareja, "Bacterial repetitive extragenic palindromic sequences are DNA targets for Insertion Sequence elements," *BMC Genomics*, vol. 7, article 62, 2006.
39. L. Fagnocchi, E. Pigozzi, V. Scarlato, and I. Delany, "In the NadR regulon, adhesins and diverse meningococcal functions are regulated in response to signals in human saliva," *Journal of Bacteriology*, vol. 194, no. 2, pp. 460–474, 2012.
40. L. A. S. Snyder, S. A. Butcher, and N. J. Saunders, "Comparative whole-genome analyses reveal over 100 putative phase-variable genes in the pathogenic *Neisseria* spp.," *Microbiology*, vol. 147, no. 8, pp. 2321–2332, 2001.
41. A. Rytkönen, B. Albiger, P. Hansson-Palo et al., "*Neisseria meningitidis* undergoes PilC phase variation and PilE sequence variation during invasive disease," *Journal of Infectious Diseases*, vol. 189, no. 3, pp. 402–409, 2004.
42. P. Martin, L. Sun, D. W. Hood, and E. R. Moxon, "Involvement of genes of genome maintenance in the regulation of phase variation frequencies in *Neisseria meningitidis*," *Microbiology*, vol. 150, no. 9, pp. 3001–3012, 2004.
43. A. R. Richardson and I. Stojiljkovic, "Mismatch repair and the regulation of phase variation in *Neisseria meningitidis*," *Molecular Microbiology*, vol. 40, no. 3, pp. 645–655, 2001.
44. L. A. S. Snyder, N. J. Saunders, and W. M. Shafer, "A putatively phase variable gene (*dca*) required for natural competence in *Neisseria gonorrhoeae* but not *Neisseria meningitidis* is located within the division cell wall (*dcw*) gene cluster," *Journal of Bacteriology*, vol. 183, no. 4, pp. 1233–1241, 2001.
45. P. van Ulsen, B. Adler, P. Fassler et al., "A novel phase-

variable autotransporter serine protease, AusI, of Neisseria meningitidis," *Microbes and Infection*, vol. 8, no. 8, pp. 2088–2097, 2006.

46. N. J. Oldfield, S. Matar, F. A. Bidmos et al., "Prevalence and phase variable expression status of two autotransporters, NalP and MspA, in carriage and disease isolates of *Neisseria meningitides*," *PLoS ONE*, vol. 8, no. 7, Article ID e69746, 2013.
47. M. J. Warren and M. P. Jennings, "Identification and characterization of *pptA*: a gene involved in the phase-variable expression of phosphorylcholine on pili of *Neisseria meningitidis*," *Infection and Immunity*, vol. 71, no. 12, pp. 6892–6898, 2003.
48. T. Davidsen, M. Bjørås, E. C. Seeberg, and T. Tønjum, "Antimutator role of DNA glycosylase MutY in pathogenic *Neisseria* species," *Journal of Bacteriology*, vol. 187, no. 8, pp. 2801–2809, 2005.
49. N. J. Saunders, A. C. Jeffries, J. F. Peden et al., "Repeat-associated phase variable genes in the complete genome sequence of *Neisseria meningitidis* strain MC58," *Molecular Microbiology*, vol. 37, no. 1, pp. 207–215, 2000.
50. M. M. E. Metruccio, E. Pigozzi, D. Roncarati et al., "A novel phase variation mechanism in the meningococcus driven by a ligand-responsive repressor and differential spacing of distal promoter elements," *PLoS Pathogens*, vol. 5, no. 12, Article ID e1000710, 2009.
51. H. L. Alexander, A. W. Rasmussen, and I. Stojiljkovic, "Identification of *Neisseria meningitidis* genetic loci involved in the modulation of phase variation frequencies," *Infection and Immunity*, vol. 72, no. 11, pp. 6743–6747, 2004.
52. A. Siddique, N. Buisine, and R. Chalmers, "The transposon-like correia elements encode numerous strong promoters and provide a potential new mechanism for phase variation in the meningococcus," *PLoS Genetics*, vol. 7, no. 1, Article ID e1001277, 2011.
53. M. Mazzone, E. De Gregorio, A. Lavitola, C. Pagliarulo, P. Alifano, and P. P. Di Nocera, "Whole-genome organization and functional properties of miniature DNA insertion sequences conserved in pathogenic *Neisseriae*," *Gene*, vol. 278, no. 1-2, pp. 211–222, 2001.
54. E. de Gregorio, C. Abrescia, M. S. Carlomagno, and P. P. di Nocera, "Asymmetrical distribution of *Neisseria* miniature insertion sequence DNA repeats among pathogenic and

nonpathogenic *Neisseria* strains," *Infection and Immunity*, vol. 71, no. 7, pp. 4217–4221, 2003.

55. R. Hilse, S. Hammerschmidt, W. Bautsch, and M. Frosch, "Site-specific insertion of IS1301 and distribution in Neisseria meningitidis strains," *Journal of Bacteriology*, vol. 178, no. 9, pp. 2527–2532, 1996.
56. E. Kugelberg, B. Gollan, C. Farrance et al., "The influence of IS 1301 in the capsule biosynthesis locus on meningococcal carriage and disease," *PLoS ONE*, vol. 5, no. 2, Article ID e9413, 2010.
57. J. Kiss, Z. Nagy, G. Tóth et al., "Transposition and target specificity of the typical IS30 family element IS1655 from Neisseria meningitidis," *Molecular Microbiology*, vol. 63, no. 6, pp. 1731–1747, 2007.
58. H. Tettelin, N. J. Saunders, J. Heidelberg et al., "Complete genome sequence of *Neisseria meningitidis* serogroup B strain MC58," *Science*, vol. 287, no. 5459, pp. 1809–1815, 2000.
59. M. Alamro, F. A. Bidmos, H. Chan et al., "Phase variation mediates reductions in expression of surface proteins during persistent meningococcal carriage," *Infection and Immunity*, vol. 82, no. 6, pp. 2472–2484, 2014.
60. M. P. Jennings, Y. N. Srikhanta, E. R. Moxon et al., "The genetic basis of the phase variation repertoire of lipopolysaccharide immunotypes in Neisseria meningitidis," *Microbiology*, vol. 145, no. 11, pp. 3013–3021, 1999.
61. P. M. Power, L. F. Roddam, M. Dieckelmann et al., "Genetic characterization of pilin glycosylation in *Neisseria meningitidis*," *Microbiology*, vol. 146, no. 4, pp. 967–979, 2000.
62. P. M. Power, L. F. Roddam, K. Rutter, S. Z. Fitzpatrick, Y. N. Srikhanta, and M. P. Jennings, "Genetic characterization of pilin glycosylation and phase variation in *Neisseria meningitidis*," *Molecular Microbiology*, vol. 49, no. 3, pp. 833–847, 2003.
63. M. M. Estabrook, N. C. Christopher, J. M. Griffiss, C. J. Baker, and R. E. Mandrell, "Sialylation and human neutrophil killing of group C *Neisseria meningitidis*," *The Journal of Infectious Diseases*, vol. 166, no. 5, pp. 1079–1088, 1992.
64. Q. Zhang, Y. Li, and C. M. Tang, "The role of the exopolyphosphatase PPX in avoidance by *Neisseria meningitidis* of complement-mediated killing," *The Journal of Biological Chemistry*, vol. 285, no. 44, pp.

34259–34268, 2010.

65. H. Claus, R. Borrow, M. Achtman et al., "Genetics of capsule O-acetylation in serogroup C, W-135 and Y meningococci," *Molecular Microbiology*, vol. 51, no. 1, pp. 227–239, 2004.
66. P. C. Fusco, E. K. Farley, C.-H. Huang, S. Moore, and F. Michon, "Protective meningococcal capsular polysaccharide epitopes and the role of O acetylation," *Clinical and Vaccine Immunology*, vol. 14, no. 5, pp. 577–584, 2007.
67. A. Batista-Duharte, B. Téllez, M. Tamayo et al., "In Silico identification of molecular mimicry between T-cell epitopes of *Neisseria meningitidis* B and the human proteome," *Revista Peruana de Medicina Experimental y Salud Publica*, vol. 30, no. 3, pp. 441–445, 2013.
68. J. Finne, U. Finne, H. Deagostini-Bazin, and C. Goridis, "Occurrence of α2-8 linked polysialosyl units in a neural cell adhesion molecule," *Biochemical and Biophysical Research Communications*, vol. 112, no. 2, pp. 482–487, 1983.
69. G. R. Moe and D. M. Granoff, "Molecular mimetics of Neisseria meningitidis serogroup B polysaccharide," *International Reviews of Immunology*, vol. 20, no. 2, pp. 201–220, 2001.
70. R. E. Mandrell and M. A. Apicella, "Lipo-oligosaccharides (LOS) of mucosal pathogens: molecular mimicry and host-modification of LOS," *Immunobiology*, vol. 187, no. 3–5, pp. 382–402, 1993.
71. A. P. Moran, M. M. Prendergast, and B. J. Appelmelk, "Molecular mimicry of host structures by bacterial lipopolysaccharides and its contribution to disease," *FEMS Immunology and Medical Microbiology*, vol. 16, no. 2, pp. 105–115, 1996.
72. H. A. Harvey, W. E. Swords, and M. A. Apicella, "The mimicry of human glycolipids and glycosphingolipids by the lipooligosaccharides of pathogenic *Neisseria* and *Haemophilus*," *Journal of Autoimmunity*, vol. 16, no. 3, pp. 257–262, 2001.
73. C. M. Tsai, "Molecular mimicry of host structures by lipooligosaccharides of Neisseria meningitidis: characterization of sialylated and nonsialylated lacto-N-neotetraose (Galbeta1-4GlcNAcbeta1-3Galbeta1-4Glc) structures in lipooligosaccharides using monoclonal antibodies and specific lectins," *Advances in Experimental Medicine and Biology*, vol. 491, pp. 525–542, 2001.
74. H.-C. Wang, T.-P. Ko, M.-L. Wu, S.-C. Ku, H.-J. Wu, and A. H.-J.

Wang, "Neisseria conserved protein DMP19 is a DNA mimic protein that prevents DNA binding to a hypothetical nitrogen-response transcription factor," *Nucleic Acids Research*, vol. 40, no. 12, pp. 5718–5730, 2012.

75. M. Coureuil, G. Mikaty, F. Miller et al., "Meningococcal type IV pili recruit the polarity complex to cross the brain endothelium," *Science*, vol. 325, no. 5936, pp. 83–87, 2009.
76. F. Miller, G. Phan, T. Brissac et al., "The hypervariable region of meningococcal major pilin PilE controls the host cell response via antigenic variation," *mBio*, vol. 5, no. 1, Article ID e01024-13, 2013.
77. M. P. Bos, B. Tefsen, P. Voet, V. Weynants, J. P. M. Van Putten, and J. Tommassen, "Function of neisserial outer membrane phospholipase A in autolysis and assessment of its vaccine potential," *Infection and Immunity*, vol. 73, no. 4, pp. 2222–2231, 2005.
78. I. W. Devoe and J. E. Gilchrist, "Release of endotoxin in the form of cell wall blebs during in vitro growth of *Neisseria meningitidis*," *Journal of Experimental Medicine*, vol. 138, no. 5, pp. 1156–1167, 1973.
79. J. S. Swartley, A. A. Marfin, S. Edupuganti et al., "Capsule switching of *Neisseria meningitidis*," *Proceedings of the National Academy of Sciences of the United States of America*, vol. 94, no. 1, pp. 271–276, 1997.
80. T. M. P. P. Castiñeiras, D. E. Barroso, J. W. Marsh et al., "Capsular switching in invasive *Neisseria meningitidis*, Brazil," *Emerging Infectious Diseases*, vol. 18, no. 8, pp. 1336–1338, 2012.
81. P. Stefanelli, C. Fazio, A. Neri, T. Sofia, and P. Mastrantonio, "First report of capsule replacement among electrophoretic type 37 *Neisseria meningitidis* strains in Italy," *Journal of Clinical Microbiology*, vol. 41, no. 12, pp. 5783–5786, 2003.
82. R. S. W. Tsang, D. K. S. Law, S. D. Tyler, G. S. Stephens, M. Bigham, and W. D. Zollinger, "Potential capsule switching from serogroup Y to B: the characterization of three such *Neisseria meningitidis* isolates causing invasive meningococcal disease in Canada," *Canadian Journal of Infectious Diseases and Medical Microbiology*, vol. 16, no. 3, pp. 171–174, 2005.
83. L. M. Hollanda, G. C. G. Cury, R. F. C. Pereira et al., "Effect of mesoporous silica under *Neisseria meningitidis* transformation process: environmental effects under meningococci transformation," *Journal of*

Nanobiotechnology, vol. 9, article 28, 2011.

84. I. B. Mattos, D. A. Alves, L. M. Hollanda, H. J. Ceragiogli, V. Baranauskas, and M. Lancellotti, "Effects of multi-walled carbon nanotubes (MWCNT) under *Neisseria meningitidis* transformation process," *Journal of Nanobiotechnology*, vol. 9, article 53, 2011.
85. U. Vogel, M.-K. Taha, J. A. Vazquez et al., "Predicted strain coverage of a meningococcal multicomponent vaccine (4CMenB) in Europe: a qualitative and quantitative assessment," *The Lancet Infectious Diseases*, vol. 13, no. 5, pp. 416–425, 2013.
86. A. J. Beddek, M.-S. Li, J. S. Kroll, T. W. Jordan, and D. R. Martin, "Evidence for capsule switching between carried and disease-causing *Neisseria meningitidis* strains," *Infection and Immunity*, vol. 77, no. 7, pp. 2989–2994, 2009.
87. L. H. Harrison, K. A. Shutt, S. E. Schmink et al., "Population structure and capsular switching of invasive *Neisseria meningitidis* isolates in the pre-meningococcal conjugate vaccine era—United States, 2000–2005," *Journal of Infectious Diseases*, vol. 201, no. 8, pp. 1208–1224, 2010.
88. P. van der Ley, L. Steeghs, H. J. Hamstra, J. Ten Hove, B. Zomer, and L. Van Alphen, "Modification of lipid a biosynthesis in *Neisseria meningitidis* lpxL mutants: influence on lipopolysaccharide structure, toxicity, and adjuvant activity," *Infection and Immunity*, vol. 69, no. 10, pp. 5981–5990, 2001.
89. J. C. Dunning Hotopp, R. Grifantini, N. Kumar et al., "Comparative genomics of *Neisseria meningitidis*: core genome, islands of horizontal transfer and pathogen-specific genes," *Microbiology*, vol. 152, no. 12, pp. 3733–3749, 2006.
90. H. Lo, C. M. Tang, and R. M. Exley, "Mechanisms of avoidance of host immunity by *Neisseria meningitidis* and its effect on vaccine development," *The Lancet Infectious Diseases*, vol. 9, no. 7, pp. 418–427, 2009.
91. M. C. Schneider, B. E. Prosser, J. J. E. Caesar et al., "*Neisseria meningitidis* recruits factor H using protein mimicry of host carbohydrates," *Nature*, vol. 458, no. 7240, pp. 890–893, 2009.
92. M. Lappann, S. Danhof, F. Guenther, S. Olivares-Florez, I. L. Mordhorst, and U. Vogel, "In vitro resistance mechanisms of *Neisseria meningitidis* against neutrophil extracellular traps," *Molecular*

Microbiology, vol. 89, no. 3, pp. 433–449, 2013.

93. M. Stork, J. Grijpstra, M. P. Bos et al., "Zinc piracy as a mechanism of Neisseria meningitidis for evasion of nutritional immunity," *PLoS Pathogens*, vol. 9, no. 10, Article ID e1003733, 2013.
94. M. Saleem, S. M. Prince, S. E. J. Rigby et al., "Use of a molecular decoy to segregate transport from antigenicity in the FrpB iron transporter from *Neisseria meningitides*," *PLoS ONE*, vol. 8, no. 2, Article ID e56746, 2013.
95. K. Hubert, M.-C. Pawlik, H. Claus, H. Jarva, S. Meri, and U. Vogel, "Opc expression, LPS immunotype switch and pilin conversion contribute to serum resistance of unencapsulated meningococci," *PLoS ONE*, vol. 7, no. 9, Article ID e45132, 2012.
96. C. M. Tsai, R. Boykins, and C. E. Frasch, "Heterogeneity and variation among *Neisseria meningitidis* lipopolysaccharides," *Journal of Bacteriology*, vol. 155, no. 2, pp. 498–504, 1983.
97. J. Chamot-Rooke, G. Mikaty, C. Malosse et al., "Posttranslational modification of pili upon cell contact triggers *N. meningitidis* dissemination," *Science*, vol. 331, no. 6018, pp. 778–782, 2011.
98. F. E.-C. Jen, C. E. Jones, J. C. Wilson, B. L. Schulz, and M. P. Jennings, "Substrate recognition of a structure motif for phosphorylcholine post-translational modification in *Neisseria meningitidis*," *Biochemical and Biophysical Research Communications*, vol. 431, no. 4, pp. 808–814, 2013.
99. E. Loh, E. Kugelberg, A. Tracy et al., "Temperature triggers immune evasion by *Neisseria meningitidis*," *Nature*, vol. 502, no. 7470, pp. 237–240, 2013.
100. M. F. Anjum, T. M. Stevanin, R. C. Read, and J. W. B. Moir, "Nitric oxide metabolism in *Neisseria meningitidis*," *Journal of Bacteriology*, vol. 184, no. 11, pp. 2987–2993, 2002.
101. P. Stefanelli, G. Colotti, A. Neri et al., "Molecular characterization of nitrite reductase gene (aniA) and gene product in *Neisseria meningitidis* isolates: is aniA essential for meningococcal survival?" *IUBMB Life*, vol. 60, no. 9, pp. 629–636, 2008.
102. H. Echenique-Rivera, A. Muzzi, E. del Tordello et al., "Transcriptome analysis of *Neisseria meningitidis* in human whole blood and mutagenesis studies identify virulence factors involved in blood

survival," *PLoS Pathogens*, vol. 7, no. 5, Article ID e1002027, 2011.

103. H. A. Hadi, K. G. Wooldridge, K. Robinson, and D. A. A. Ala'Aldeen, "Identification and characterization of App: an immunogenic autotransporter protein of *Neisseria meningitidis*," *Molecular Microbiology*, vol. 41, no. 3, pp. 611–623, 2001.
104. J. Arenas, S. Cano, R. Nijland et al., "The meningococcal autotransporter AutA is implicated in autoaggregation and biofilm formation," *Environmental Microbiology*, vol. 17, no. 4, pp. 1321–1337, 2015.
105. N. H. Chen, R. M. Couñago, K. Y. Djoko et al., "A glutathione-dependent detoxification system is required for formaldehyde resistance and optimal survival of neisseria meningitidis in biofilms," *Antioxidants and Redox Signaling*, vol. 18, no. 7, pp. 743–755, 2013.
106. R. B. Neil and M. A. Apicella, "Role of HrpA in biofilm formation of *Neisseria meningitidis* and regulation of the hrpBAS transcripts," *Infection and Immunity*, vol. 77, no. 6, pp. 2285–2293, 2009.
107. A. J. Potter, S. P. Kidd, M. P. Jennings, and A. G. McEwan, "Evidence for distinctive mechanisms of S-nitrosoglutathione metabolism by AdhC in two closely related species, *Neisseria gonorrhoeae* and *Neisseria meningitidis*," *Infection and Immunity*, vol. 75, no. 3, pp. 1534–1536, 2007.
108. K. Yi, A. W. Rasmussen, S. K. Gudlavalleti, D. S. Stephens, and I. Stojiljkovic, "Biofilm formation by *Neisseria meningitidis*," *Infection and Immunity*, vol. 72, no. 10, pp. 6132–6138, 2004.
109. S. Agarwal, S. Vasudhev, R. B. DeOliveira, and S. Ram, "Inhibition of the classical pathway of complement by meningococcal capsular polysaccharides," *The Journal of Immunology*, vol. 193, no. 4, pp. 1855–1863, 2014.
110. A. C. Pridmore, D. H. Wyllie, F. Abdillahi et al., "A lipopolysaccharide-deficient mutant of *Neisseria meningitidis* elicits attenuated cytokine release by human macrophages and signals via toll-like receptor (TLR) 2 but not via TLR_4/MD2," *The Journal of Infectious Diseases*, vol. 183, no. 1, pp. 89–96, 2001.
111. L. M. Willis and C. Whitfield, "KpsC and KpsS are retaining 3-deoxy-D-manno-oct-2- ulosonic acid (Kdo) transferases involved in synthesis of bacterial capsules," *Proceedings of the National Academy of Sciences*

of the United States of America, vol. 110, no. 51, pp. 20753–20758, 2013.

112. T. R. Sampson and D. S. Weiss, "CRISPR-Cas systems: new players in gene regulation and bacterial physiology," *Frontiers in Cellular and Infection Microbiology*, vol. 4, p. 37, 2014.
113. K. M. Esvelt, P. Mali, J. L. Braff, M. Moosburner, S. J. Yaung, and G. M. Church, "Orthogonal Cas9 proteins for RNA-guided gene regulation and editing," *Nature Methods*, vol. 10, no. 11, pp. 1116–1123, 2013.
114. E. Kugelberg, B. Gollan, and C. M. Tang, "Mechanisms in *Neisseria meningitidis* for resistance against complement-mediated killing," *Vaccine*, vol. 26, supplement 8, pp. I34–I39, 2008.
115. A. E. Deghmane, D. Giorgini, M. Larribe, J.-M. Alonso, and M.-K. Taha, "Down-regulation of pili and capsule of *Neisseria meningitidis* upon contact with epithelial cells is mediated by CrgA regulatory protein," *Molecular Microbiology*, vol. 43, no. 6, pp. 1555–1564, 2002.
116. R. Ieva, C. Alaimo, I. Delany, G. Spohn, R. Rappuoli, and V. Scarlato, "CrgA is an inducible LysR-type regulator of *Neisseria meningitidis*, acting both as a repressor and as an activator of gene transcription," *Journal of Bacteriology*, vol. 187, no. 10, pp. 3421–3430, 2005.
117. K. O. Johswich, J. Zhou, D. K. S. Law et al., "Invasive potential of nonencapsulated disease isolates of *Neisseria meningitides*," *Infection and Immunity*, vol. 80, no. 7, pp. 2346–2353, 2012.
118. M. J. Uria, Q. Zhang, Y. Li et al., "A generic mechanism in *Neisseria meningitidis* for enhanced resistance against bactericidal antibodies," *Journal of Experimental Medicine*, vol. 205, no. 6, pp. 1423–1434, 2008.
119. T. M. Stevanin, J. R. Laver, R. K. Poole, J. W. B. Moir, and R. C. Read, "Metabolism of nitric oxide by *Neisseria meningitidis* modifies release of NO-regulated cytokines and chemokines by human macrophages," *Microbes and Infection*, vol. 9, no. 8, pp. 981–987, 2007.
120. T. M. Stevanin, J. W. B. Moir, and R. C. Read, "Nitric oxide detoxification systems enhance survival of *Neisseria meningitidis* in human macrophages and in nasopharyngeal mucosa," *Infection and Immunity*, vol. 73, no. 6, pp. 3322–3329, 2005.

121. T. Davidsen, E. K. Amundsen, E. A. Rødland, and T. Tønjum, "DNA repair profiles of disease-associated isolates of *Neisseria meningitidis*," *FEMS Immunology and Medical Microbiology*, vol. 49, no. 2, pp. 243–251, 2007.

122. A. K. Criss and H. S. Seifert, "A bacterial siren song: intimate interactions between *Neisseria* and neutrophils," *Nature Reviews Microbiology*, vol. 10, no. 3, pp. 178–190, 2012.

123. S. Schielke, C. Spatz, R. F. Schwarz et al., "Characterization of FarR as a highly specialized, growth phase-dependent transcriptional regulator in *Neisseria meningitidis*," *International Journal of Medical Microbiology*, vol. 301, no. 4, pp. 325–333, 2011.

124. J. Kortekaas, S. A. Müller, P. Ringler et al., "Immunogenicity and structural characterisation of an in vitro folded meningococcal siderophore receptor (FrpB, FetA)," *Microbes and Infection*, vol. 8, no. 8, pp. 2145–2153, 2006.

125. J. Kortekaas, A. Pettersson, J. van der Biezen et al., "Shielding of immunogenic domains in *Neisseria meningitidis* FrpB (FetA) by the major variable region," *Vaccine*, vol. 25, no. 1, pp. 72–84, 2007.

126. K. O. Mironov, A. E. Platonov, I. S. Koroleva et al., "Monitoring for *Neisseria meningitidis* species using sequencing of variable fragments of surface proteins FetA and PorA genes," *Zhurnal Mikrobiologii, Epidemiologii, i Immunobiologii*, no. 3, pp. 23–27, 2009.

127. J.A.WelschandS.Ram,"FactorHand*Neisserialpathogenesis*," *Vaccine*, vol. 26, no. 8, pp. I40–I45, 2008.

128. P. van der Ley, J. van der Biezen, R. Sutmuller, P. Hoogerhout, and J. T. Poolman, "Sequence variability of FrpB, a major iron-regulated outer-membrane protein in the pathogenic neisseriae," *Microbiology*, vol. 142, no. 11, pp. 3269–3274, 1996.

129. I. Delany, R. Rappuoli, and V. Scarlato, "Fur functions as an activator and as a repressor of putative virulence genes in *Neisseria meningitidis*," *Molecular Microbiology*, vol. 52, no. 4, pp. 1081–1090, 2004.

130. H. Takahashi, K. Hirose, and H. Watanabe, "Necessity of meningococcal γ-glutamyl aminopeptidase for *Neisseria meningitidis* growth in rat cerebrospinal fluid (CSF) and CSF-like medium," *Journal of Bacteriology*, vol. 186, no. 1, pp. 244–247, 2004.

131. A. W. Rasmussen, H. L. Alexander, D. Perkins-Balding, W. M. Shafer, and I. Stojiljkovic, "Resistance of *Neisseria meningitidis* to the toxic effects of heme iron and other hydrophobic agents requires expression of *ght*," *Journal of Bacteriology*, vol. 187, no. 15, pp. 5214–5223, 2005.
132. F. Putker, A. Grutsch, J. Tommassen, and M. P. Bos, "Ght protein of neisseria meningitidis is involved in the regulation of lipopolysaccharide biosynthesis," *Journal of Bacteriology*, vol. 196, no. 4, pp. 780–789, 2014.
133. R. Colicchio, S. Ricci, F. Lamberti et al., "The meningococcal ABC-type L-glutamate transporter GltT is necessary for the development of experimental meningitis in mice," *Infection and Immunity*, vol. 77, no. 9, pp. 3578–3587, 2009.
134. A. Talà, C. Monaco, K. Nagorska et al., "Glutamate utilization promotes meningococcal survival in vivo through avoidance of the neutrophil oxidative burst," *Molecular Microbiology*, vol. 81, no. 5, pp. 1330–1342, 2011.
135. C. S. Hong, W. Hashimoto, A. M. Fialho, T. K. Das Gupta, and A. M. Chakrabarty, "Disrupting the entry barrier and attacking brain tumors: the role of the *Neisseria* H.8 epitope and the Laz protein," *Cell Cycle*, vol. 5, no. 15, pp. 1633–1641, 2006.
136. W. Baehr, E. C. Gotschlich, and P. J. Hitchcock, "The virulence-associated gonococcal H.8 gene encodes 14 tandemly repeated pentapeptides," *Molecular Microbiology*, vol. 3, no. 1, pp. 49–55, 1989.
137. J. R. Black, W. J. Black, and J. G. Cannon, "Neisserial antigen H.8 is immunogenic in patients with disseminated gonococcal and meningococcal infections," *Journal of Infectious Diseases*, vol. 151, no. 4, pp. 650–657, 1985.
138. T. D. Ray, L. A. Lewis, S. Gulati, P. A. Rice, and S. Ram, "Novel blocking human IgG directed against the pentapeptide repeat motifs of *Neisseria meningitidis* Lip/H.8 and Laz lipoproteins," *Journal of Immunology*, vol. 186, no. 8, pp. 4881–4894, 2011.
139. I. Tauseef, O. B. Harrison, K. G. Wooldridge et al., "Influence of the combination and phase variation status of the haemoglobin receptors HmbR and HpuAB on meningococcal virulence," *Microbiology*, vol. 157, no. 5, pp. 1446–1456, 2011.

140. J. Lucidarme, J. Findlow, H. Chan et al., "The distribution and 'in vivo' phase variation status of haemoglobin receptors in invasive meningococcal serogroup B disease: genotypic and phenotypic analysis," *PLoS ONE*, vol. 8, no. 9, Article ID e76932, 2013.

141. H. Omer, G. Rose, K. A. Jolley et al., "Genotypic and phenotypic modifications of *Neisseria meningitidis* after an accidental human passage," *PLoS ONE*, vol. 6, no. 2, Article ID e17145, 2011.

142. L. Fantappiè, M. M. E. Metruccio, K. L. Seib et al., "The RNA chaperone Hfq is involved in stress response and virulence in *Neisseria meningitidis* and is a pleiotropic regulator of protein expression," *Infection and Immunity*, vol. 77, no. 5, pp. 1842–1853, 2009.

143. S. Vitovski, R. C. Read, and J. R. Sayers, "Invasive isolates of *Neisseria meningitidis* possess enhanced immunoglobulin A1 protease activity compared to colonizing strains," *The FASEB Journal*, vol. 13, no. 2, pp. 331–337, 1999.

144. S. A. Spence, V. L. Clark, and V. M. Isabella, "The role of catalase in gonococcal resistance to peroxynitrite," *Microbiology*, vol. 158, no. 2, pp. 560–570, 2012.

145. H.-J. Wu, K. L. Seib, J. L. Edwards, M. A. Apicella, A. G. McEwan, and M. P. Jennings, "Azurin of pathogenic *Neisseria* spp. is involved in defense against hydrogen peroxide and survival within cervical epithelial cells," *Infection and Immunity*, vol. 73, no. 12, pp. 8444–8448, 2005.

146. V. Roussel-Jazédé, I. Jongerius, M. P. Bos, J. Tommassen, and P. Van Ulsen, "NalP-mediated proteolytic release of lactoferrin-binding protein B from the meningococcal cell surface," *Infection and Immunity*, vol. 78, no. 7, pp. 3083–3089, 2010.

147. R. Exley, M. Winterbotham, Q. Zhang, V. Pelicic, and C. M. Tang, "A Conserved Major Pilin subunit in Neisseria meningitidis," http://www.meningitis.org/assets/x/52309.

148. S. A. Brown, K. L. Palmer, and M. Whiteley, "Revisiting the host as a growth medium," *Nature Reviews Microbiology*, vol. 6, no. 9, pp. 657–666, 2008.

149. S. Piek, Z. Wang, J. Ganguly et al., "The role of oxidoreductases in determining the function of the neisserial lipid a phosphoethanolamine transferase required for resistance to polymyxin," *PLoS ONE*, vol. 9,

no. 9, Article ID e106513, 2014.

150. R. S. W. Tsang, D. K. S. Law, C.-M. Tsai, and L.-K. Ng, "Detection of the 1st gene in different serogroups and LOS immunotypes of *Neisseria meningitidis*," *FEMS Microbiology Letters*, vol. 199, no. 2, pp. 203–206, 2001.
151. N. J. Griffiths, D. J. Hill, E. Borodina et al., "Meningococcal surface fibril (Msf) binds to activated vitronectin and inhibits the terminal complement pathway to increase serum resistance," *Molecular Microbiology*, vol. 82, no. 5, pp. 1129–1149, 2011.
152. M. K. Bielecka, N. Devos, M. Gilbert et al., "Recombinant protein truncation strategy for inducing bactericidal antibodies to the macrophage infectivity potentiator protein of *Neisseria meningitidis* and circumventing potential cross-reactivity with human FK506-binding proteins," *Infection and Immunity*, vol. 83, no. 2, pp. 730–742, 2015.
153. N. Tsolakos, C. Brookes, S. Taylor et al., "Identification of vaccine antigens using integrated proteomic analyses of surface immunogens from serogroup B *Neisseria meningitidis*," *Journal of Proteomics*, vol. 101, pp. 63–76, 2014.
154. Y.-L. Tzeng, A. Datta, K. Ambrose et al., "The MisR/MisS two-component regulatory system influences inner core structure and immunotype of lipooligosaccharide in *Neisseria meningitidis*," *The Journal of Biological Chemistry*, vol. 279, no. 33, pp. 35053–35062, 2004.
155. J. Adu-Bobie, P. Lupetti, B. Brunelli et al., "GNA33 of *Neisseria meningitidis* is a lipoprotein required for cell separation, membrane architecture, and virulence," *Infection and Immunity*, vol. 72, no. 4, pp. 1914–1919, 2004.
156. K. L. Seib, H.-J. Tseng, A. G. McEwan, M. A. Apicella, and M. P. Jennings, "Defenses against oxidative stress in *Neisseria gonorrhoeae* and *Neisseria meningitidis*: distinctive systems for different lifestyles," *The Journal of Infectious Diseases*, vol. 190, no. 1, pp. 136–147, 2004.
157. M. Antoine, A. Gand, S. Boschi-Muller, and G. Branlant, "Characterization of the amino acids from *Neisseria meningitidis* MsrA involved in the chemical catalysis of the methionine sulfoxide reduction step," *The Journal of Biological Chemistry*, vol. 281, no. 51, pp. 39062–39070, 2006.

158. Y.-L. Tzeng, K. D. Ambrose, S. Zughaier et al., "Cationic antimicrobial peptide resistance in *Neisseria meningitidis*," *Journal of Bacteriology*, vol. 187, no. 15, pp. 5387–5396, 2005.
159. R. M. Delahay, B. D. Robertson, J. T. Balthazar, W. M. Shafer, and C. A. Ison, "Involvement of the gonococcal MtrE protein in the resistance of *Neisseria gonorrhoeae* to toxic hydrophobic agents," *Microbiology*, vol. 143, no. 7, pp. 2127–2133, 1997.
160. M. Comanducci, S. Bambini, B. Brunelli et al., "NadA, a novel vaccine candidate of *Neisseria meningitidis*," *The Journal of Experimental Medicine*, vol. 195, no. 11, pp. 1445–1454, 2002.
161. B. Capecchi, J. Adu-Bobie, F. di Marcello et al., "*Neisseria meningitidis* NadA is a new invasin which promotes bacterial adhesion to and penetration into human epithelial cells," *Molecular Microbiology*, vol. 55, no. 3, pp. 687–698, 2005.
162. C. Mazzon, B. Baldani-Guerra, P. Cecchini et al., "IFN-γ and R-848 dependent activation of human monocyte-derived dendritic cells by *Neisseria meningitidis* adhesin A," *The Journal of Immunology*, vol. 179, no. 6, pp. 3904–3916, 2007.
163. G. Bozza, M. Capitani, P. Montanari et al., "Role of ARF6, Rab11 and external Hsp90 in the trafficking and recycling of recombinant-soluble neisseria meningitidis adhesin A (rNadA) in human epithelial cells," *PLoS ONE*, vol. 9, no. 10, Article ID e110047, 2014.
164. V. Nägele, J. Heesemann, S. Schielke, L. F. Jiménez-Soto, O. Kurzai, and N. Ackermann, "*Neisseria meningitidis* adhesin NadA targets beta1 integrins: functional similarity to *Yersinia* invasion," *The Journal of Biological Chemistry*, vol. 286, no. 23, pp. 20536–20546, 2011.
165. S. Franzoso, C. Mazzon, M. Sztukowska et al., "Human monocytes/macrophages are a target of *Neisseria meningitidis* Adhesin A (NadA)," *Journal of Leukocyte Biology*, vol. 83, no. 5, pp. 1100–1110, 2008.
166. P. van Ulsen, L. van Alphen, J. ten Hove, F. Fransen, P. van der Ley, and J. Tommassen, "A Neisserial autotransporter NalP modulating the processing of other autotransporters," *Molecular Microbiology*, vol. 50, no. 3, pp. 1017–1030, 2003.
167. S. Bambini, J. Piet, A. Muzzi et al., "An analysis of the sequence variability of meningococcal fHbp, NadA and NHBA over a 50-year period in the Netherlands," *PLoS ONE*, vol. 8, no. 5, Article ID e65043,

2013.

168. I. R. Peak, Y. N. Srikhanta, V. E. Weynants, C. Feron, J. T. Poolman, and M. P. Jennings, "Evaluation of truncated NhhA protein as a candidate meningococcal vaccine antigen," *PLoS ONE*, vol. 8, no. 9, Article ID e72003, 2013.

169. H. Sjölinder and A.-B. Jonsson, "Imaging of disease dynamics during meningococcal sepsis," *PLoS ONE*, vol. 2, no. 2, article e241, 2007.

170. K. R. Barth, V. M. Isabella, and V. L. Clark, "Biochemical and genomic analysis of the denitrification pathway within the genus *Neisseria*," *Microbiology*, vol. 155, no. 12, pp. 4093–4103, 2009.

171. M. Aspholm, F. E. Aas, O. B. Harrison et al., "Structural alterations in a component of cytochrome c oxidase and molecular evolution of pathogenic *Neisseria* in humans," *PLoS Pathogens*, vol. 6, no. 8, Article ID e1001055, pp. 55–56, 2010.

172. J. R. Laver, T. M. Stevanin, S. L. Messenger et al., "Bacterial nitric oxide detoxification prevents host cell S-nitrosothiol formation: a novel mechanism of bacterial pathogenesis," *The FASEB Journal*, vol. 24, no. 1, pp. 286–295, 2010.

173. S. Chandra, D. Singh, and T. R. Singh, "Prediction and characterization of T-cell epitopes for epitope vaccine design from outer membrane protein of *Neisseria meningitidis* serogroup B," *Bioinformation*, vol. 5, no. 4, pp. 155–161, 2010.

174. S. Giuntini, D. M. Vu, and D. M. Granoff, "FH-dependent complement evasion by disease-causing meningococcal strains with absent fHbp genes or frameshift mutations," *Vaccine*, vol. 31, no. 38, pp. 4192–4199, 2013.

175. L. A. Lewis, D. M. Vu, D. M. Granoff, and S. Ram, "Inhibition of the alternative pathway of nonhuman infant complement by porin B2 contributes to virulence of Neisseria meningitidis in the infant rat model," *Infection and Immunity*, vol. 82, no. 6, pp. 2574–2584, 2014.

176. J. D. Rock, M. J. Thomson, R. C. Read, and J. W. B. Moir, "Regulation of denitrification genes in *Neisseria meningitidis* by nitric oxide and the repressor NsrR," *Journal of Bacteriology*, vol. 189, no. 3, pp. 1138–1144, 2007.

177. K. Heurlier, M. J. Thomson, N. Aziz, and J. W. B. Moir, "The nitric oxide (NO)-sensing repressor NsrR of *Neisseria meningitidis* has a compact

regulon of genes involved in NO synthesis and detoxification," *Journal of Bacteriology*, vol. 190, no. 7, pp. 2488–2495, 2008.

178. H. J. Lee, B. Rakić, M. Gilbert, W. W. Wakarchuk, S. G. Withers, and N. C. J. Strynadka, "Structural and kinetic characterizations of the polysialic acid O-acetyltransferase OatWY from *Neisseria meningitidis*," *Journal of Biological Chemistry*, vol. 284, no. 36, pp. 24501–24511, 2009.
179. M. Sadarangani, A. J. Pollard, and S. D. Gray-Owen, "Opa proteins and CEACAMs: pathways of immune engagement for pathogenic Neisseria," *FEMS Microbiology Reviews*, vol. 35, no. 3, pp. 498–514, 2011.
180. M. J. Callaghan, C. Buckee, N. D. McCarthy et al., "Opa protein repertoires of disease-causing and carried meningococci," *Journal of Clinical Microbiology*, vol. 46, no. 9, pp. 3033–3041, 2008.
181. E. Carbonnelle, D. J. Hill, P. Morand et al., "Meningococcal interactions with the host," *Vaccine*, vol. 27, supplement 2, pp. B78–B89, 2009.
182. K. O. Johswich, S. E. McCaw, E. Islam et al., "*In vivo* adaptation and persistence of *Neisseria meningitidis* within the nasopharyngeal mucosa," *PLoS Pathogens*, vol. 9, no. 7, Article ID e1003509, 2013.
183. H. A. Rowe, N. J. Griffiths, D. J. Hill, and M. Virji, "Co-ordinate action of bacterial adhesins and human carcinoembryonic antigen receptors in enhanced cellular invasion by capsulate serum resistant *Neisseria meningitidis*," *Cellular Microbiology*, vol. 9, no. 1, pp. 154–168, 2007.
184. K. Wolff and A. Stern, "Identification and characterization of specific sequences encoding pathogenicity associated proteins in the genome of commensal *Neisseria* species," *FEMS Microbiology Letters*, vol. 125, no. 2-3, pp. 255–263, 1995.
185. M. M. Eason and X. Fan, "The role and regulation of catalase in respiratory tract opportunistic bacterial pathogens," *Microbial Pathogenesis*, vol. 74, pp. 50–58, 2014.
186. R. Ieva, D. Roncarati, M. M. E. Metruccio, K. L. Seib, V. Scarlato, and I. Delany, "OxyR tightly regulates catalase expression in *Neisseria meningitidis* through both repression and activation mechanisms," *Molecular Microbiology*, vol. 70, no. 5, pp. 1152–1165, 2008.
187. C. Dehio, S. D. Gray-Owen, and T. F. Meyer, "Host cell invasion by pathogenic *Neisseriae*," *Sub-Cellular Biochemistry*, vol. 33, pp. 61–

96, 2000.

188. J. P. Dillard and K. T. Hackett, "Mutations affecting peptidoglycan acetylation in *Neisseria gonorrhoeae* and *Neisseria meningitidis*," *Infection and Immunity*, vol. 73, no. 9, pp. 5697–5705, 2005.

189. I. Linhartova, M. Basler, J. Ichikawa et al., "Meningococcal adhesion suppresses proapoptotic gene expression and promotes expression of genes supporting early embryonic and cytoprotective signaling of human endothelial cells," *FEMS Microbiology Letters*, vol. 263, no. 1, pp. 109–118, 2006.

190. H. Takahashi, T. Yanagisawa, K. S. Kim, S. Yokoyama, and M. Ohnishi, "Meningococcal pilV potentiates *Neisseria meningitidis* type IV pilus-mediated internalization into human endothelial and epithelial cells," *Infection and Immunity*, vol. 80, no. 12, pp. 4154–4166, 2012.

191. S. C. Bernard, N. Simpson, O. Join-Lambert et al., "Pathogenic *Neisseria meningitidis* utilizes CD147 for vascular colonization," *Nature Medicine*, vol. 20, no. 7, pp. 725–731, 2014.

192. S. V. Balasingham, R. F. Collins, R. Assalkhou et al., "Interactions between the lipoprotein PilP and the secretin PilQ in *Neisseria meningitidis*," *Journal of Bacteriology*, vol. 189, no. 15, pp. 5716–5727, 2007.

193. S. K. Dash, M. Sharma, S. Khare, and A. Kumar, "rmpM Gene as a genetic marker for human bacterial meningitis," *Cellular and Molecular Biology*, vol. 58, no. 1, pp. 26–30, 2012.

194. A. Hey, M.-S. Li, M. J. Hudson, P. R. Langford, and J. Simon Krolla, "Transcriptional profiling of *Neisseria meningitidis* interacting with human epithelial cells in a long-term in vitro colonization model," *Infection and Immunity*, vol. 81, no. 11, pp. 4149–4159, 2013.

195. R. Grifantini, E. Frigimelica, I. Delany et al., "Characterization of a novel *Neisseria meningitidis* Fur and iron-regulated operon required for protection from oxidative stress: utility of DNA microarray in the assignment of the biological role of hypothetical genes," *Molecular Microbiology*, vol. 54, no. 4, pp. 962–979, 2004.

196. R. Gasparini, D. Amicizia, P. L. Lai, and D. Panatto, "*Neisseria meningitidis*: pathogenetic mechanisms to overcome the human immune defences," *Journal of Preventive Medicine and Hygiene*, vol. 53, no. 2, pp. 50–55, 2012.

197. A. Unkmeir, U. Kämmerer, A. Stade et al., "Lipooligosaccharide and polysaccharide capsule: virulence factors of *Neisseria meningitidis* that determine meningococcal interaction with human dendritic cells," *Infection and Immunity*, vol. 70, no. 5, pp. 2454–2462, 2002.

198. M. R. Spinosa, C. Progida, A. Talà, L. Cogli, P. Alifano, and C. Bucci, "The *Neisseria meningitidis* capsule is important for intracellular survival in human cells," *Infection and Immunity*, vol. 75, no. 7, pp. 3594–3603, 2007.

199. A. Talà, L. Cogli, M. D. Stefano et al., "Serogroup-specific interaction of *Neisseria meningitidis* capsular polysaccharide with host cell microtubules and effects on tubulin polymerization," *Infection and Immunity*, vol. 82, no. 1, pp. 265–274, 2014.

200. D. J. Hill, N. J. Griffiths, E. Borodina, and M. Virji, "Cellular and molecular biology of *Neisseria meningitidis* colonization and invasive disease," *Clinical Science*, vol. 118, no. 9, pp. 547–564, 2010.

201. M. Drogari-Apiranthitou, E. J. Kuijper, N. Dekker, and J. Dankert, "Complement activation and formation of the membrane attack complex on serogroup B *Neisseria meningitidis* in the presence or absence of serum bactericidal activity," *Infection and Immunity*, vol. 70, no. 7, pp. 3752–3758, 2002.

202. S. Ram, F. G. Mackinnon, S. Gulati et al., "The contrasting mechanisms of serum resistance of *Neisseria gonorrhoeae* and group B *Neisseria meningitidis*," *Molecular Immunology*, vol. 36, no. 13-14, pp. 915–928, 1999.

203. M. J. Simões, M. Cunha, F. Almeida, C. Furtado, and L. Brum, "Molecular surveillance of *Neisseria meningitidis* capsular switching in Portugal, 2002–2006," *Epidemiology and Infection*, vol. 137, no. 2, pp. 161–165, 2009.

204. L. Craig, N. Volkmann, A. S. Arvai et al., "Type IV pilus structure by cryo-electron microscopy and crystallography: implications for pilus assembly and functions," *Molecular Cell*, vol. 23, no. 5, pp. 651–662, 2006.

205. M. Kirchner and T. F. Meyer, "The PilC adhesin of the Neisseria type IV pilus—binding specificities and new insights into the nature of the host cell receptor," *Molecular Microbiology*, vol. 56, no. 4, pp. 945–957, 2005.

206. C. J. Orihuela, J. Mahdavi, J. Thornton et al., "Laminin receptor

initiates bacterial contact with the blood brain barrier in experimental meningitis models," *Journal of Clinical Investigation*, vol. 119, no. 6, pp. 1638–1646, 2009.

207. J. L. Diaz, M. Virji, and J. E. Heckels, "Structural and antigenic differences between two types of meningococcal pili," *FEMS Microbiology Letters*, vol. 21, no. 2, pp. 181–184, 1984.
208. R. A. Helm and H. S. Seifert, "Frequency and rate of pilin antigenic variation of *Neisseria meningitidis*," *Journal of Bacteriology*, vol. 192, no. 14, pp. 3822–3823, 2010.
209. M. E. Wörmann, C. L. Horien, J. S. Bennett et al., "Sequence, distribution and chromosomal context of class I and class II pilin genes of *Neisseria meningitidis* identified in whole genome sequences," *BMC Genomics*, vol. 15, no. 1, article 253, 2014.
210. E. Carbonnelle, S. Helaine, X. Nassif, and V. Pelicic, "A systematic genetic analysis in *Neisseria meningitidis* defines the Pil proteins required for assembly, functionality, stabilization and export of type IV pili," *Molecular Microbiology*, vol. 61, no. 6, pp. 1510–1522, 2006.
211. M. Kirchner, D. Heuer, and T. F. Meyer, "CD46-independent binding of neisserial type IV pili and the major pilus adhesin, PilC, to human epithelial cells," *Infection and Immunity*, vol. 73, no. 5, pp. 3072–3082, 2005.
212. M. Coureuil, H. Lécuyer, M. G. H. Scott et al., "Meningococcus hijacks a *β*2-adrenoceptor/*β*-arrestin pathway to cross brain microvasculature endothelium," *Cell*, vol. 143, no. 7, pp. 1149–1160, 2010.
213. T. Brissac, G. Mikaty, G. Duménil, M. Coureuil, and X. Nassif, "The meningococcal minor pilin PilX is responsible for type IV pilus conformational changes associated with signaling to endothelial cells," *Infection and Immunity*, vol. 80, no. 9, pp. 3297–3306, 2012.
214. M. Virji, S. M. Watt, S. Barker, K. Makepeace, and R. Doyonnas, "The N-domain of the human CD66a adhesion molecule is a target for Opa proteins of *Neisseria meningitidis* and *Neisseria gonorrhoeae*," *Molecular Microbiology*, vol. 22, no. 5, pp. 929–939, 1996.
215. E. R. Watkins, Y. H. Grad, S. Gupta, and C. O. Buckee, "Contrasting within- and between-host immune selection shapes *Neisseria* Opa repertoires," *Scientific Reports*, vol. 4, Article ID e6554, 2014.
216. R. Leuzzi, L. Serino, D. Serruto, and M. Pizza, "Newly described

surface structures and adhesion strategies of the pathogenic Neisseriae," in *Neisseria: Molecular Mechanisms of Pathogenesis*, Caister Academic Press, Norfolk, UK, 2010.

217. J. Pohlner, R. Halter, and T. F. Meyer, "*Neisseria gonorrhoeae* IgA protease. Secretion and implications for pathogenesis," *Antonie van Leeuwenhoek*, vol. 53, no. 6, pp. 479–484, 1987.
218. D. P. J. Turner, A. G. Marietou, L. Johnston et al., "Characterization of MspA, an immunogenic autotransporter protein that mediates adhesion to epithelial and endothelial cells in Neisseria meningitidis," *Infection and Immunity*, vol. 74, no. 5, pp. 2957–2964, 2006.
219. M. Sjölinder, G. Altenbacher, M. Hagner, W. Sun, S. Schedin-Weiss, and H. Sjölinder, "Meningococcal outer membrane protein NhhA triggers apoptosis in macrophages," *PLoS ONE*, vol. 7, no. 1, Article ID e29586, 2012.
220. V. Nägele, J. Heesemann, S. Schielke, L. F. Jiménez-Soto, O. Kurzai, and N. Ackermann, "*Neisseria meningitidis* adhesin NadA targets β1 integrins: functional similarity to *Yersinia* invasion," *The Journal of Biological Chemistry*, vol. 286, no. 23, pp. 20536–20546, 2011.
221. F. S. Archibald and M. N. Duong, "Superoxide dismutase and oxygen toxicity defenses in the genus Neisseria," *Infection and Immunity*, vol. 51, no. 2, pp. 631–641, 1986.
222. S. Sainsbury, J. Ren, J. E. Nettleship, N. J. Saunders, D. I. Stuart, and R. J. Owens, "The structure of a reduced form of OxyR from *Neisseria meningitidis*," *BMC Structural Biology*, vol. 10, article 10, 2010.
223. K. E. Wilks, K. L. R. Dunn, J. L. Farrant et al., "Periplasmic superoxide dismutase in meningococcal pathogenicity," *Infection and Immunity*, vol. 66, no. 1, pp. 213–217, 1998.
224. C. Schoen, L. Kischkies, J. Elias, and B. J. Ampattu, "Metabolism and virulence in *Neisseria meningitidis*," *Frontiers in Cellular and Infection Microbiology*, vol. 4, p. 114, 2014.
225. J. A. Gómez, M. T. Criado, and C. M. Ferreirós, "Cooperation between the components of the meningococcal transferrin receptor, TbpA and TbpB, in the uptake of transferrin iron by the 37-kDa ferric-binding protein (FbpA)," *Research in Microbiology*, vol. 149, no. 6, pp. 381–387, 1998.
226. K. L. R. Dunn, J. L. Farrant, P. R. Langford, and J. S. Kroll, "Bacterial [Cu,Zn]-cofactored superoxide dismutase protects opsonized,

encapsulated Neisseria meningitidis from phagocytosis by human monocytes/macrophages," *Infection and Immunity*, vol. 71, no. 3, pp. 1604–1607, 2003.

227. R. M. Exley, J. Shaw, E. Mowe et al., "Available carbon source influences the resistance of *Neisseria meningitidis* against complement," *Journal of Experimental Medicine*, vol. 201, no. 10, pp. 1637–1645, 2005.
228. D. Serruto and C. L. Galeotti, "The signal peptide sequence of a lytic transglycosylase of *Neisseria meningitidis* is involved in regulation of gene expression," *Microbiology*, vol. 150, no. 5, pp. 1427–1437, 2004.
229. H. Takahashi, K. S. Kim, and H. Watanabe, "Meningococcal internalization into human endothelial and epithelial cells is triggered by the influx of extracellular L-glutamate via GltT L-glutamate ABC transporter in neisseria meningitidis," *Infection and Immunity*, vol. 79, no. 1, pp. 380–382, 2011.
230. M. K. Pangburn, V. P. Ferreira, and C. Cortes, "Discrimination between host and pathogens by the complement system," *Vaccine*, vol. 26, no. 8, supplement, pp. I15–I21, 2008.
231. I. H. Lepow, L. Pillemer, M. D. Schoenberg et al., "The properdin system and immunity. X. Characterization of partially purified human properdin," *Journal of Immunology*, vol. 83, pp. 428–436, 1959.
232. N. A. Cunliffe, N. Snowden, E. M. Dunbar, and M. R. Haeney, "Recurrent meningococcal septicaemia and properdin deficiency," *Journal of Infection*, vol. 31, no. 1, pp. 67–68, 1995.
233. D. Panatto, D. Amicizia, P. L. Lai, and R. Gasparini, "*Neisseria meningitidis* B vaccines," *Expert Review of Vaccines*, vol. 10, no. 9, pp. 1337–1351, 2011.
234. R. Gasparini, D. Amicizia, A. Domnich, P. L. Lai, and D. Panatto, "*Neisseria meningitidis* B vaccines: recent advances and possible immunization policies," *Expert Review of Vaccines*, vol. 13, no. 3, pp. 345–364, 2014.
235. D. Panatto, D. Amicizia, P. L. Lai, M. L. Cristina, A. Domnich, and R. Gasparini, "New versus old meningococcal Group B vaccines: how the new ones may benefit infants & toddlers," *Indian Journal of Medical Research*, vol. 138, pp. 835–846, 2013.
236. D. Male, J. Brostoff, D. B. Roth et al., *Immunology*, Mosby Elsevier, Toronto, Canada, 2006.
237. A. Gasparoni, L. Ciardelli, A. Avanzini et al., "Age-related changes

in intracellular Th1/Th2 cytokine production, immunoproliferative T lymphocyte response and natural killer cell activity in newborns, children and adults," *Biology of the Neonate*, vol. 84, no. 4, pp. 297–303, 2003.

238. C. Härtel, N. Adam, T. Strunk, P. Temming, M. Müller-Steinhardt, and C. Schultz, "Cytokine responses correlate differentially with age in infancy and early childhood," *Clinical and Experimental Immunology*, vol. 142, no. 3, pp. 446–453, 2005.
239. B. Morein, G. Blomqvist, and K. Hu, "Immune responsiveness in the neonatal period," *Journal of Comparative Pathology*, vol. 137, no. 1, pp. S27–S31, 2007.
240. I. Goldschneider, E. C. Gotschlich, and M. S. Artenstein, "Human immunity to the meningococcus. II. Development of natural immunity," *Journal of Experimental Medicine*, vol. 129, no. 6, pp. 1327–1348, 1969.
241. S. Flexner, "The results of the serum treatment in thirteen hundred cases of epidemic meningitis," *The Journal of Experimental Medicine*, vol. 17, no. 5, pp. 553–576, 1913.
242. V. Davenport, E. Groves, C. G. Hobbs, N. A. Williams, and R. S. Heyderman, "Regulation of Th-1 T cell-dominated immunity to *Neisseria meningitidis* within the human mucosa," *Cellular Microbiology*, vol. 9, no. 4, pp. 1050–1061, 2007.
243. N. A. Korzhueva, D. I. Rybalko, V. P. Naglenko, and A. P. Zinchenko, "Specific immunological reactivity in generalized forms of meningococcal infection in children," *Zhurnal Mikrobiologii Epidemiologii i Immunobiologii*, no. 8, pp. 110–114, 1984.
244. S. Raziuddin, M. E.-H. El-Awad, and N. A. Mir, "Bacterial meningitis: T cell activation and immunoregulatory $CD4^+$ T cell subset alteration," *Journal of Allergy and Clinical Immunology*, vol. 87, no. 6, pp. 1115–1120, 1991.
245. A. J. Pollard, R. Galassini, E. M. Rouppe van der Voort et al., "Cellular immune responses to *Neisseria meningitidis* in children," *Infection and Immunity*, vol. 67, no. 5, pp. 2452–2463, 1999.
246. A. C. Cruz, B. S. Figueredo, S. L. Souza, M. C. Rebelo, A. G. Carvalho, and L. G. Milagres, "Antibody and memory $CD4^+$ T-cell responses after meningococcal disease," *Vaccine*, vol. 32, no. 40, pp. 5145–5148, 2014.

247. V. Davenport, T. Guthrie, J. Findlow, R. Borrow, N. A. Williams, and R. S. Heyderman, "Evidence for naturally acquired T cell-mediated mucosal immunity to *Neisseria meningitidis*," *The Journal of Immunology*, vol. 171, no. 8, pp. 4263–4270, 2003.
248. E. C. Gotschlich, T. Y. Liu, and M. S. Artenstein, "Human immunity to the meningococcus. 3. Preparation and immunochemical properties of the group A, group B, and group C meningococcal polysaccharides," *The Journal of Experimental Medicine*, vol. 129, no. 6, pp. 1349–1365, 1969.
249. E. C. Gotschlich, I. Goldschneider, and M. S. Artenstein, "Human immunity to the meningococcus. IV. Immunogenicity of group A and group C meningococcal polysaccharides in human volunteers," *Journal of Experimental Medicine*, vol. 129, no. 6, pp. 1367–1384, 1969.
250. E. C. Gotschlich, I. Goldschneider, and M. S. Artenstein, "Meningococcal infections 2. Field trial of group C meningococcal polysaccharide vaccine in 1969-70," *Bulletin of the World Health Organization*, vol. 45, Article ID 196970, pp. 279–282, 1971.
251. M. S. Artenstein, "Control of meningococcal meningitis with meningococcal vaccines," *Yale Journal of Biology and Medicine*, vol. 48, no. 3, pp. 197–200, 1975.
252. J. F. Brundage, M. A. K. Ryan, B. H. Feighner, and F. J. Erdtmann, "Meningococcal disease among United States military service members in relation to routine uses of vaccines with different serogroup-specific components, 1964–1998," *Clinical Infectious Diseases*, vol. 35, no. 11, pp. 1376–1381, 2002.
253. R. Biselli, A. Fattorossi, P. M. Matricardi, R. Nisini, T. Stroffolini, and R. D'Amelio, "Dramatic reduction of meningococcal meningitis among military recruits in Italy after introduction of specific vaccination," *Vaccine*, vol. 11, no. 5, pp. 578–581, 1993.
254. World Health Organization, "International travel and health, 'Meningococcal disease'," http://www.who.int/ith/diseases/meningococcal/en/.
255. K. E. Stein, "Thymus-independent and thymus-dependent responses to polysaccharide antigens," *Journal of Infectious Diseases*, vol. 165, supplement 1, pp. S49–S52, 1992.
256. J. I. Ward and K. M. Zangwill, "Haemophilus influenzae vaccines," in *Vaccines*, WB Saunders, Philadelphia, Pa, USA, 1999.

257. P. J. Baker, "T cell regulation of the antibody response to bacterial polysaccharide antigens: an examination of some general characteristics and their implications," *Journal of Infectious Diseases*, vol. 165, supplement, pp. S44–S48, 1992.

258. M. A. Breukels, A. Zandvoort, G. P. J. M. van den Dobbelsteen et al., "Pneumococcal conjugate vaccines overcome splenic dependency of antibody response to pneumococcal polysaccharides," *Infection and Immunity*, vol. 69, no. 12, pp. 7583–7587, 2001.

259. C. A. Siegrist, "Vaccine immunology," http://www.who.int/immunization/documents/Elsevier_Vaccine_immunology.pdf.

260. European Medicines Agency, "Nimenrix," http://www.ema.europa.eu/docs/en_GB/document_library/EPAR_-_Public_assessment_report/human/002226/WC500127664.pdf.

261. Centers for Disease Control and Prevention, "Prevention and control of meningococcal disease: recommendations of the advisory committee on immunization practices (ACIP)," *Morbidity and Mortality Weekly Report*, vol. 62, no. 2, 2013, http://www.cdc.gov/mmwr/pdf/rr/rr6202.pdf.

262. European Medicines Agency, "Menveo," http://www.ema.europa.eu/docs/en_GB/document_library/EPAR_-_Public_assessment_report/human/001095/WC500090150.pdf.

263. D. Serruto, T. Spadafina, L. Ciucchi et al., "*Neisseria meningitidis* GNA2132, a heparin-binding protein that induces protective immunity in humans," *Proceedings of the National Academy of Sciences of the United States of America*, vol. 107, no. 8, pp. 3770–3775, 2010.

264. J. A. Welsch, G. R. Moe, R. Rossi, J. Adu-Bobie, R. Rappuoli, and D. M. Granoff, "Antibody to genome-derived neisserial antigen 2132, a Neisseria meningitidis candidate vaccine, confers protection against bacteremia in the absence of complement-mediated bactericidal activity," *The Journal of Infectious Diseases*, vol. 188, no. 11, pp. 1730–1740, 2003.

265. J. S. Plested and D. M. Granoff, "Vaccine-induced opsonophagocytic immunity to *Neisseria meningitidis* group B," *Clinical and Vaccine Immunology*, vol. 15, no. 5, pp. 799–804, 2008.

266. G. Madico, J. A. Welsch, L. A. Lewis et al., "The meningococcal vaccine candidate GNA1870 binds the complement regulatory protein

factor H and enhances serum resistance," *The Journal of Immunology*, vol. 177, no. 1, pp. 501–510, 2006.

267. M. C. Schneider, R. M. Exley, H. Chan et al., "Functional significance of factor H binding to *Neisseria meningitidis*," *The Journal of Immunology*, vol. 176, no. 12, pp. 7566–7575, 2006.
268. T. Vesikari, S. Esposito, R. Prymula et al., "Immunogenicity and safety of an investigational multicomponent, recombinant, meningococcal serogroup B vaccine (4CMenB) administered concomitantly with routine infant and child vaccinations: results of two randomised trials," *The Lancet*, vol. 381, no. 9869, pp. 825–835, 2013.
269. D. Toneatto, P. Oster, A. C. W. DeBoer et al., "Early clinical experience with a candidate meningococcal B recombinant vaccine (rMenB) in healthy adults," *Human Vaccines*, vol. 7, no. 7, pp. 781–791, 2011.
270. European Medicines Agency, *Bexsero*, European Medicines Agency, London, UK, 2013, http://www.ema.europa.eu/ema/index.jsp?curl=pages/medicines/human/medicines/002333/human_med_001614.jsp&mid=WC0b01ac058001d124.
271. Agenzia Italiana del Farmaco, Determina 29 luglio 2013. Classificazione del medicinale per uso umano "Bexsero", ai sensi dell'art 8, comma 10, della legge 24 dicembre 1993, n. 537 (Determina 698/2013). Gazzetta Ufficiale—Serie Generale—no. 194 del 20 agosto 2013.
272. Food and Drug Administration, "Trumenba," http://www.fda.gov/downloads/BiologicsBloodVaccines/Vaccines/ApprovedProducts/UCM421139.pdf.
273. L. D. Fletcher, L. Bernfield, V. Barniak et al., "Vaccine Potential of the *Neisseria meningitidis* 2086 Lipoprotein," *Infection and Immunity*, vol. 72, no. 4, pp. 2088–2100, 2004.
274. D. Zhu, Y. Zhang, V. Barniak, L. Bernfield, A. Howell, and G. Zlotnick, "Evaluation of recombinant lipidated P2086 protein as a vaccine candidate for group B *Neisseria meningitidis* in a murine nasal challenge model," *Infection and Immunity*, vol. 73, no. 10, pp. 6838–6845, 2005.
275. D. Zhu, V. Barniak, Y. Zhang, B. Green, and G. Zlotnick, "Intranasal immunization of mice with recombinant lipidated P2086 protein reduces nasal colonization of group B *Neisseria meningitidis*," *Vaccine*, vol. 24, no. 26, pp. 5420–5425, 2006.
276. A. S. Anderson, L. Hao, Q. Jiang et al., "Potential impact of the bivalent

rLP2806vaccine on *Neisseria meningitidis* carriage and invasive serogroup B disease," *Human Vaccines and Immunotherapeutics*, vol. 9, no. 3, pp. 471–479, 2013.

277. F. Martinon-Torres, F. Gimenez-Sanchez, E. Bernaola-Iturbe, J. Diez-Domingo, Q. Jiang, and J. L. Perez, "A randomized, phase 1/2 trial of the safety, tolerability, and immunogenicity of bivalent rLP2086 meningococcal B vaccine in healthy infants," *Vaccine*, vol. 32, no. 40, pp. 5206–5211, 2014.

278. H. S. Marshall, P. C. Richmond, M. D. Nissen et al., "A phase 2 open-label safety and immunogenicity study of a meningococcal B bivalent rLP2086 vaccine in healthy adults," *Vaccine*, vol. 31, no. 12, pp. 1569–1575, 2013.

279. S. L. Harris, D. Zhu, E. Murphy et al., "Preclinical evidence for the potential of a bivalent fHBP vaccine to prevent *Neisseria meningitidis* serogroup C disease," *Human Vaccines*, vol. 7, supplement 1, pp. 68–74, 2011.

280. M. Konar, D. M. Granoff, and P. T. Beernink, "Importance of inhibition of binding of complement factor H for serum bactericidal antibody responses to meningococcal factor H-binding protein vaccines," *The Journal of Infectious Diseases*, vol. 208, no. 4, pp. 627–636, 2013.

281. L. Lemée, E. Hong, M. Etienne et al., "Genetic diversity and levels of expression of factor H binding protein among carriage isolates of *Neisseria meningitidis*," *PLoS ONE*, vol. 9, no. 9, Article ID e107240, 2014.

282. M. Delgado, D. Yero, O. Niebla et al., "Lipoprotein NMB0928 from *Neisseria meningitidis* serogroup B as a novel vaccine candidate," *Vaccine*, vol. 25, no. 50, pp. 8420–8431, 2007.

283. C.-A. Hsu, W.-R. Lin, J.-C. Li et al., "Immunoproteomic identification of the hypothetical protein NMB1468 as a novel lipoprotein ubiquitous in *Neisseria meningitidis* with vaccine potential," *Proteomics*, vol. 8, no. 10, pp. 2115–2125, 2008.

284. A. Pettersson, J. Kortekaas, V. E. Weynants et al., "Vaccine potential of the *Neisseria meningitidis* lactoferrin-binding proteins LbpA and LbpB," *Vaccine*, vol. 24, no. 17, pp. 3545–3557, 2006.

285. M.-C. Hung, J. E. Heckels, and M. Christodoulides, "The adhesin complex protein (ACP) of Neisseria meningitidis is a new adhesin with vaccine potential," *mBio*, vol. 4, no. 2, 2013.

286. D. Martin, B. R. Brodeur, J. Hamel et al., "Candidate *Neisseria meningitidis* NspA vaccine," *Journal of Biotechnology*, vol. 83, no. 1-2, pp. 27–31, 2000.

287. S. Ying, J. He, M. Yu et al., "Recombinant Neisseria surface protein A is a potential vaccine candidate against *Neisseria meningitides* serogroup B," *Molecular Medicine Reports*, vol. 10, no. 3, pp. 1619–1625, 2014.

288. F. Haghi, S. N. Peerayeh, S. D. Siadat, and H. Zeighami, "Recombinant outer membrane secretin $PilQ_{406-770}$ as a vaccine candidate for serogroup B *Neisseria meningitidis*," *Vaccine*, vol. 30, no. 9, pp. 1710–1714, 2012.

289. S. K. Gupta, S. Smita, A. N. Sarangi, M. Srivastava, B. A. Akhoon, and Q. Rahman, "*In silico* $CD4^+$ T-cell epitope prediction and HLA distribution analysis for the potential proteins of *Neisseria meningitidis* Serogroup B—A clue for vaccine development," *Vaccine*, vol. 28, no. 43, pp. 7092–7097, 2010.

290. T. Fiebig, F. Berti, F. Freiberger et al., "Functional expression of the capsule polymerase of *Neisseria meningitidis* serogroup X: a new perspective for vaccine development," *Glycobiology*, vol. 24, no. 2, pp. 150–158, 2014.

291. J. Holst, P. Oster, R. Arnold et al., "Vaccines against meningococcal serogroup B disease containing outer membrane vesicles (OMV) lessons from past programs and implications for the future," *Human Vaccines and Immunotherapeutics*, vol. 9, no. 6, pp. 1241–1253, 2013.

292. J. Newcombe, T. A. Mendum, C.-P. Ren, and J. McFadden, "Identification of the immunoproteome of the meningococcus by cell surface immunoprecipitation and MS," *Microbiology*, vol. 160, no. 2, pp. 429–438, 2014.

293. E. Altindis, R. Cozzi, B. Di Palo et al., "Protectome analysis: a new selective bioinformatics tool for bacterial vaccine candidate discovery," *Molecular & Cellular Proteomics*, vol. 14, no. 2, pp. 418–429, 2015.

294. M. Christodoulides, "*Neisseria* proteomics for antigen discovery and vaccine development," *Expert Review of Proteomics*, vol. 11, no. 5, pp. 573–591, 2014.

295. N. E. Rosenstein, B. A. Perkins, D. S. Stephens et al., "The changing epidemiology of meningococcal disease in the United States, 1992–1996," *The Journal of Infectious Diseases*, vol. 180, no. 6, pp. 1894–1901, 1999.

296. E. Miller, D. Salisbury, and M. Ramsay, "Planning, registration, and implementation of an immunisation campaign against meningococcal serogroup C disease in the UK: a success story," *Vaccine*, vol. 20, supplement 1, pp. S58–S67, 2001.

297. R. Gasparini and D. Panatto, "Meningococcal glycoconjugate vaccines," *Human Vaccines*, vol. 7, no. 2, pp. 170–182, 2011.

298. F. M. La Force, K. Konde, S. Viviani, and M.-P. Préziosi, "The meningitis vaccine project," *Vaccine*, vol. 25, supplement 1, pp. 97–100, 2007.

299. F. Micoli, M. R. Romano, M. Tontini et al., "Development of a glycoconjugate vaccine to prevent meningitis in Africa caused by meningococcal serogroup X," *Proceedings of the National Academy of Sciences of the United States of America*, vol. 110, no. 47, pp. 19077–19082, 2013.

300. R. Rappuoli, "The challenge of developing universal vaccines," *F1000 Medicine Reports*, vol. 3, article 16, 2011.

301. C. L. Trotter, N. J. Gay, and W. J. Edmunds, "Dynamic models of meningococcal carriage, disease, and the impact of serogroup C conjugate vaccination," *American Journal of Epidemiology*, vol. 162, no. 1, pp. 89–100, 2005.

302. H. Campbell, R. Borrow, D. Salisbury, and E. Miller, "Meningococcal C conjugate vaccine: the experience in England and Wales," *Vaccine*, vol. 27, supplement 2, pp. B20–B29, 2009.

303. R. Gasparini, M. Comanducci, D. Amicizia et al., "Molecular and serological diversity of *Neisseria meningitidis* carrier strains isolated from Italian students aged 14 to 22 years," *Journal of Clinical Microbiology*, vol. 52, no. 6, pp. 1901–1910, 2014.

304. Public Health Agency of Canada, *Neisseria SPP*, Public Health Agency of Canada, Ottawa, Canada, 2011, http://www.phac-aspc.gc.ca/lab-bio/res/psds-ftss/neisseria-eng.php.

8

Cervical Cancer Cells Express Markers Associated with Immunosurveillance

Adriana Gutiérrez-Hoya[1,2], Octavio Zerecero-Carreón[1], Arturo Valle-Mendiola[1], Martha Moreno-Lafont[3], Rubén López-Santiago[3], Benny Weiss-Steider[1] and Isabel Soto-Cruz[1]

[1]Laboratorio de Oncología Molecular, FES Zaragoza, Universidad Nacional Autónoma de México, Batalla 5 de mayo s/n Col. Ejército de Oriente, CP 09230 Ciudad de México, Mexico
[2]Cátedras CONACyT, CONACYT, Avenida Insurgentes Sur 1582, Benito Juárez, Crédito Constructor, 03940 Ciudad de México, Mexico
[3]Departamento de Inmunología, Escuela Nacional de Ciencias Biológicas, Instituto Politécnico Nacional, Prolongación de Carpio y Plan de Ayala S/N. Colonia Santos Tomás, Miguel Hidalgo, 11340 Ciudad de México, Mexico

ABSTRACT

Cervical cancer is the second most frequent cancer in women in Mexico, and its development depends on the presence of human papillomaviruses in the uterine cervix. These oncogenic viruses transform cells where the control

Citation: Cervical Cancer Cells Express Markers Associated with Immunosurveillance. Adriana Gutiérrez-Hoya, Octavio Zerecero-Carreón, Arturo Valle-Mendiola, Martha Moreno-Lafont, Rubén López-Santiago, Benny Weiss-Steider, and Isabel Soto-Cruz. Hindawi Journal of Immunology Research Volume 2019, Article ID 1242979, 10 pages https://doi.org/10.1155/2019/1242979

over cell cycle disappears, and the capacity to induce apoptosis is absent. On the other hand, some mutations confer to the transformed cells the ability to evade recognition by the immune system. The expression of markers of the immune system such as CD95, MICA/B, CD39, CD73, NKp30, NKp46, CD44, CD24, NKG2A, and CTLA-4 was analysed by flow cytometry on cervical cancer cells INBL (HPV 18, stage IVB), HeLa (HPV 18), CaSki (HPV 16), and C33A (HPV-). Our results showed the presence of atypical markers on cervical cancer cells; some of them are molecules involved in tumour cell recognition such as MICA/B and CD95. Other markers associated with immune system escape, such as CD39, CD73, and CTLA-4, were also present. Furthermore, we found that some cervical cancer cells expressed typical markers of NK cells like NKp30, NKp46, NKG2A, and KIR3DL1. It is not clear whether these molecules confer any gain to the tumour cells or if they represent a disadvantage, but we hypothesise that these molecules that are present in cervical cancer cells allow them to mimic in front of the immune system.

INTRODUCTION

One of the main strategies of the transformed cells to establish premalignant and malignant lesions is the modulation of the microenvironment and the evasion of the immune system. Our research group determined that cervical cancer cells express diverse molecules that play a central role in cell proliferation and that could serve as biomarkers of the disease. One of these molecules is the IL-2 receptor (IL-2R) which is present in cervical cancer cells. These cells express the α, β, and γ subunits of IL-2R; the addition of exogenous IL-2 induces the interaction with its receptor to initiate the JAK/STAT signalling pathway essential for proliferation, similar to that activated in lymphocytes [1, 2]. These tumour cells also express constitutively active JAK3 and STAT5 proteins [3]. Another set of molecules involved in tumour cell growth are proteins that belong to the epidermal growth factor receptor (EGFR) family. Data from our working group show that EGFR and HER2 are also present in cervical cancer lines and that they could be involved in the pathogenesis and progression of cervical lesions, due to a direct association of EGFR expression and HPV infection [4]. However, tumour cells are also able of expressing atypical receptors along with their respective ligands to induce tumour proliferation. For example, a small subpopulation of cervical cancer cells express the NKG2D receptor on the cell surface and secrete the ligands MICA/B and thereby induce tumour proliferation and probably use these molecules as an immunological escape mechanism [5]. The presence of these tumour markers that can be used by cervical cancer cells to induce their proliferation or, furthermore, an evasion of the immune system, led us to evaluate the presence of other typical and atypical cell markers to elucidate their role in the development of cervical cancer and consider their use as suitable tumour markers.

MATERIALS AND METHODS

Cell Lines and Antibodies

Biological materials, the HeLa (HPV 18), CaSki (HPV 16), and C33A (HPV-) cell lines, were obtained from the American Type Culture Collection (ATCC). INBL, an HPV 18 cell line derived from an invasive stage IVB squamous cell carcinoma, was established at the Cell Differentiation Laboratory of FES Zaragoza [6]. Cells were cultured in RPMI-1640 medium (Microlab, Mexico) supplemented with 5% foetal bovine serum (FBS, Invitrogen, USA). All cultures were maintained in an incubator at 37°C, 5% CO_2, and saturated humidity.

Analysis of Cell Surface and Intracellular Markers of Tumor Cells

Approximately 1×10^6 cervical cancer cells per condition were seeded onto Petri dishes. Cells were incubated for 35 min at 4°C in staining buffer (PBS, 0.5% BSA, and 0.01% sodium azide). The cells were stained with specific antibodies against the following surface markers: anti-NKp30-PE, anti-CD95-APC, anti-CD39-FITC, anti-CD-73-PE-Cy7, anti-KIR3DL1-APC, anti-NKp46-APC, anti-CD25-PE, anti-CD25-PerCP-Cy5.5, anti-CD44-FITC, and anti-CD24-PE (Becton Dickinson, USA); anti-NKG2A-PerCP, anti-CD158e1-APC (R&D, Minneapolis, MN); and anti-CD158b2-FITC (BioLegend). Cells were washed and fixed with 1% paraformaldehyde for 20 minutes. For permeabilisation, cells were treated with Cytofix/Cytoperm buffer for 20 minutes (Becton Dickinson, USA). For intracellular staining, anti-human CTLA-4-APC, anti-CD95-PE-Cy7 (Becton Dickinson, USA), and anti-NKG2A-PerCP (R&D, Minneapolis, MN) antibodies were used. Cell samples were incubated for 35 minutes at 4°C in the dark. Finally, cells were washed and resuspended in staining buffer. The phenotype of tumor cells was characterized by flow cytometry on a FACSAria II cytometer (Becton Dickinson, USA). Data were analysed using Summit 4.4 software.

RESULTS AND DISCUSSION

Activating Ligands of the Immune Response Are Differentially Expressed by Cervical Cancer Cell Lines

We analysed the expression of MICA/B and CD95, markers of the immune system, in cervical cancer cell lines infected with HPV18 (INBL and HeLa), HPV16 (CaSki), and HPV-negative (C33A). MICA or MICB is not expressed on the cell surface of healthy cells; these molecules are known to activate cytotoxic cells, including NK cells and CD8 T cells. We previously demonstrated that cervical cancer cells express the NKG2D receptor on the surface membrane, and the interaction of NKG2D with MICA/B induced the proliferation on tumour cells [5]. Our results indicate that MICA/B expression in cervical cancer cells is low (4-17%, Figure 1(a)); however, it is present in all cell lines. These results are consistent with those of other research groups that demonstrated that MICA or MICB is present in most tumour samples analysed, with predominantly intracellular localisation and only occasional cell surface localisation. Nevertheless, they report a similar profile for many epithelial cells in normal tissues [7]. It is not clear why normal tissues do express MICA or MICB, but our results with cervical cancer cells suggest that it is a strategy to evade the immune system. Our workgroup demonstrated that a small subpopulation of the cervical cancer cell lines bears the NKG2D receptor, and by interacting with its ligands (MICA/B), tumour proliferation is induced [5]. Thus, the presence of these molecules might help tumour cells to proliferate and evade the immune system.

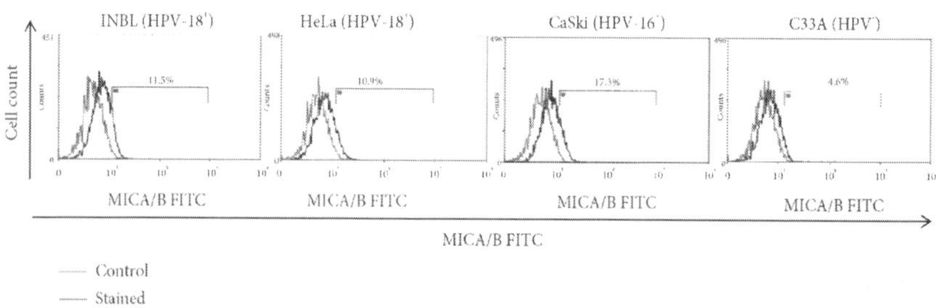

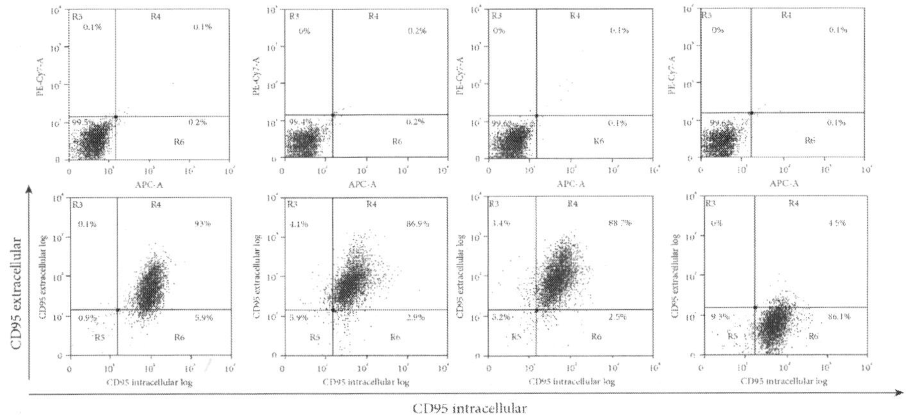

(b)

Figure 1. Presence of activating markers of the immune response in cervical cancer cells. (a) Expression of MICA/B on the INBL, HeLa, CaSki, and C33A cells was determined using anti-MICA/B antibodies conjugated with FITC. In the histograms, the grey lines represent the staining controls and the black lines represent the stained cells. (b) Expression of CD95 extracellular and intracellular on the INBL, HeLa, CaSki, and C33A cells was determined using anti-CD95 antibodies conjugated with PE-Cy7 (for extracellular expression) and APC (for intracellular expression); the upper dot plots show the staining controls for each cell line, and the lower dot plots show the cells stained for CD95.

CD95 (Fas/APO-1) and its ligand, CD95L, are very important for inducing apoptosis to maintain immune homeostasis and for the elimination of virus-infected cells, damaged cells, and cancer cells. We analysed whether the cervical cancer cells expressed CD95 on the cell membrane, and our results demonstrated that cervical cancer cells positive for HPV (INBL, HeLa, and CaSki) show high levels of CD95 (86-93%). C33A cells, negative for HPV, express shallow levels of CD95 (4.5%; Figure 1(b)). To further explore the presence of CD95 in C33A cells, we evaluated the presence of intracellular CD95. Our results indicate that all cervical cancer cell lines, including C33A, express high intracellular levels of CD95 (90-99%; Figure 1(b)). Filippova et al. showed that the E6 protein of HPV-16 could bind to the death domains of FAS (FADD) and thereby degrade them. The loss of FADD prevents the activation of the apoptotic cascade and with this the elimination of the tumour cells [8]. Additional experimental studies have referred to the presence of CD95L with a poor prognosis in cervical cancer [9, 10], which was initially attributed as a mechanism for immune escape. Recently, some studies report that several tumour cells can use the CD95/CD95L pathway to induce survival and proliferation following the activation of signalling pathways such as NF-κB, Erk1/2, JNK, and Jun [11]. Altogether, results suggest that CD95 could be considered as an important marker that might be involved in tumour growth and survival rather than in tumor elimination.

Inhibitory Markers of the Immune Response Expressed by Cervical Cancer Cell Lines

Tumour cells have different strategies to evade the immune system; one strategy is the secretion of immunomodulatory cytokines such as IL-10 and TGFβ. Another approach is the expression of molecules that suppress the immune response such as CTLA-4, CD25, CD39, and CD73. These molecules are important markers in Treg cells and contribute by inhibiting the activation of dendritic cells by joining CTLA-4/CD80 and CD86 and by deprivation of ATP, a role played by the CD39 and CD73 ectonucleases or for the elimination of IL-2 by CD25. CD39 hydrolyses extracellular ATP and adenosine diphosphate (ADP) into adenosine monophosphate (AMP), and then AMP is degraded into adenosine by the CD73 ecto-5′-nucleotidase. Finally, this adenosine binds to A2A receptors. Adenosine causes the accumulation of intracellular cAMP preventing TCR-triggered CD25 upregulation and inhibiting activation of T lymphocyte proliferation and the secretion of inflammatory cytokines [12]. We evaluated the presence of markers that suppress the immune response such as CD73 and CD39. The results showed that cervical cancer cells express CD73 and CD39 (Figures 2(a) and 2(b)). These data are consistent with the expression of CD73 which plays a significant role in the production of adenosine that suppresses antitumour effector cells, as well as with the fact that cervical cancer cells that are HPV-positive express a higher percentage suggesting that HPV infection contributes to overexpression of CD73 [13]. However, the role of CD39 in cervical cancer is poorly understood. Studies have shown that tumour cells like B lymphoma cells, melanoma cells, B cell chronic lymphocytic leukaemia cells, and ovarian cancer cells overexpress CD39 and thus activate the adenosinergic pathway to inhibit the proliferation of CD4 and CD8 T cells and the cytotoxic effector CD8 T cells in a CD39 adenosine-dependent manner [14]. These data suggest that overexpression of CD39 in cervical cancer cell lines is important for tumour survival. We observed that the expression of CD39 is higher in HPV-positive cells (74-88%) versus HPV-negative cervical tumour cells (17%); this observation points out the importance of the presence of the virus to favour the expression of this molecule.

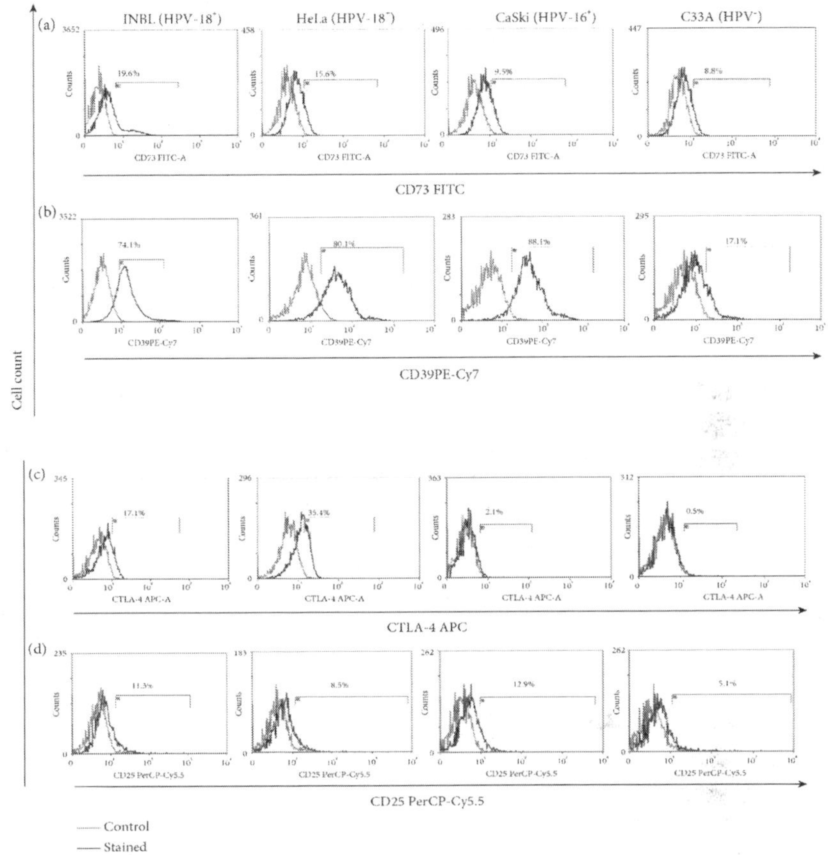

Figure 2. Inhibitory markers of the immune response present in cervical cancer cells. (a) The expression of CD73 on INBL, HeLa, CaSki, and C33A cells. (b) The expression of CD39 on INBL, HeLa, CaSki, and C33A cells. (c) The expression of CTLA-4 on INBL, HeLa, CaSki, and C33A cells. (d) The expression of CD25 on INBL, HeLa, CaSki, and C33A cells. In the histograms, the grey lines represent the staining controls and the black represent the stained cells.

CTLA-4 plays a crucial role in the regulation of the immune system. We determined its presence on the cell surface of cervical cancer cell lines. The results showed a very low or null presence of CTLA-4 on the cell surface (data not shown). Therefore, we evaluated the intracellular form of this protein. The results revealed that all cervical cell lines express CTLA-4 (0.5-35%; Figure 2(c)). These results are similar to those reported by Contardi et al., who observed a weak presence of CTLA-4 in HeLa and C33A cells; they also evaluated the presence of this molecule in a large number of cell lines derived from a variety of malignant human solid tumours including breast carcinoma, melanoma, neuroblastoma, rhabdomyosarcoma, and osteosarcoma. This working group found that the treatment of cells expressing CTLA-4 with recombinant forms of the costimulatory molecules CD80 and CD86 induced apoptosis associated with the activation of caspase-8 and caspase-3 [15]. However, Chen et al. showed that the coculture of dendritic cells activated with breast cancer cells positive for CTLA-4 decreases the expression of costimulatory molecules. Moreover, the suppressed dendritic cells further inhibited proliferation of $CD4^{+}/CD8^{+}$ T cells, which implies the role of CTLA-4 in the immunosuppression of the immune system [16]. We demonstrated the presence of these molecules; however, at this point, the role of CTLA-4 in cervical cancer cells remains to be determined. Additionally, further research is needed to conclude whether these inhibitory markers of the immune response are important in the tumour microenvironment.

CD25 is a critical immunoregulatory molecule constitutively expressed on Treg cells for deprivation of IL-2 to induce apoptosis and suppress effector T cells [17]. CD25, the α-chain of IL-2R, is essential to form the high-affinity receptor ($Kd = 10-11$ M) [18]. Previously, our workgroup demonstrated that some HPV 18^{+} cervical cancer cells express the β and γ chain of the IL-2 receptor [1, 2]. We evaluated the presence of CD25 in HPV-positive and HPV-negative cervical cancer cells; the results showed a low percentage of CD25-positive cells (5-12%; Figure 2(d)). The role of these molecules remains elusive; however, several reports exist about the role of Treg associated with a prognosis in cancer. Nevertheless, only Loose et al. report the expression of CD25 in squamous cell carcinoma of the head and neck (SCCHN), but they did not find any relationship between the CD25 expression on tumour cells with a prognosis in SCCHN patients [19]. We do not know if the presence of this marker confers the tumour cells a strategy to evade the immune response due to IL-2 depletion. But previous reports of our work group show that low doses of IL-2 can be used by cervical tumour cells to proliferate [3].

Epithelial-Mesenchymal Transition Markers or Cancer Stem Cell Population

Metastasis is the primary cause of death in patients with cervical cancer; evidence exists that metastasis develops due to the presence of tumour stem cells, which show slow replication, are resistant to chemotherapy, and have a metastatic capacity. One potential marker for the cancer stem-like cell subpopulation is CD117/c-kit, a tyrosine kinase receptor associated with cancer progression and stem cell maintenance. Stimulation of CD117 by its ligand, stem cell factor (SCF; kit ligand), activates several signalling pathways driving proliferation, survival, and migration [20]. The importance of c-kit is evident in renal carcinoma, ovarian cancer, melanoma, acute myeloid leukaemia, and gastrointestinal tumour amongst others [21, 22]. Therefore, we decided to evaluate the presence of c-kit in cervical cancer cell lines. The results showed that only a small proportion of the cells express this marker (1.8-2.9%; Figure 3(a)), but this percentage is similar in cells positive for HVP 18 and 16 or negative for HPV; however, these results suggest that they could represent a subpopulation of cancer stem cells (CSC). Other markers associated with stem cells and the epithelial-mesenchymal transition (EMT) are CD44 and CD24; these markers have been best characterized in breast cancer, where the subpopulation of CSC expresses low levels of stable heat antigen (CD24) and high levels of the hyaluronic receptor (CD44). This cell subpopulation has the capacity of self-renewal, to initiate tumour formation, and is intrinsically resistant to chemotherapy. However, although these markers associate with metastasis, they are not sufficient to predict the capacity of metastasis [23, 24]. On the other hand, Liu et al. showed that HPV16-positive cervical cancer cells SiHa $CD44^{+}/CD24^{+}$ are resistant to cellular apoptosis induced by irradiation therapy and that these cells possess characteristics of stem cells such as potent tumorigenicity and colony formation capacity and have high *in vitro* balling ability in suspension culture [25]. We determined the expression of CD44 and CD24 in the different cervical cancer cell lines; the results demonstrate that HeLa, INBL, and CaSki cells express high levels of both molecules (86-95%). However, HPV-negative C33A cells do not express CD44, but they do express CD24 by more than 80% (Figure 3(b)). Taking these results into consideration, we hypothesise that the presence of the virus may control the overexpression of CD44.

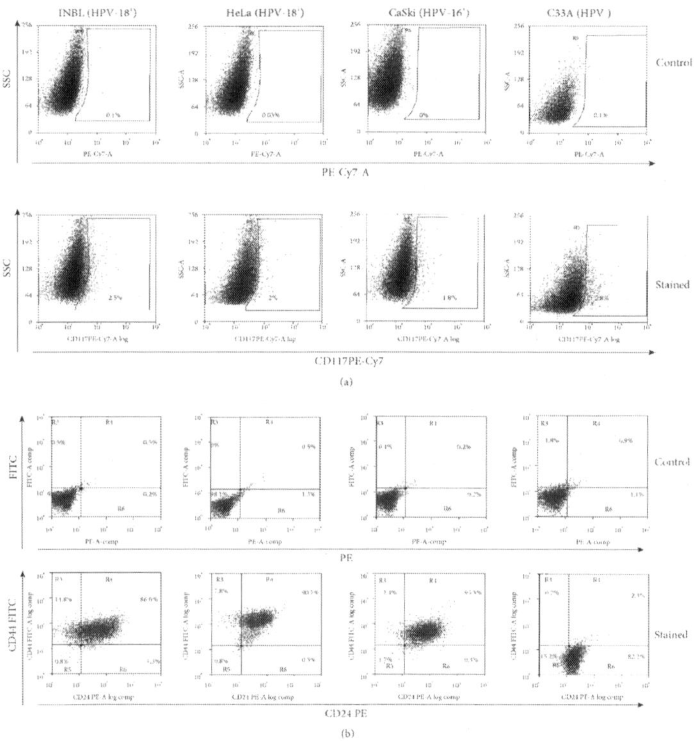

Figure 3. Markers of stem cells and epithelial-mesenchymal transition present in cervical cancer cells. (a) Expression of CD117 (c-kit) on INBL, HeLa, CaSki, and C33A cells was determined using anti-CD117 antibodies conjugated with PE-Cy7. The upper graphics indicating granularity versus fluorescence in PE-Cy7 show the staining controls for each cell line, and the lower graphics indicating granularity versus CD117 PE-Cy7 show the positive cells stained for CD117. (b) Expression of CD24 and CD44 on the INBL, HeLa, CaSki, and C33A cells was determined using anti-CD24 antibodies conjugated with PE and anti-CD44 antibodies conjugated with FITC. The upper dot plots show the staining controls for each cell line, and the lower dot plots show the double-positive cells stained for CD24 and CD44. For C33A cells, only CD24-positive cells were found.

A large number of cancers overexpress CD44; however, its role in cervical cancer is unknown. CD44 has a pleiotropic role and is involved in the induction of EMT altering the cytoskeleton, promoting drug resistance, and activating antiapoptotic mechanisms. At present, anti-CD44 antibodies are evaluated in preclinical trials to know if blocking this molecule could serve as a therapy for cancer [26].

NK Cell Markers in Cervical Cancer Cells

We previously reported that a small subpopulation of cervical cancer cells expresses the NKG2D receptor, a receptor believed to be exclusive of cytotoxic cells. Also, we demonstrated that tumour cells secreted NKG2D ligands MICA/B to induce their proliferation [5]. Later, Tang and collaborators found that NKG2D expression in tumour cells contributes to the immunological escape by the acute myeloid leukaemia cells [27]. With this information in mind, we evaluated the presence of molecules, which are characteristic of cytotoxic cells, in cervical cancer cells. Interestingly, we found that small subpopulations of tumour cells expressed atypical markers, for example, in all HPV-positive cervical cancer cells, we found a small subpopulation that expresses intracellular NKG2A (1-6%). However, a bigger subpopulation (36.4%) of the INBL cell line expresses the NKG2A receptor (Figure 4(a)). The NKG2A receptor plays a vital role in immunosurveillance by binding to HLA-E complexes; it is an inhibitory receptor signalling through two inhibitory ITIM motifs [28]. Some studies showed that blocking the interaction between NKG2A and HLA-E (NKG2A ligand) can regain the activity of NK cells, suggesting that NKG2A blockade has the potential to restore immunity against tumours [29, 30]. At this point, the function of NKG2A in cancer cells is not clear, but in cervical cancer cells, the intracellular expression is high. One possible explanation is that tumour cells could secrete NKG2A to interact with HLA-E expressed by the same tumour cells, and this could activate a survival signal. It is also possible that soluble NKG2A could interact with HLA-A-expressing cells in the immune system blocking their activation; however, this remains to be determined. After evidencing the presence of NKG2A, a molecule with inhibitory functions typically expressed by NK cells, we search for other molecules that interfere with the activation of NK cells. As shown in Figure 4(b), approximately 23% of INBL and 14% of HeLa cells express the inhibitory molecule KIR3DL1, whereas HeLa, CaSki, and C33A express a low percentage of this molecule. The presence of KIR3DL1 on NK cells inhibits the lysis of target cells bearing HLA-Bw4. Also, small populations of $\gamma\delta$ T cells, $CD4^+$ T cells, and $CD8^+$ T cells express KIR3DL1 to control its cytolytic ability, cytokine production, and proliferation. Interestingly, Chwae and coworkers showed that the expression of KIR3DL1 could protect these cells from activation-induced cell death following stimulation, at least in part by preventing the upregulation of the Fas ligand [31, 32]. In this context, our results show that HPV-positive cervical cancer cells express a high percentage of CD95 and INBL cells express KIR3DL1;

therefore, it is possible that these molecules could inhibit the induction of apoptosis by the CD95 pathway. We evaluated the expression of CD158b (KIR2DL2/DL3) with typically inhibitory functions expressed by NK cells [33]. There are no references on the appearance of this molecule in tumour cells, and there are few references on its expression on cells of the immune system and its association with cancer. Guerra et al. showed the presence of this molecule in infiltrates of lymphocytes in renal carcinomas and how the lythic activity of autologous tumour cells increased with the addition of the anti-CD158b antibody, indicating that this molecule participates in the tumour immune response [34]. Our results show that INBL cells express CD158b (16%; Figure 4(c)), while in the other cell lines, this molecule is not present. We do not know the advantages conferred by the expression of CD158b to these tumour cells, and further research is necessary to analyse if the expression of KIR favours the evasion of the immune system.

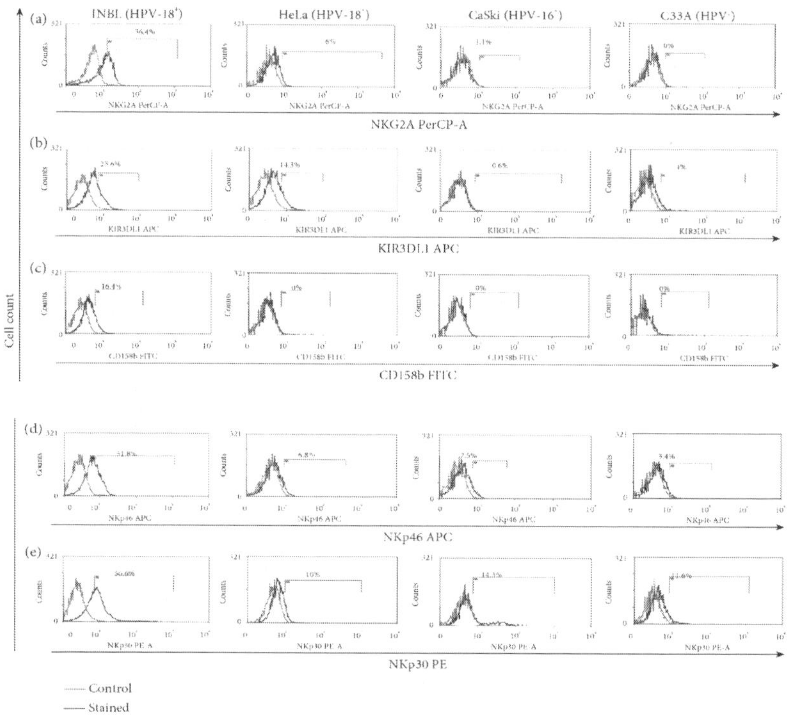

Figure 4. Cervical cancer cells express NK cell markers. (a) Intracellular expression of NKG2A on INBL, HeLa, CaSki, and C33A cells. (b) Expression of KIR3DL1 on INBL, HeLa, CaSki, and C33A cells. (c) Expression of NKp46 on INBL, HeLa, CaSki, and C33A cells. (d) Expression of NKp30 on INBL, HeLa,

CaSki, and C33A cells. (e) Expression of CD158b on INBL, HeLa, CaSki, and C33A cells. In the histograms, the grey lines represent the staining controls and the black lines represent the stained cells.

To further evaluate the presence of activating markers of the immune system in cervical cancer cells, we analysed whether tumour cells express the activating receptors of NK cells. These receptors belong to the family of Ig-like molecules named natural cytotoxicity receptors (NCRs) which include NKp30, NKp44, and NKp46 molecules. We evaluated the expression of NKp30 and NKp46 in cervical cancer cells. As shown in Figure 4(d), INBL cells express NKp46 (51%) while a low percentage of HeLa (6.8%), CaSki (7.5%), and C33A (3.4%) express this molecule. NKp46 is a specific NK killer receptor that recognises some influenza hemagglutinins and unknown tumour ligands. Cagnano and coworkers showed that malignant melanocytes but not normal melanocytes express ligands to NKp46; these data are consistent with those demonstrated by Elboim et al. where tumour growth of cells expressing high levels of the NKp46 ligands depends upon the presence of the NKp46 receptor [35, 36]. Altogether, these results suggest that tumour cells expressing NKp46 could secrete ligands for this receptor and thereby induce tumour growth. Further research is needed to describe if the stimulation of NKp46-positive cervical tumour cells with its ligand could induce its survival or proliferation.

Finally, we determined the presence of the receptor NKp30, an activator of the immune response in NK cells. Our results show that all cervical cancer cell lines express this receptor; however, INBL cells express a higher percentage of this molecule (56%; Figure 4(e)) while HeLa, CasKi, and C33A express a lower percentage (10-14.3%). NKp30, a natural cytotoxicity receptor on NK cells, is involved in multiple processes, such as allowing NK cells to interact with dendritic cells, modulating the maturation or elimination of dendritic cells, and the maturation of NK cells [37]. One of the ligands for NKp30 is galectin-3; interestingly, cervical tumour cells secrete different galectins, including galectin-3, which participates in the induction of proliferation, metastasis, angiogenesis, and resistance to chemo/radiotherapy. Galectin-3 is associated with the inhibition of apoptosis and the immune response. The concentration of galectin-3 is related to the stage of cervical cancer [38–40]. Since we observed that NKp30 is present in all cervical cancer cell lines, and higher expression in INBL cells, we propose that tumour cells can produce galectin-3 to interact with the NKp30 receptor and maybe promote metastasis, proliferation, angiogenesis, and chemoresistance. However, this requires further investigation to analyse the function of this signalling pathway in tumour cells.

Our results show that there is a large variety of subpopulations of tumour cells (Figure 5), some that bear activating markers of the immune response. Evidence exists that cervical tumour cells have strategies to evade the activation of the immune response; thus, it is possible that these molecules instead of activating the immune response promote proliferation, survival, and metastasis. Another set of tumour cells expressing markers of the immune response induce tolerance or lack of activation of the immune system; also, there are small subpopulations that express stem cell and epithelial-mesenchymal transition markers. However, these small subpopulations of tumour cells could be responsible for relapses and metastases. Finally, in this article, we show the presence of typical NK cell markers, molecules that in NK cells promote their inhibition or activation; however, at this point, the effect of these markers in tumour cells is not clear. Further research is needed to conclude whether these molecules contribute to tumour proliferation or to the promotion of metastasis and, furthermore, to determine if tumour cells use these molecules as camouflage to prevent the activation of the immune system.

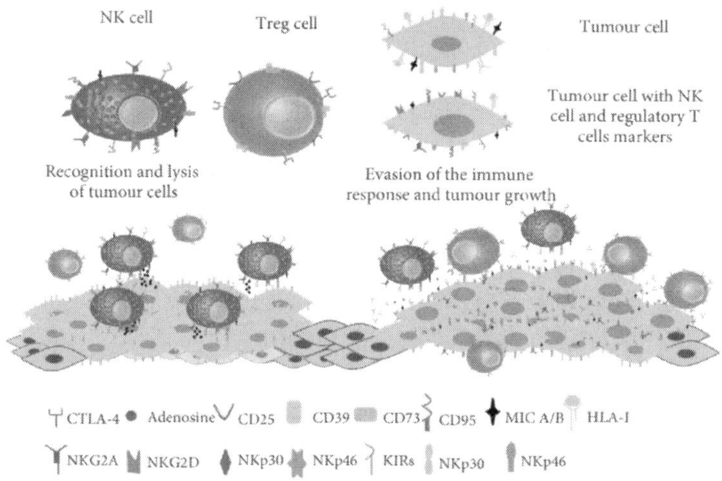

Figure 5. Cancer cells evade the immune system. The tumour cells express markers of the immune system such as CD25, CTLA-4, CD39, and CD73 that help to regulate the tumour microenvironment. Also, these cells express typical markers of NK cells, apparently as a camouflage system to counteract the effect of increased expression of molecules that induce cytotoxicity such as CD95.

The analysis of all these molecules provides a better understanding of tumour cell biology and opens the possibility of analysing more signalling pathways related to proliferation, metastasis, or resistance to cell death that could be therapeutic targets. Then, it is necessary to evaluate the presence of each of these molecules in early lesions of the cervix, carcinoma in situ, and invasive cancer, to know whether they have clinical and medical relevance and, further, to determine if they can be functional biomarkers.

CONCLUSION

The tumour cells have developed a series of mechanisms that allow them to express molecules that promote their proliferation and enhance its survival capacity. Some of these molecules are typical of their cell lineage, and others are not present in normal tissues of the same origin. Our data indicate that the cervical tumour cells express markers associated with activation of the immune system (CD95 and MICA/B), as well as markers related to inhibition of the immune system (CD73, CD39, CTLA-4, and CD25). However, there are very small subpopulations that express stem cell and epithelial-mesenchymal transition markers. Additionally, some subpopulations of cervical tumour cells express typical receptors of NK cells, activating inhibitor markers. At this point, we do not know if these molecules allow cervical cancer cells to proliferate and enhance its survival capacity. Altogether, our results suggest that these molecules could provide cervical cancer cells with a useful strategy to mimic the immune cell and therefore the ability to evade the immune response for the development of tumour masses.

ACKNOWLEDGMENTS

This work was supported by grants from the DGAPA-UNAM (PAPIIT IN222915 and IN222218) and from the National Council of Science and Technology (CONACYT, project number CB 253262).

REFERENCES

1. Rangel-Corona R., Corona-Ortega T., Soto-Cruz I., et al. Evidence that cervical cancer cells secrete IL-2, which becomes an autocrine growth factor. *Cytokine*. 2010;50(3):273–277. doi: 10.1016/j.cyto.2010.02.013.
2. Rocha-Zavaleta L., Huitron C., Caceres-Cortés J. R., et al. Interleukin-2 (IL-2) receptor-$\beta\gamma$ signalling is activated by c-Kit in the absence of IL-2, or by exogenous IL-2 via JAK3/STAT5 in human papillomavirus-associated cervical cancer. *Cellular Signalling*. 2004;16(11):1239–1247. doi: 10.1016/j.cellsig.2004.03.011.
3. Valle-Mendiola A., Weiss-Steider B., Rocha-Zavaleta L., Soto-Cruz I. IL-2 enhances cervical cancer cells proliferation and JAK3/STAT5 phosphorylation at low doses, while at high doses IL-2 has opposite effects. *Cancer Investigation*. 2014;32(4):115–125. doi: 10.3109/07357907.2014.883526.
4. Soto-Cruz I., Valle-Mendiola A., Martinez-Castro J., Zerecero-Carreon O., Bustos-Rodriguez R., Weiss-Steider B. An inactive receptor for epidermal growth factor (EGFR) interacts with HER-2 to induce proliferation in cervical cancer cell lines: p06-159. *The FEBS Journal*. 2012;279:p. 184.
5. Weiss-Steider B., Soto-Cruz I., Martinez-Campos C. A., Mendoza-Rincon J. F. Expression of MICA, MICB and NKG2D in human leukemic myelomonocytic and cervical cancer cells. *Journal of Experimental & Clinical Cancer Research*. 2011;30(1):p. 37. doi: 10.1186/1756-9966-30-37.
6. Cáceres-Cortes J. R., Alvarado-Moreno J. A., Waga K., et al. Implication of tyrosine kinase receptor and steel factor in cell density-dependent growth in cervical cancers and leukemias. *Cancer Research*. 2001;61(16):6281–6289.
7. Ghadially H., Brown L., Lloyd C., et al. MHC class I chain-related protein A and B (MICA and MICB) are predominantly expressed intracellularly in tumour and normal tissue. *British Journal of Cancer*. 2017;116(9):1208–1217. doi: 10.1038/bjc.2017.79.
8. Filippova M., Parkhurst L., Duerksen-Hughes P. J. The human papillomavirus 16 E6 protein binds to Fas-associated death domain and protects cells from Fas-triggered apoptosis. *Journal of*

Biological Chemistry. 2004;279(24):25729–25744. doi: 10.1074/jbc.M401172200.

9. Munakata S., Watanabe O., Ohashi K., Morino H. Expression of Fas ligand and bcl-2 in cervical carcinoma and their prognostic significance. *American Journal of Clinical Pathology*. 2005;123(6):879–885. doi: 10.1309/0773N4Q3GFP34J5V.
10. Barnhart B. C., Legembre P., Pietras E., Bubici C., Franzoso G., Peter M. E. CD95 ligand induces motility and invasiveness of apoptosis-resistant tumor cells. *The EMBO Journal*. 2004;23(15):3175–3185. doi: 10.1038/sj.emboj.7600325.
11. Chen L., Park S. M., Tumanov A. V., et al. CD95 promotes tumour growth. *Nature*. 2010;465(7297):492–496. doi: 10.1038/nature09075.
12. Bastid J., Cottalorda-Regairaz A., Alberici G., Bonnefoy N., Eliaou J. F., Bensussan A. ENTPD1/CD39 is a promising therapeutic target in oncology. *Oncogene*. 2013;32(14):1743–1751. doi: 10.1038/onc.2012.269.
13. Mora-García M. L., Ávila-Ibarra L. R., García-Rocha R., et al. Cervical cancer cells suppress effector functions of cytotoxic T cells through the adenosinergic pathway. *Cellular Immunology*. 2017;320:46–55. doi: 10.1016/j.cellimm.2017.09.002.
14. Bastid J., Regairaz A., Bonnefoy N., et al. Inhibition of CD39 enzymatic function at the surface of tumor cells alleviates their immunosuppressive activity. *Cancer Immunology Research*. 2015;3(3):254–265. doi: 10.1158/2326-6066.CIR-14-0018.
15. Contardi E., Palmisano G. L., Tazzari P. L., et al. CTLA-4 is constitutively expressed on tumor cells and can trigger apoptosis upon ligand interaction. *International Journal of Cancer*. 2005;117(4):538–550. doi: 10.1002/ijc.21155.
16. Chen X., Shao Q., Hao S., et al. CTLA-4 positive breast cancer cells suppress dendritic cells maturation and function. *Oncotarget*. 2017;8(8):13703–13715. doi: 10.18632/oncotarget.14626.
17. Vignali D. A. A., Collison L. W., Workman C. J. How regulatory T cells work. *Nature Reviews Immunology*. 2008;8(7):523–532. doi: 10.1038/nri2343.
18. Benjamin D., Rosolen A., Wormsley S. B., DeBault L. E., Colamonici O. R. Expression of low-, intermediate-, and high-affinity IL-2 receptors on B cell lines derived from patients

with undifferentiated lymphoma of Burkitt's and non-Burkitt's types. *Cellular Immunology*. 1990;129(1):112–124. doi: 10.1016/0008-8749(90)90191-S.

19. Loose D., Signore A., Bonanno E., et al. Prognostic value of CD25 expression on lymphocytes and tumor cells in squamous-cell carcinoma of the head and neck. *Cancer Biotherapy and Radiopharmaceuticals*. 2008;23(1):25–33. doi: 10.1089/cbr.2007.0373.
20. Foster B., Zaidi D., Young T., Mobley M., Kerr B. CD117/c-kit in cancer stem cell-mediated progression and therapeutic resistance. *Biomedicine*. 2018;6(1):p. 31. doi: 10.3390/biomedicines6010031.
21. Pastò A., Bellio C., Pilotto G., et al. Cancer stem cells from epithelial ovarian cancer patients privilege oxidative phosphorylation, and resist glucose deprivation. *Oncotarget*. 2014;5(12):4305–4319. doi: 10.18632/oncotarget.2010.
22. Zhao F., Chen Y., Wu Q., Wang Z., Lu J. Prognostic value of CD117 in cancer: a meta-analysis. *International Journal of Clinical and Experimental Pathology*. 2014;7(3):1012–1021.
23. Li J., Zhou B. P. Activation of β-catenin and Akt pathways by Twist are critical for the maintenance of EMT associated cancer stem cell-like characters. *BMC Cancer*. 2011;11(1):p. 49. doi: 10.1186/1471-2407-11-49.
24. Sheridan C., Kishimoto H., Fuchs R. K., et al. $CD44^+/CD24^-$breast cancer cells exhibit enhanced invasive properties: an early step necessary for metastasis. *Breast Cancer Research*. 2006;8(5, article R59) doi: 10.1186/bcr1610.
25. Liu H., Wang Y. J., Bian L., Fang Z. H., Zhang Q. Y., Cheng J. X. $CD44^+/CD24^+$ cervical cancer cells resist radiotherapy and exhibit properties of cancer stem cells. *European Review for Medical and Pharmacological Sciences*. 2016;20(9):1745–1754.
26. Chen C., Zhao S., Karnad A., Freeman J. W. The biology and role of CD44 in cancer progression: therapeutic implications. *Journal of Hematology & Oncology*. 2018;11(1):p. 64. doi: 10.1186/s13045-018-0605-5.
27. Tang M., Acheampong D. O., Wang Y., Xie W., Wang M., Zhang J. Tumoral NKG2D alters cell cycle of acute myeloid leukemic cells and reduces NK cell-mediated immune surveillance. *Immunologic*

Research. 2016;64(3):754–764. doi: 10.1007/s12026-015-8769-3.

28. Kaiser B. K., Pizarro J. C., Kerns J., Strong R. K. Structural basis for NKG2A/CD94 recognition of HLA-E. *Proceedings of the National Academy of Sciences of the United States of America*. 2008;105(18):6696–6701. doi: 10.1073/pnas.0802736105.
29. McWilliams E., Mele J. M., Fiazuddin F., Cheney C., Muthusamy N., Awan F. T. Targeting the tumor evasion interaction of NKG2A and its ligand HLA-E increases natural-killer cell activity in chronic lymphocytic leukemia. *Blood*. 2015;126(23):p. 1289.
30. Sun C., Xu J., Huang Q., et al. High NKG2A expression contributes to NK cell exhaustion and predicts a poor prognosis of patients with liver cancer. *OncoImmunology*. 2017;6(1, article e1264562) doi: 10.1080/2162402X.2016.1264562.
31. Chwae Y. J., Chang M. J., Park S. M., et al. Molecular mechanism of the activation-induced cell death inhibition mediated by a p70 inhibitory killer cell Ig-like receptor in Jurkat T cells. *The Journal of Immunology*. 2002;169(7):3726–3735. doi: 10.4049/jimmunol.169.7.3726.
32. O'Connor G. M., McVicar D. W. The yin-yang of KIR3DL1/S1: molecular mechanisms and cellular function. *Critical Reviews in Immunology*. 2013;33(3):203–218. doi: 10.1615/CritRevImmunol.2013007409.
33. Farag S. S., Fehniger T. A., Ruggeri L., Velardi A., Caligiuri M. A. Natural killer cell receptors: new biology and insights into the graft-versus-leukemia effect. *Blood*. 2002;100(6):1935–1947. doi: 10.1182/blood-2002-02-0350.
34. Guerra N., Guillard M., Angevin E., et al. Killer inhibitory receptor (CD158b) modulates the lytic activity of tumor-specific T lymphocytes infiltrating renal cell carcinomas. *Blood*. 2000;95(9):2883–2889.
35. Cagnano E., Hershkovitz O., Zilka A., et al. Expression of ligands to NKp46 in benign and malignant melanocytes. *Journal of Investigative Dermatology*. 2008;128(4):972–979. doi: 10.1038/sj.jid.5701111.
36. Elboim M., Gazit R., Gur C., Ghadially H., Betser-Cohen G., Mandelboim O. Tumor immunoediting by NKp46. *The Journal of Immunology*. 2010;184(10):5637–5644. doi: 10.4049/jimmunol.0901644.

37. Hsieh C. L., Nagasaki K., Martinez O. M., Krams S. M. NKp30 is a functional activation receptor on a subset of rat natural killer cells. *European Journal of Immunology*. 2006;36(8):2170–2180. doi: 10.1002/eji.200635982.
38. Wang L., Zhao Y., Wang Y., Wu X. The role of galectins in cervical cancer biology and progression. *BioMed Research International*. 2018;2018:11. doi: 10.1155/2018/2175927.2175927
39. Punt S., Thijssen V. L., Vrolijk J., de Kroom C.D., Gorter A., Jordanova E. S. Galectin-1, -3 and -9 expression and clinical significance in squamous cervical cancer. *PlosOne*. 2015;10(6, article e0129119) doi: 10.1002/eji.200635982.
40. Wang W., Guo H., Geng J., et al. Tumor-released galectin-3, a soluble inhibitory ligand of human NKp30, plays an important role in tumor escape from NK cell attack. *Journal of Biological Chemistry*. 2014;289(48):33311–33319. doi: 10.1074/jbc.M114.603464.

9

Colors and Some Morphological Traits as Defensive Mechanisms in Anurans

Luís Felipe Toledo[1] and Célio F. B. Haddad[2]

[1]Museu de Zoologia "Prof. Dr. Adão José Cardoso", Instituto de Biologia, Universidade Estadual de Campinas, Rua Albert Einstein s/n, CP 6109, 13083-863 Campinas, SP, Brazil
[2]Departamento de Zoologia, Instituto de Biociências, Universidade Estadual Paulista, CP 199, 13506-970 Rio Claro, SP, Brazil

ABSTRACT

Anurans may be brightly colored or completely cryptic. Generally, in the former situation, we are dealing with aposematism, and the latter is an example of camouflage. However, these are only simple views of what such colorations really mean and which defensive strategy is implied. For instance, a brightly colored frog may be part of a mimicry ring, which could be either Batesian, Müllerian, or Browerian. These are only examples of the diversity of color-usage systems as defensive strategies. Unfortunately, reports on

Citation: Colors and Some Morphological Traits as Defensive Mechanisms in Anurans. Luıs Felipe Toledo ´ and Celio F. B. Haddad. Hindawi Publishing Corporation International Journal of Zoology Volume 2009, Article ID 910892, 12 pages doi:10.1155/2009/910892

the use of colors as defensive mechanisms are widespread in the available literature, and the possible functions are rarely mentioned. Therefore, we reviewed the literature and added new data to this subject. Then, we the use of colors (as defensive mechanism) into categories. Mimicry was divided into the subcategories camouflage, homotypy, and nondeceitful homotypy, and these groups were also subcategorized. Dissuasive coloration was divided into behavioral display of colors, polymorphism, and polyphenism. Aposematism was treated apart, but aposematic colorations may be present in other defensive strategies. Finally, we propose functions and forms of evolution for some color systems in post-metamorphic anurans and hope that this review can be the basis for future research, even on other animal groups.

INTRODUCTION

Anuran coloration results from natural selection acting simultaneously on different aspects of natural history, such as protection against solar radiation, thermoregulation, osmoregulation, nitrogen metabolism (e.g., [1–4]), reproductive biology (e.g., [5]), and defensive strategy (e.g., [6]).

Anurans are remarkable for their color patterns, which may range from a uniform black dorsum, as in *Cycloramphus boraceiensis* (Cycloramphidae), to bright blue, orange, red, yellow, and green in the same individual, as in *Agalychnis callidryas* (Hylidae). In the context of defensive strategies, these two distinct situations may be related to two different types of primary defenses, which operate regardless of whether or not a predator is in the vicinity [7]. Both situations: crypsis and aposematism, as generally understood, help anurans to survive in nature. The former may improve chances of avoiding detection, and the latter may signalize that the individual is dangerous in one or several ways, where being toxic is the most common [7]. Besides these, there are several other defensive strategies involving coloration, such as body rising, mimicry, or flash color (e.g., [6, 8, 9]).

The presence of these and other defensive colorations in anurans has been published for several species in a fragmented way, and has never been reviewed. Herein, in order to organize current knowledge and ground future research, we reviewed this subject, added new data, and provided information about the evolution of color usage in anurans, with special reference to Neotropical species.

MATERIALS AND METHODS

We reviewed the literature by searching for coloration-related defensive strategies in anurans. Both natural and experimental observations were considered. Additional data were obtained during several field expeditions in Brazil from 1972 to 2007, mainly in the Cerrado and Atlantic Forest domains in the southeast.

All amphibian scientific names follow Frost [10], and the specific nomenclature for coloration categories was adapted from Pasteur [11]. It has been suggested that the terms "dupe" and "selective agent" be preferentially used instead of "predator," due to their broader meaning [11, 12]. In other words, an anuran's coloration is not always selected or maintained to warn or avoid a predator; the "signal receiver" [13] may often be another anuran (e.g., in intraspecific communication) or their prey (i.e., the anuran may benefit from not being perceived by its prey). However, in the present paper, we choose to use "predator" to designate any animal that could be a potential risk to the anuran because we are analyzing coloration as a defensive strategy. For the same reason, we will not cover aggressive and reproductive mimicry in this paper.

RESULTS

The use of colors can be divided into three major categories: mimicry, deceptive coloration, and aposematism (Table 1).

Table 1. Major categories and subcategories of colors used as defensive strategies in post-metamorphic anurans.

Major	2nd major	3rd major	Subcategories
		Eucrypsis	Visible color spectrum mimesis
	Camouflage		Nonvisible color spectrum mimesis
		Mimesis	Cryptic mimesis
			Phaneric mimesis
Mimicry		Concrete homotypy	Batesian mimicry*
	Homotypy		Browerian mimicry*
		Abstract homotypy	Definable model
			Model not definable*
	Nondeceitful homotypy		Müllerian mimicry*
	Arithmetic mimicry		

Deceptive coloration Polymorphism* Polyphenism	Behavioral display of colors*
Aposematism*	

for where aposematic coloration may or must be present

Mimicry

Mimicry is generally considered as per the Batesian mimicry concept, in which a nontoxic (or, otherwise, dangerous, e.g., the species may bite) species mimics a dangerous model species (generally toxic). However, Batesian mimicry is one of several types of mimicry into which anurans may be included (see Table 1). Mimicry (as presently defined) occurs when a group of organisms, the mimics, has evolved to share common perceived characteristics with another group of organisms, the models, through the selective action of a predator. Collectively, they are known as a mimicry complex. Mimicry was herein divided into three major groups: camouflage, homotypy, and nondeceitful homotypy (see what follows).

Camouflage

Camouflage may be defined as the resemblance of an animal with a part of the environment [7], especially as viewed by the predator at the time and place in which the prey is most vulnerable to predation (Endler [19]).

In post-metamorphic anurans, camouflage may be optical, chemical (e.g., production of floral, leaf-like, and ammonia odors), or acoustic (may occur, e.g., when frogs stop calling in the presence of a predator; e.g., [20]). In what follows, we will give special reference to optical camouflage, which implies avoidance of detection by possible predators.

(1) *Eucrypsis*Definition: homochromy (imitation of reflected light) which is acting alone. The model is undefined, that is, it is the background.

(a) *Visible Color Spectrum Mimesis*Many frogs are cryptic with the substrate they use, and there are a great variety of backgrounds and mimic frogs. As substrates, anurans may use rocks with lichens, tree trunks, leaves, forest litter, as well as mossy and rocky fields, for example. For any such substrates, there are mimic frogs that live there (Figure 1). Eucrypsis may be strengthened or weakened by a predator's angle of vision; that is, the anuran

may be more cryptic from a lateral view than from a dorsal view (Figures 1(f) and 1(e), resp.).The more distant predator is from the site occupied by an anuran, the higher may be the crypsis benefits. For instance, it is easy to find a *Dendropsophus nanus* in its reproductive site when we are close to it. However, from a certain distance, the colors and shape of this hylid get mixed with the general view of the area. The size and colors of *D. nanus* are generally similar to part of the foliage, and it is hard to distinguish them from yellowish leaves (Figures 1(k), 1(l)).

Figure 1. Different situations of cryptic mimicry in anurans: (a) *Dendropsophus acreanus* (Hylidae) on a tree trunk, (b) *Bokermannohyla alvarengai* (Hylidae) on a rock with lichens, (c) *Ischnocnema guentheri* (Brachycephalidae) on litter, (d) green morph of *Hypsiboas prasinus* (Hylidae) perched on a leaf, ((e)-(f), resp.) dorsal and lateral views of *Theloderma horridum* (Rhacophoridae) on a tree, (g) *Lankanectes corrugatus* (Nyctibatrachidae) in a lotic water body, (h) *Hylodes asper* (Hylodidae) on a wet rock, (i) *Proceratophrys boiei* (Cycloramphidae) on litter, (j) *Scythrophrys sawayae* (Leptodactylidae) on a dead leaf, (k) an adult male of *Dendropsophus nanus* (Hylidae) of practically the same size and shape as the leaves and (l) its reproductive site.

(b) *Nonvisible Color Spectrum Mimesis* Some studies have shown that several anuran species, from different families, may show a

pronounced rise in reflectance in the infrared part of the spectrum (e.g., [21–23]). Furthermore, some predators of anurans, such as pit vipers, may be able to perceive infrared wavebands [24]. Others, such as rodents, other amphibians, lizards, and mainly birds, have tetrachromatic color vision (including ultraviolet cones), and may, therefore, perceive ultraviolet wavelengths (e.g., [25]). Therefore, it is possible that infrared reflectance may have evolved for anurans to remain cryptic in the foliage (which also has the same infrared reflectance property), even against those predators who are able to detect infrared or ultraviolet spectra (e.g., [22, 26]).

(2) *Mimesis* Definition: it is homomorphy (imitation of morphology) and/or homokinemy (imitation of movements), in addition to homochromy (defined before). The model is defined, that is, it is an object.

(a) *Cryptic Mimesis*It holds when the model is a dominant element of the mimic's environment, such as green or brown leaves, sticks, rocks, lichens, and mosses. Many examples may be cited, but to mention some, we may refer to species of genera *Proceratophrys* and *Scythrophrys* which resemble fallen leaves (Figures 1(i), 1(j), resp.), as well as some species of *Theloderma* and *Bokermannohyla* that resemble mosses and lichens, respectively (Figure 1(b)).The phyllomedusines of genus *Phasmahyla*, when walking, sway slightly as if caught by the movement of the wind (= homokinemy). Indeed, their genus name *Phasmahyla* was coined in allusion to the similarity of their moving style with the walking-sticks of order Phasmatodea [27]. The illusion created by this behavior is increased with the leaf-like color pattern of this species.

(b) *Phaneric Mimesis* It holds when the model is an isolated and conspicuous inanimate element of the mimic's environment, such as animal droppings or rocks (when there are few rocks in the environment). As examples, we may cite some *Theloderma* spp. and *Dendropsophus marmoratus*, which resemble bird droppings, and many species of *Rhinella* and *Bufo* that resemble stones on the floor.

(3) *Factors of Camouflage Enhancement* Both eucrypsis and mimesis imply camouflage, which could be strengthened by (a)

countershading, (b) disruptive coloration, (c) shadow camouflage, (d) wetting, and (e) integumentary structures.

(a) *Countershading* Countershading occurs when the anuran's pigmentation is darker dorsally and lighter ventrally. This transition may be gradual or abrupt, which could involve different camouflage strategies (see [28, 29]).Two main functions have been attributed to countershading. (i) it is believed to have the effect of reducing conspicuous shadows cast on the ventral region of an animal's body. In essence, the distribution of light on objects lit from above will cause unequal reflection of light by a solid body of uniform color. Such shadows could provide predators with visual cues to a prey's shape and projection. Countershading, therefore, reduces the ease with which prey is detected by potential predators by counterbalancing the effects of shadowing. This effect occurs mainly in animals that present a gradual transition of colors [28], and should be properly named "self-shadow concealment" [29]. (ii) Countershading coloration would work via background matching; that is, when seen from the top, the dorsum (if cryptic colored) blends into the background below, which could be a body of water (when the anuran is swimming or floating) or dry-to-flooded ground (when the anuran is on the floor or perched in the vegetation). When seen from below, the lighter ventral area blends into the sun or moonlight (see [29]). This second explanation generally occurs with an abrupt transition of colors and seems to be the more adequate for anurans.Countershading could result from selective pressures other than predation avoidance. For example, the dorsal surface must be protected against the damaging properties of UV light and/or abrasion Kiltie [30, 31], and the ventral side does not need to be pigmented, as pigmentation may be costly [29]. Therefore, the occurrence of countershading may result from multiple factors.This system is so widespread among aquatic and terrestrial fauna that several authors have stated that it is perhaps the most universal feature of animal coloration (see [29] and references therein). Likewise, it is present in several anuran species and might work against terrestrial predators, for example, which may be on the ground when the anuran is perched on a tree branch. In aquatic species, such as the pipids (e.g., *Pipa* spp. and *Xenopus* spp.), the system may work as described for fishes.

(b) *Disruptive Coloration* Disruptive coloration is a color pattern that breaks the appearance of body form. Several anuran species have dorsal lines and/or blotches that may be considered constitutive of disruptive coloration, breaking the general outline of the body. Some species may enhance their camouflage by having high-contrast lines on the edges of colored patterns (see [32]). In addition to this, several species present lateral lines that cross the eyes, breaking their rounded shape (Figure 1(j)). These are the most common forms of disruptive coloration in anurans.A possible variation is the presence of aposematic coloration (see what follows for an explanation of aposematism) simultaneously with disruptive coloration, depending on the predator and/or brightness of the night. This may occur because the colorful stripes and/or blotches of an aposematic anuran (Figure 2(a)) may become a disruptive pattern when seen by a color-blind predator, or by a color vision predator hunting on dark nights (Figure 2(b); see also [33, 34]).Recently, it has been suggested and/or demonstrated that disruptive coloration is advantageous compared to simple eucrypsis (see [35, 36], and references therein).

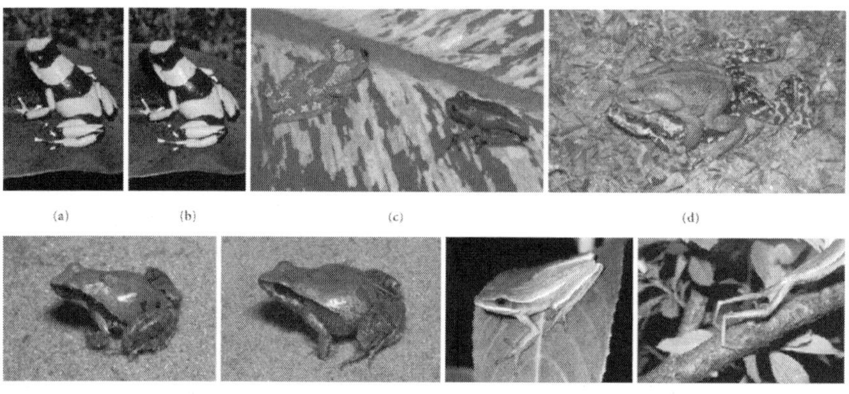

(a) (b) (c) (d) (e) (f) (g) (h)

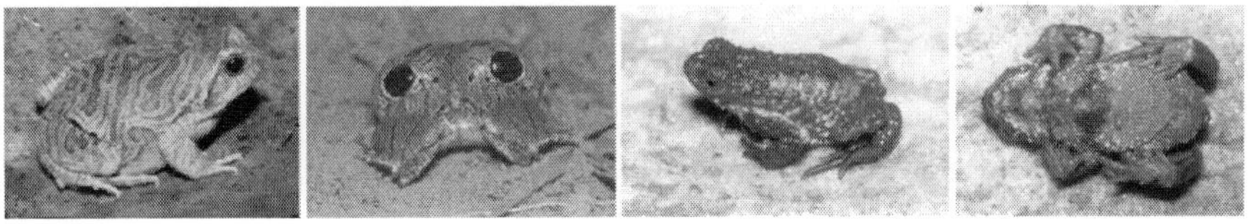

Figure 2. (a) An aposematic *Oophaga lehmanni* (Dendrobatidae) (b) may have disruptive coloration on dark nights or when searched for by a color-blind predator. (c) An adult and a juvenile *Hypsiboas semilineatus* (Hylidae), (d) an amplectant pair of *Rhinella icterica* (Bufonidae), and ((e)-(f)) two morphs of *Physalaemus cuvieri* (Leiuperidae) are examples of polyphenism. ((g)-(h)) Flash color exhibited by *Hypsiboas caingua* (Hylidae), ((i)-(j)) deimatic behavior by *Eupemphix nattereri* (Leiuperidae), and ((k)–(l)) thanatosis by *Melanophryniscus moreirae* (Bufonidae).

(c) *Shadow Camouflage*Anurans may rest in areas with combined spots of sunlight and shade, making it difficult to recognize the animals on the substrate. If part of the anuran is exposed to sunlight and the other part is in shadow, such light play may enhance the anuran's disruptive pattern (e.g., [32]).

(d) *Wetting*Some individuals may remain in lotic water bodies, covered by a film of water or by water drops. This situation may enhance an animal's crypsis against terrestrial predators, by creating reflected shiny spots on its dorsum matching the shiny spots in the water or substrate, for example, rocks (Figures 1(g)-1(h)).

(e) *Integumentary Structures*Some integumentary structures seem to be associated with disruptive outlines and thereby aid in concealment. Such structures include small, irregular ridges, supraciliary processes (e.g., species of *Proceratophrys* and *Ceratophrys*), scalloped folds on the outer edges of limbs (e.g., *Cruziohyla craspedopus*), and calcars (e.g., several *Hypsiboas* spp.). Dorsal glands may also enhance crypsis by promoting resemblance with lichens (e.g., *Bokermannohyla alvarengai*, *Itapotihyla langsdorffii*, and *Scinax nebulosus*) or mosses (e.g., *Theloderma corticale* and other Vietnamese mossy frogs) (Figure 1).

Homotypy (Assimilation to Another Type)

It acts differently from camouflage: in homotypy animals can be perceived by the predator. Homotypy involves the mimetic imitation of another object (which can be the same or another species, an object of the environment, or an undefined model), being, therefore, included as mimicry.

(1) *Concrete Homotypy (Actual Model).* The model is definite or an existing species (or cluster of similar species).

(a) *Batesian Mimicry*The concept of Batesian mimicry [37] involves the success of a specific coloration against a predator's experience; that is, predators learn to avoid unpalatable species, which are identified by their aposematic coloration (which does not always mean bright colors: e.g., [11, 38]), odor, sound, or another signal. Thus, a mimic could obtain protection by resembling the unpalatable or less palatable model. For anurans, there are few cases described wherein some palatable frogs may mimic some poison frogs (Table 2).Batesian mimicry involves the predator's ability to learn or have innate knowledge. Several predators, such as invertebrates in general, are not as well-endowed in terms of sight and memory as are mammals, and, therefore, may not have been the promoters of selective pressures for the evolution and persistence of Batesian mimicry (see also [11]).A Batesian mimic does not necessarily need to be identical to its model. Sometimes, it may exhibit intermediate resemblances to two (or more) models. In this manner, the mimic may escape from some predators that avoid one model and from others that avoid the other model. This dual mimicry system has been proposed for coral snake mimics [39] and may be present in anurans.An intriguing situation is the Batesian mimicry proposed for the *Leptodactylus lineatus* and *Allobates femoralis* complex. This complex would actually represent a case of Müllerian mimicry (where both species are unpalatable; see further explanations below), instead of Batesian mimicry as initially proposed [16], as *L. lineatus* seems to be a noxious species as well. Upon handling and fixing individuals of *L. lineatus*, they exude a great amount of milky, probably noxious secretions (C. F. B. Haddad, personal observation). However, some tests performed on *A. femoralis* indicated that this species is not toxic (see [40], and references therein). If *A. femoralis* is not toxic and *L. lineatus* is, this would

be once again a case of Batesian mimicry where *L. lineatus* is the model. If both species are discovered to be nontoxic, it would be a case of Arithmetic mimicry (where both species are palatable; see further explanations below). Therefore, the relationship of this complex remains unresolved and requires further research.

Table 2. Occurrence of Batesian and Müllerian mimicry in anurans, and distribution overlap between species.

Mimic	Model	Mimetism type	Sympatric species	Source
Aromobatidae	Dendrobatidae			
Allobates zaparo	*Ameerega bilinguis*	Batesian	Yes	[14]
Allobates zaparo	*Ameerega parvula*	Batesian	Yes	[14]
Brachycephalidae	Dendrobatidae			
Pristimantis gaigeae	*Phyllobates lugubris*	Batesian	Yes	[15]
Pristimantis gaigeae	*Phyllobates auro-taenia*	Batesian	Yes	[15]
Leptodactylidae	Aromobatidae			
Leptodactylus linea-tus[**]	*Allobates femoralis*[**]	Batesian	Yes	[16]
Dendrobatidae	Dendrobatidae			
Ranitomeya imitator	*Ranitomeya fantas-ticus*	Müllerian	Yes	[9]
Ranitomeya imitator	*Ranitomeya variabilis*	Müllerian	Yes	[9]
Ranitomeya imitator	*Ranitomeya ventri-maculata*	Müllerian	Yes	[9]
Mantellidae	Mantellidae			
Mantella laevigata[*]	*Mantella manery*[*]	Müllerian	Yes	[17]
Mantella baroni[*]	*Mantella madagas-cariensis*[*]	Müllerian	Yes	[17]
Mantella pulchra[*]	*Mantella cowanii*[*]	Müllerian	Yes	[18]
Dendrobatidae	Dendrobatidae			
Phyllobates auro-taenia[*]	*Phyllobates lugubris*[*]	Müllerian	No	Present study

*These cases, definition of the mimic and model is not possible.

**Further studies are needed in this case: see text.

(b) *Browerian Mimicry* When individuals within a species differ in palatability to predators, the more palatable individuals (mimics) will gain benefits from those less palatable (models). The models can be of the same or opposite sex to the mimics. Albeit never reported that this type of mimicry may be present in aromobatids, bufonids, dendrobatids, mantellids, and myobatrachids, at least. Individuals of the same noxious species of these families (cited above) acquire the alkaloids contained in their noxious secretions from dietary arthropods (e.g., [41–43], and references therein). Therefore, if for any reason a group of individuals within a species either does not feed on arthropods that carry the alkaloids, feeds on them but at a lower rate/proportion, or does not sequester these components, they will be less poisonous than the others (e.g., [44–47]).Indeed, there are reports that show spatial (geographic) and temporal (seasonal) variation in the alkaloid profiles of poison frogs [48], which may support the Browerian mimicry theory for anurans.

(2) *Abstract Homotypy (Virtual Model)* When the model is not an actual species, homotypy is an abstract.

(a) *Definable Model* It occurs when the model looks like a general type of organism, or part or indirect vestiges of another organism, but is not identifiable at species level. For example, the deimatic eyespots present on the back of leiuperids, which could resemble snakes' eyes (see, e.g., [6]), but not of a specific species of snake. Other examples are frogs that rest on leaves and look like bird droppings (e.g., *Dendropsophus marmoratus* and *Theloderma* spp.). Besides this, the presence of calcars on many species of rainforest tree frogs, and their absence in other anurans, may suggest that they might serve as points of water runoff, mimicking drip tips of leaves [49, page 371].

(b) *Model Not Definable* Occurs when the model is not identifiable at all, but a frightening or cryptic form is conjured up. This seems to be the case for the leg interweaving behavior described elsewhere (see [50]).

Nondeceitful Homotypy

This category was created (*sensu* [11]) basically for the inclusion of Müllerian mimicry [51, 52] and Arithmetic mimicry [53]. In both cases, the predator is not deceived, that is, the predator can recognize whether the anuran is palatable or unpalatable. In this category, both the model and the mimic are either unpalatable and/or dangerous in some way (Müllerian mimicry), or palatable (arithmetic mimicry).

These mimicry complexes are formed by species with similar (or even exactly the same) color pattern. However, it is sometimes hard to distinguish real nondeceitful homotypy from a possible phylogenetic influence; that is, closely related species, such as two species of *Dendrobates*, may resemble one another due to a simplesiomorphy (sharing of an ancestral character) and not because of coloration convergence (homoplasy). Whether a convergence or a simplesiomorphy, the resemblance would benefit the two (or more) species involved.

(a) *Müllerian Mimicry* Müllerian mimicry involves the mutual benefit of two dangerous (e.g., toxic, noxious, unpalatable, able to harm the aggressor by any means, such as biting and scratching, etc.) species by sharing similar phenotypes [51, 52]. To our knowledge, there are only few cases of Müllerian mimicry in anurans (see Table 2); however, other possible mimicry relationships may be suggested based on the current reports. We may join two mimicry complexes of *Mantella* (*M. baroni/madagascariensis* and *M. pulchra/cowanii*) into a larger group of mimicry species, as all individuals are sympatric and exhibit homochromy and homomorphy [17, 18]. Furthermore, from the evidence of Batesian mimicry by *Pristimantis gaigeae* with both *Phyllobates aurotaenia* and *P. lugubris* as models, we may suggest another Müllerian mimicry system. This system would be comprised of *P. aurotaenia* and *P. lugubris*. These two species are not sympatric, however, chronosympatry (presence of model and mimic in the same time and place) is not a rule for the existence of mimicry [54, 55]; for instance, their past distributions could have overlapped (see also [15]), and/or the present distribution of selective agents (predators) may overlap with the distribution of both species (including Batesian mimic *P. gaigeae*).Furthermore, it is possible that other Müllerian mimicry systems exist that are not based on coloration but still comprise visual mimetism. For example, if two different-colored species, or different morphs of the same species, present the same toxic substances, they could be chemical mimics. Therefore, based on properties other than

color, such as body shape and brightness, they could be part of a Müllerian mimicry ring (see also [56]). This would partly explain the several cases of polymorphism among aposematic species (see below).

(b) *Arithmetic Mimicry.* Similar sympatric edible species share the burden of predation in proportion to their relative frequencies. In other words, the higher the abundance of a certain (color) morph in the predator foraging area, the lower the chances of an individual prey being preyed upon. In this case, predator learning (ontogenetic or inherited) is irrelevant. To our knowledge, arithmetic mimicry has never been reported for anurans; however, it may be a very widespread phenomenon involving several sympatric (or even syntopic) similar (e.g., in simultaneous homochromy, homomorphy, and homokinemy) species. As examples of species pairs (or more), there are syntopic *Leptodactylus furnarius* and *L. jolyi*; *Dendropsophus nanus*, *D. sanborni*, and *D. minutus*; *Proceratophrys appendiculata*, *P. boiei*, and *P. melanopogon*; as well as several species of *Ischnocnema*, all of them found chronosympatrically in the Neotropical region.Furthermore, two species with differences in coloration (for instance), but of similar sizes (for instance), may also be arithmetic mimics. Both of these edible species sharing a predator foraging area may be equally nutritive; therefore, provided that the predator can perceive that they are both nutritive, they will be nutritional arithmetic mimics, benefited by the saturating theory (see [56]).

Deceptive Coloration

Behavioral Display of Colors

Many anurans may exhibit deceptive coloration, either intentionally or not, by performing movements (e.g., flash color behavior) or by adopting specific postures (e.g., body raising).

A fleeing anuran may escape from predators by displaying a flash of aposematic color(s), generally followed by remaining motionless. This is known as flash color behavior. This coloration is only visible when the anuran is moving (generally while jumping), and concealed during resting posture (Figures 2(g)-2(h)). Flash color is widespread among anurans, occurring in several species and families. In the Neotropics, it is very common in species of genera *Scinax* and *Hypsiboas* (Hylidae).

Flash color behavior may serve to disorientate and confuse an attacking predator [7] and/or warn predators of the presence of toxins [57]. These two distinct functions, and the implied predator responses, may vary among individuals or species of predator. In other words, some predators may be warned, while some may get confused. In this latter case, the predator may lose the anuran: the flash behavior may precede the motionless behavior, creating a prey search image that quickly disappears [7, 58]. Also, flash color behavior may halt the predator's attack for an instant and, thus, provide extra time for the frog to escape [59].

Another possibility of color display is body raising. For example, in some leiuperids (e.g., *Eupemphix nattereri*, *Physalaemus deimaticus*, and *Pleurodema brachyops*), when individuals lift their hindparts, they exhibit a pair of black eyespots (Figures 2(i)-2(j)). Similarly to flash color behavior, this exhibition may be viewed as an aposematic signal (as these species have large poison glands below the eye-spots) or may be a deceptive signal. In this latter case, the predator may have the illusion of facing a different (and larger) animal. Therefore, it would be better classified as an abstract homotypy where the model is definable (see above).

Polymorphism

Polymorphism in anurans is characterized by the presence of fixed chromatic phenotypes within or between populations. The individuals seem unable to change their color, so there must be genetic control involved. Polymorphism may benefit the anuran in such a way that one or more of the phenotypes are not included in the predator's search image. Several species are known to present different chromatic morphotypes, and such polymorphism may occur in three ways as follows.

(a) *between Individuals, Independent of Sex and Life Stage* Two morphotypes: for example, some adult individuals of *Physalaemus cuvieri* (Leiuperidae) and *Paratelmatobius* spp. (Leptodactylidae) are green, while others are brown (Figures 2(e)-2(f)). Another type of color dimorphism is exhibited by individuals of *Leptodactylus fuscus* (Leptodactylidae), which may or may not have a dorso-vertebral white line.More than two morphotypes: several species of cryptic genera *Eleutherodactylus* and *Ischnocnema* (Brachycephalidae) exhibit many different color morphotypes [60]. Even in aposematic species, this sort of polymorphism

can also be observed; for example, in *Anomaloglossus beebei* (Aromobatidae), there are five color morphotypes [61], and in *Oophaga pumilio* (Dendrobatidae), there are at least 15 color morphotypes (see, e.g., [5, 23]). Multiple aposematic phenotypes are also observed in other dendrobatids [62] and African mantellids (see, e.g., [63]). Multiple aposematic phenotypes may add the benefits of aposematism to the benefit regarding predator search image (see before).

(b) *between Life Stages: Juveniles Different from Adults* This is the case of *Hypsiboas geographicus*, *H. semilineatus*, *H. lundii*, *H. pardalis*, and *H. raniceps* (Hylidae), for instance, where juveniles have a totally different coloration from adults, independent of sex (Figure 2(c)).

(c) *between Sexes: Males Different from Females*As examples, we may cite *Rhinella icterica* (Bufonidae), where males are light yellowish and females are black and white (Figure 2(d)); *Leptopelis vermiculatus* (Arthroleptidae), where males are green and females are brown; and most contrasting is *Incilius periglenes* (Bufonidae), where males are bright orange and females are black, red, and yellow.

Polyphenism

Polyphenism is the ability to generate different phenotypes, by color changing in this case, by the same individual. Polyphenism may be a better term to describe this phenomenon than polymorphism, which generally implies a stronger genetic element for each particular appearance [64].

Many anurans can change their dorsal coloration by rearranging their chromatophores, which involves sophisticated physiological control of skin structures. There is a continuous gradient of color change timing in anurans: the change may occur instantaneously, or may take a few minutes, hours, days, or even weeks to occur.

Some species may change their color very quickly. We placed one individual of *Bokermannohyla circumdata* (Hylidae) in a dark place (under tree bark) and left another individual exposed to the laboratory light. After 15 minutes, they were completely different from their initial, dark reddish-brownish coloration. The first individual (kept in the dark) was much darker, almost black, whereas the second individual (exposed) was almost white. A similar polyphenism was described for *Bokermannohyla alvarengai*, but in this case, the color change was studied with regard to physiological adjustments for temperature and water loss control [4].

The dorsal coloration of *Scinax fuscomarginatus* (Hylidae) males is yellowish during the night (during reproductive activity) and grayish or brownish during the day. This may be related to the specific site they use: during the night, they remain perched on yellowish grass vegetation [65], while during the day they can be found in dark sites, such as tree holes, under tree bark, and in the middle of clumps of grass. Conversely, this color change may be due to testosterone amount during reproductive activity (calling), as the yellowish males also have an odor which is distinct from the one they have during the daytime resting period. Additionally, individuals killed while still yellowish left a yellow tinge in the preservative liquid (formalin, alcohol). We recorded the same phenomenon for *Scinax fuscovarius* and *S. hayii*. Furthermore, some phyllomedusines can change from purplish during night activity to greenish during daytime resting. Such polyphenism was observed in *Phasmahyla cochranae*, *P. guttata*, and *P. jandaia*. This phenomenon can occasionally be observed in *Phyllomedusa azurea*, *P. megacephala*, and *P. rohdei* (see colored pictures in [66]).

We observed a seasonal polyphenism in *Hypsiboas prasinus* (Hylidae), in which a higher presence of green individuals occurred mainly in the hot, rainy season of the year, whereas a higher presence of brown individuals occurred in the cold, dry season of the year. This pattern overlaps with the frequency of green and brown leaves in the Mesophytic Semideciduos Forest where this species dwells [67]: the peak of leaf fall precedes the peak of brown morphs and the peak of leaf flushing precedes the peak of green morphs (Figure 3). Seasonal color changing has also been observed in *Pseudacris regilla* (Hylidae), and has been considered a response to divergent selection for crypsis in a heterogeneous, seasonally variable environment [68]. This is likely to be an explanation for the polymorphism of *H. prasinus* as well.

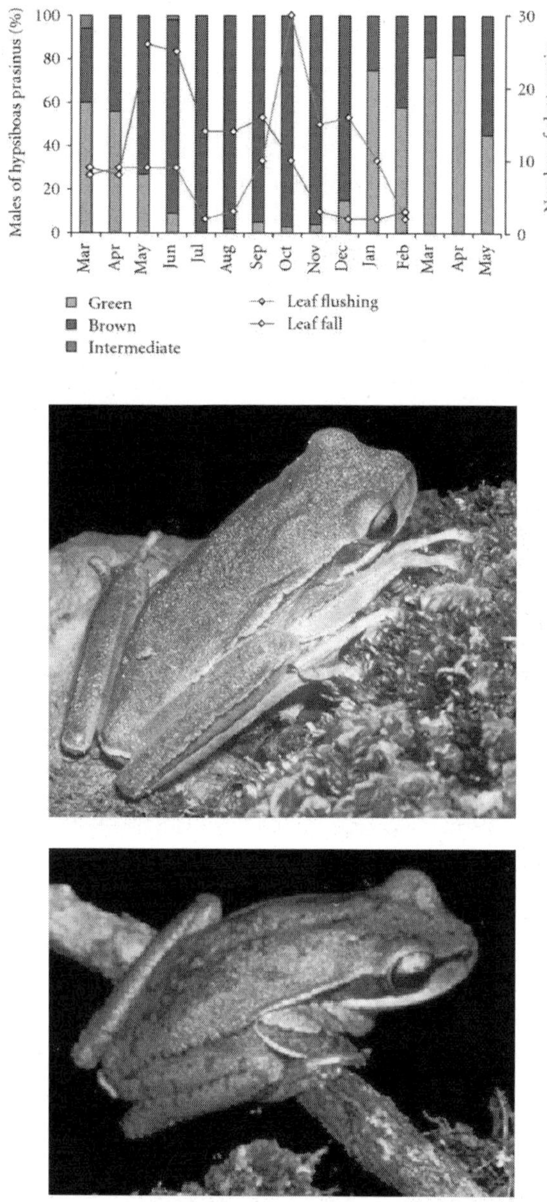

Figure 3. Seasonal variation (March 1988 to May 1989) in the dorsal coloration of reproductive males of *Hypsiboas prasinus* (Hylidae), and number of plants that exhibited leaf fall (brown lines) and leaf flushing (green lines), in Serra do Japi, Jundiaí, São Paulo, Brazil. Brown bars represent brown morphs (upper picture), green bars represent green morphs (lower picture), and olive

bars represent intermediate (olive) coloration morphs. Plant phenology data were obtained from Morellato et al. [67].

Polyphenism may be advantageous over polymorphism because the anuran may select a substrate and then adjust its color pattern. Polymorphic anurans may find adequate substrates to fit their general coloration, and such are not necessarily hard to find, but polyphenic species may have a wider range of substrates that they can use.

Aposematism

Aposematic coloration has also been referred to as sematic, conspicuous, or warning coloration. Aposematism is the presence of contrasting and conspicuous coloration that is generally related to the presence of skin toxins in the individuals [7, 58]. Furthermore, it may also signalize that the anuran is dangerous or unpleasant in any other way. For example, aposematic anurans may bite, exhibit spine aggression, and/or produce loud defensive screams.

Aposematic coloration is generally bright red, orange, yellow, and/or blue on a dark (generally black), contrasting background. This aposematic coloration is most commonly widespread over the entire body, such as in species of *Allobates* (Aromobatidae), *Ameerega* and *Dendrobates* (Dendrobatidae), *Mantella* (Mantellidae), *Atelopus* (Bufonidae), and *Brachycephalus* (Brachycephalidae).

Aposematic coloration is often confined to parts that are usually concealed when the frog is in its resting posture (e.g., some leiuperids (*Eupemphix nattereri* and *Physalaemus* spp.), bombinatorids (*Bombina* spp.), hylids (*Phyllomedusa* spp. and *Hypsiboas* of the *marmoratus* group), bufonids (*Melanophryniscus* spp.), and leptodactylids (*Leptodactylus* of the *pentadactylus* group)). In this case, the bright coloration is generally present on the back of the thighs, the underside of the body, the limbs, feet, and hands, and is displayed by specific posturing such as the unken reflex [69] or body raising [65].

In species of *Ceratoprhys* and *Hemiphractus*, aposematic coloration may be present on the lips or on the tongue, respectively, which are displayed when the individuals perform the mouth-gapping display or emit defensive screams. In *Phyllomedusa* spp., the bright, contrasting coloration may be displayed intentionally (e.g., during foot flagging) or when the individuals walk on tree branches. The foot signaling of *Atelopus zeteki*, in addition to being an intraspecific communication, may be a visual display, including aposematic coloration (bright yellow contrasting with black stripes) for potential predators [70]. In *Ameerega flavopicta*, there seems to be a relationship between parental care and body raising [71]. Several species have a conspicuously colored belly (or other underside parts; e.g., *Paratelmatobius* spp., *Leptodactylus pustulatus*, or *Melanophryniscus* spp.). When facing a predator, they might not present any specific behavior to display these colors. However, they may display thanatosis (a widespread behavior in frogs) and, during handling by the predator, the anurans may be turned upside down and would thus display the aposematic coloration (Figures 2(k)-2(l)). Species of *Oreophrynella*, after displaying "balling behavior" while fleeing, may stop belly-side-up, exposing their bright, contrasting ventral coloration [72].

Several factors are involved in the evolution of aposematism, such as unpalatability, honest signals, relative predator-prey abundance-dependence, and kin selection (review in [73]). In anurans, aposematism has evolved multiple times (e.g., [74–76]). However, some of these authors have based their results on the hypothesis of evolutionary relationships that have been recently modified or complemented [40, 77]. Therefore, a new overview of these evolutionary approaches to defense is needed, because their assumptions may be modified. For example, Summers and Clough [74] did their study based on the assumption that family Dendrobatidae (as considered at that time) was monophyletic, and that the toxic aposematic dendrobatids clade was also monophyletic. However, this apparent monophyly has been rejected [75, 76], the former dendrobatid family has been divided into two sister families (Dendrobatidae and Aromobatidae), and the toxic species are divided into several genera (e.g., *Adelphobates*, *Allobates*, *Ameerega*, *Dendrobates*, *Epipedobates*, *Hyloxalus*, *Myniobates*, and *Oophaga*) in both families [40].

Although there may be changes to the phylogenetic hypotheses, it is still possible that aposematic coloration has evolved in tandem with toxicity in anurans of the Bufonidae, Dendrobatidae, Aromobatidae, and Mantellidae families, as previously proposed (e.g., [43, 74]). "If differences among species in dietary preferences or predatory capabilities are heritable, then natural selection could act to favor brighter coloration in species that consistently have preferences for or access to prey with more or more potent toxins" [74]. Speculations apart, the most evident fact is that anuran aposematism has evolved by means of multiple convergent radiations, within the order, its families, and/or its genera (e.g., [43, 75, 76, 78]).

DISCUSSION

As we observed above, anuran coloration may provide protection against predators by providing concealment (e.g., camouflage, homotypy, and arithmetic mimicry) by alerting the predator about a possible hidden danger or unpleasant characteristic (aposematism), or by deceiving predators (deceptive coloration and some cases of mimicry). These strategies may act on the two first stages of predation as reviewed by Endler [79]: detection and identification.

The different types of defensive coloration presented in this study may have been selected differently throughout anuran evolution (see Table 3). Cryptic coloration (e.g., camouflage) may have been selected for by predation pressures imposed by predators upon anurans (defensive strategies), by the predatory behavior of anurans upon their prey (feeding strategies), or by both. All deceptive coloration categories (polyphenism, polymorphism, and behavioral display of colors) are directly related with predator search image. Therefore, the pressures that may have promoted them are tightly associated with the predator's vision and cerebral capability. Aposematic coloration may have multiple factors involved in its evolution. It may have evolved by means of sexual selection and/or after the acquisition of any dangerous or unpleasant defensive strategy. In the former case, in anurans, such strategies can be biting, spine aggression, defensive screams, unpalatability (in terms of bad taste), or, most commonly, noxious skin secretions. Therefore, we would not expect to find a harmless aposematic anuran, except for those involved in mimicry rings (mainly Batesian and Browerian). Indeed, as far as we know, every aposematic anuran (except for some mimics) produces harmful defenses, such as poison (e.g., some species of Mantellidae, Dendrobatidae, and Brachycephalidae: [40, 43, 80]), or aggressive defenses (e.g., some species of Ceratophryidae and Leptodactylidae: L. F. Toledo, I. Sazima, and C. F. B. Haddad, unpublished data). Therefore, multiple selection factors may be synergistically involved in the origin and evolution of anuran coloration.

Table 3. Characteristics, benefits acquired, and constraints involved in the evolution of different classes of anuran coloration.

Coloration classes	Main characteristic	Benefits acquired	Evolutionary constraints involved
Cryptic coloration	Background matching	Detection avoidance (by prey and predators)	Predator and prey search image
Aposematic coloration	Background contrasting	Predator avoidance	Predator learning, presence of harmful defenses*, sexual selection
		Reproductive success	
Deceptive coloration	Dissuasive coloration	Predator avoidance	Predator search image, sexual selection
		Reproductive success	

be an exception in cases of Browerian and Batesian mimicry.

Use of colors by anurans as a defensive strategy is a very large field of knowledge, and quite unexplored up to the present moment. Now, with this review, we attempt to organize part of our knowledge, generating at least a standardization of the nomenclature that may be applied to anuran coloration as regarding defensive strategy. Also, we added some information and insights on the relationships between predation and the defensive mechanisms of post-metamorphic anurans. From this point forward, we recommend some lines of research which may (i) complement this and other recent reviews with more observational information (which is truly lacking at present), (ii) focus on specific defensive strategies against predators and reports of predator-prey interactions, (iii) complete a broader, meta-analysis of predator-prey interactions, and (iv) advance further in the understanding of the evolution (including phylogenetic approaches) of defensive strategies and their relationships with present and past predators.

ACKNOWLEDGMENTS

Ivan Sazima, Anne D'Heursel, Itamar Martins, Ricardo Sawaya, Rogério Bastos, and two anonymous referees made valuable comments during early versions of the manuscript. André Antunes, Cynthia Prado, Daniel Loebmann, Juliana Zina, José Pombal Jr., Luís Giasson, Olívia Araújo, and Rodrigo Lingnau helped during field expeditions. Julián Faivovich and Patrícia Morellato helped with some references. Peter Janzen, Glenn Tattersall, André Antunes, Ricardo Sawaya, Zoltan Takacs, and Jeet Sukumaran provided the pictures of *Lankanectes corrugatus*, *Bokermannohyla alvarengai*, *Eleutherodactylus guentheri*, *Physalaemus nattereri*, *Oophaga lehmanni*, and *Theloderma horridum*, respectively. FAPESP and CNPq supported the Laboratório de Herpetologia, Departamento de Zoologia, UNESP, Rio Claro, São Paulo, Brazil. The authors also thank CAPES (PRODOC), FAPESP (JP proc. no. 2008/50325-5 and 2008/52847-9), Idea Wild, and Neotropical Grassland Conservancy and Fauna Pro Assessoria e Consultoria Ambiental for grants, scholarships, equipment donation, and supporting some of the expeditions.

REFERENCES

1. F. Kobelt and K. E. Linsenmair, "Adaptations of the reed frog *Hyperolius viridiflavus* (Amphibia, Anura, Hyperoliidae) to its arid environment. 1. The skin of *Hyperolius viridiflavus nitidulus* in wet and dry season conditions," *Oecologia*, vol. 68, pp. 533–541, 1986.
2. R. Schmuek and K. E. Linsenmair, "Adaptations of the reed frog *Hyperolius viriditlavus* (Amphibia, Anura, Hyperoliidae) to its arid environment. III. Aspects of nitrogen metabolism and osmoregulation in the reed frog, *Hyperolius viriditlavus taeniatus*, with special reference to the role of iridophores," *Oecologia*, vol. 75, no. 3, pp. 354–361, 1988.
3. R. Kaul and V. H. Shoemaker, "Control of thermoregulatory evaporation in the waterproof treefrog *Chiromantis xerampelina*," *Journal of Comparative Physiology B*, vol. 158, no. 6, pp. 643–649, 1989.
4. G. J. Tattersall, P. C. Eterovick, and D. V. de Andrade, "Tribute to R. G. Boutilier: skin colour and body temperature changes in basking *Bokermannohyla alvarengai* (Bokermann 1956)," *Journal of Experimental Biology*, vol. 209, no. 7, pp. 1185–1196, 2006.
5. A. Siddiqi, T. W. Cronin, E. R. Loew, M. Vorobyev, and K. Summers, "Interspecific and intraspecific views of color signals in the strawberry poison frog *Dendrobates pumilio*," *The Journal of Experimental Biology*, vol. 207, no. 14, pp. 2471–2485, 2004.
6. I. Sazima and U. Caramaschi, "Descrição de *Physalaemus deimaticus*, sp. n., e observações sobre comportamento deimático em *P. nattereri* (Steindn.)—Anura, Leptodactylidae," *Revista de Biologia*, vol. 13, no. 1–4, pp. 91–101, 1988.
7. M. Edmunds, *Defence in Animals: A Survey of Anti-Predator Defences*, Longman, New York, NY, USA, 1974.
8. M. Martins, "Deimatic behavior in *Pleurodema brachyops*," *Journal of Herpetology*, vol. 23, no. 3, pp. 305–307, 1989.
9. R. Symula, R. Schulte, and K. Summers, "Molecular phylogenetic evidence for a mimetic radiation in Peruvian poison frogs supports a Müllerian mimicry hypothesis," *Proceedings of the Royal Society B*, vol. 268, no. 1484, pp. 2415–2421, 2001.
10. D. R. Frost, "Amphibian Species of the World 5.2, an Online Reference," American Museum of Natural History, New York, NY, USA, 2008, http://research.amnh.org/herpetology/amphibia/index.php.

11. G. Pasteur, "A classificatory review of mimicry systems," *Annual Review of Ecology and Systematics*, vol. 13, pp. 169–199, 1982.
12. A. Starrett, "Adaptive resemblance: a unifying concept for mimicry and crypsis," *Biological Journal of the Linnean Society*, vol. 48, no. 4, pp. 299–317, 1993.
13. W. Wickler, "Mimicry and the evolution of animal communication," *Nature*, vol. 208, no. 5010, pp. 519–521, 1965.
14. C. R. Darst and M. E. Cummings, "Predator learning favours mimicry of a less-toxic model in poison frogs," *Nature*, vol. 440, no. 7081, pp. 208–211, 2006.
15. C. W. Myers and J. W. Daly, "Dart-poison frogs," *Scienctific American*, vol. 248, no. 2, pp. 120–133, 1983.
16. C. E. Nelson and G. A. Miller, "A possible case of mimicry in frogs," *Herpetological Review*, vol. 3, p. 109, 1971.
17. H.-C. Schaefer, M. Vences, and M. Veith, "Molecular phylogeny of Malagasy poison frogs, genus *Mantella* (Anura: Mantellidae): homoplastic evolution of colour pattern in aposematic amphibians," *Organisms Diversity & Evolution*, vol. 2, no. 2, pp. 97–105, 2002.
18. F. Andreone, "Syntopy of *Mantella cowani* Boulenger and *Mantella madagascariensis* (Grandidier) in central-eastern Madagascar, with notes on the colouration in the genus *Mantella* (Anura: Mantellidae)," *Bollettino del Museo Regionale di Scienze Naturali*, vol. 10, no. 2, pp. 421–450, 1992.
19. J. A. Endler, "A predator's view of animal color patterns," *Evolutionary Biology*, vol. 11, pp. 319–364, 1978.
20. M. D. Tuttle, L. K. Taft, and M. J. Ryan, "Acoustical location of calling frogs by *Philander Opossums*," *Biotropica*, vol. 13, no. 3, pp. 233–234, 1981.
21. P. A. Schwalm, P. H. Starrett, and R. W. McDiarmid, "Infrared reflectance in leaf-sitting neotropical frogs," *Science*, vol. 196, no. 4295, pp. 1225–1227, 1977.
22. S. B. Emerson, T. A. Cooper, and J. R. Ehleringer, "Convergence in reflectance spectra among treefrogs," *Functional Ecology*, vol. 4, no. 1, pp. 47–51, 1990.
23. K. Summers, T. W. Cronin, and T. Kennedy, "Variation in spectral reflectance among populations of *Dendrobates pumilio*, the strawberry poison frog, in the Bocas del Toro Archipelago, Panama," *Journal of*

Biogeography, vol. 30, no. 1, pp. 35–53, 2003.

24. C. H. Deiques, "The development of the pit organ of *Bothrops jararaca* and *Crotalus durissus terrificus* (Serpentes,Viperidae): support for the monophyly of the subfamily Crotalinae," *Acta Zoologica*, vol. 83, no. 3, pp. 175–182, 2002.
25. J. Honkavaara, M. Koivula, E. Korpimäki, H. Siitari, and J. Viitala, "Ultraviolet vision and foraging in terrestrial vertebrates," *Oikos*, vol. 98, no. 3, pp. 505–511, 2002.
26. J. A. Endler and P. W. Mielke Jr., "Comparing entire colour patterns as birds see them," *Biological Journal of the Linnean Society*, vol. 86, no. 4, pp. 405–431, 2005.
27. C. A. G. Cruz, "Sobre as relações intergenéricas de Phyllomedusinae da floresta atlântica (Amphibia, Anura, Hylidae)," *Revista Brasileira de Biologia*, vol. 50, pp. 709–726, 1990.
28. A. H. Thayer, "The law which underlies protective coloration," *The Auk*, vol. 13, pp. 124–127, 1896.
29. G. D. Ruxton, M. P. Speed, and D. J. Kelly, "What, if anything, is the adaptive function of countershading?" *Animal Behaviour*, vol. 68, no. 3, pp. 445–451, 2004.
30. R. A. Kiltie, "Countershading: universally deceptive or deceptively universal," *Trends in Ecology and Evolution*, vol. 3, pp. 21–23, 1988.
31. S. Braude, D. Ciszek, N. E. Berg, and N. Shefferly, "The ontogeny and distribution of countershading in colonies of the naked mole-rat (*Heterocephalus glaber*)," *Journal of Zoology*, vol. 253, no. 3, pp. 351–357, 2001.
32. D. Osorio and M. V. Srinivasan, "Camouflage by edge enhancement in animal coloration patterns and its implications for visual mechanisms," *Proceedings of the Royal Society B*, vol. 244, no. 1310, pp. 81–85, 1991.
33. B. H. Brattstrom, "The coral snake 'Mimic' problem and protective coloration," *Evolution*, vol. 9, no. 2, pp. 217–219, 1955.
34. E. D. Brodie Jr. and M. S. Tumbarello, "The antipredator functions of *Dendrobates auratus* (Amphibia, Anura, Dendrobatidae) skin secretion in regard to a snake predator (*Thamnophis*)," *Journal of Herpetology*, vol. 12, no. 2, pp. 264–265, 1978.

35. I. C. Cuthill, M. Stevens, J. Sheppard, T. Haddocks, C. A. Párraga, and T. S. Troscianko, "Disruptive coloration and background pattern matching," *Nature*, vol. 434, no. 7029, pp. 72–74, 2005.
36. J. A. Endler, "Disruptive and cryptic coloration," *Proceedings of the Royal Society B*, vol. 273, no. 1600, pp. 2425–2426, 2006.
37. H. W. Bates, "Contributions to an insect fauna of the Amazon valley," *Transactions of the Linnean Society of London*, vol. 23, pp. 495–566, 1862.
38. W. Wüster, C. S. E. Allum, I. B. Bjargardóttir et al., "Do aposematism and Batesian mimicry require bright colours? A test, using European viper markings," *Proceedings of the Royal Society B*, vol. 271, no. 1556, pp. 2495–2499, 2004.
39. F. H. Pough, "Mimicry of vertebrates: are the rules different?" *The American Naturalists*, vol. 131, supplement 1, pp. S67–S102, 1988.
40. T. Grant, D. R. Frost, J. P. Caldwell et al., "Phylogenetic systematics of dart-poison frogs and their relatives (Amphibia: Athesphatanura: Dendrobatidae)," *Bulletin of the American Museum of Natural History*, vol. 299, pp. 1–262, 2006.
41. J. W. Daly, "Thirty years of discovering arthropod alkaloids in amphibian skin," *Journal of Natural Products*, vol. 61, no. 1, pp. 162–172, 1998.
42. R. A. Saporito, H. M. Garraffo, M. A. Donnelly, A. L. Edwards, J. T. Longino, and J. W. Daly, "Formicine ants: an arthropod source for the pumiliotoxin alkaloids of dendrobatid poison frogs," *Proceedings of the National Academy of Sciences of the United States of America*, vol. 101, no. 21, pp. 8045–8050, 2004.
43. V. C. Clark, C. J. Raxworthy, V. Rakotomalala, P. Sierwald, and B. L. Fisher, "Convergent evolution of chemical defense in poison frogs and arthropod prey between Madagascar and the Neotropics," *Proceedings of the National Academy of Sciences of the United States of America*, vol. 102, no. 33, pp. 11617–11622, 2005.
44. J. W. Daly, S. I. Secunda, H. M. Garraffo, T. F. Spande, A. Wisnieski, and J. F. Cover Jr., "An uptake system for dietary alkaloids in poison frogs (Dendrobatidae)," *Toxicon*, vol. 32, no. 6, pp. 657–663, 1994.
45. J. W. Daly, W. L. Padgett, R. L. Saunders, and J. F. Cover Jr., "Absence of tetrodotoxins in a captive-raised riparian frog, *Atelopus varius*," *Toxicon*, vol. 35, no. 5, pp. 705–709, 1997.

46. J. W. Daly, H. M. Garraffo, G. S. E. Hall, and J. F. Cover Jr., "Absence of skin alkaloids in captive-raised Madagascan mantelline frogs (*Mantella*) and sequestration of dietary alkaloids," *Toxicon*, vol. 35, no. 7, pp. 1131–1135, 1997.
47. J. P. Caldwell, "The evolution of myrmecophagy and its correlates in poison frogs (Family Dendrobatidae)," *Journal of Zoology*, vol. 240, no. 1, pp. 75–101, 1996.
48. R. A. Saporito, M. A. Donnelly, H. M. Garraffo, T. F. Spande, and J. W. Daly, "Geographic and seasonal variation in alkaloid-based chemical defenses of *Dendrobates pumilio* from Bocas del Toro, Panama," *Journal of Chemical Ecology*, vol. 32, no. 4, pp. 795–814, 2006.
49. W. E. Duellman and L. Trueb, *Biology of Amphibians*, The Johns Hopkins University Press, London, UK, 2nd edition, 1994.
50. A. Channing and K. Howell, "*Phlyctimantis keithae* (Wot-wot)," *Defensive Behavior Herpetological Review*, vol. 34, pp. 52–53, 2003.
51. F. Müller, "Über die Vortheile der Mimicry bei Schmetterlingen," *Zoologischer Anzeiger*, vol. 1, pp. 54–55, 1878.
52. F. Müller, "Ituna and Thyridia: a remarkable case of mimicry in butterflies," *Proceedings of the Entomological Society of London*, vol. 1879, pp. 20–29, 1879.
53. V. G. L. van Someren and T. H. E. Jackson, "Some comments on protective resemblance amongst African Lepidoptera (Rhopalocera)," *Journal of the Lepidopterists' Society*, vol. 13, pp. 121–150, 1959.
54. M. Rothschild, "Is the buff ermine a mimic of the white ermine?" *Proceedings of the Royal Entomological Society of London A*, vol. 38, pp. 159–164, 1963.
55. M. Rothschild, "The mimicrats must move with the times," *Biological Journal of the Linnean Society*, vol. 16, no. 1, pp. 21–23, 1981.
56. J. R. G. Turner and M. P. Speed, "How weird can mimicry get?" *Evolutionary Ecology*, vol. 13, no. 7-8, pp. 807–827, 1999.
57. M. C. Dickerson, *The Frog Book*, Doubleday, Page, New York, NY, USA, 1906.
58. R. C. Stebbins and N. W. Cohen, *A Natural History of Amphibians*, Princeton University Press, Princeton, NJ, USA, 1995.
59. T. Caro, *Antipredator Defenses in Birds and Mammals*, The University of Chicago Press, London, UK, 2005.

60. E. A. Hoffman and M. S. Blouin, "A review of colour and pattern polymorphisms in anurans," *Biological Journal of the Linnean Society*, vol. 70, no. 4, pp. 633–665, 2000.

61. G. R. Bourne, "Color pattern, predator avoidance, and foraging behavior in the golden frog *Colostethus beebei* (Anura: Dendrobatidae)," *Herpetological Review*, vol. 32, no. 4, pp. 225–228, 2001.

62. J. L. Roberts, J. L. Brown, R. Schulte, W. Arizabal, and K. Summers, "Rapid diversification of colouration among populations of a poison frog isolated on sky peninsulas in the central cordilleras of Peru," *Journal of Biogeography*, vol. 34, no. 3, pp. 417–426, 2006.

63. J. W. Daly, M. A. Andriamaharavo, M. Andriantsiferana, and C. W. Myers, "Madagascan poison frogs (*Mantella*) and their skin alkaloids," *American Museum Novitates*, vol. 3177, pp. 1–34, 1996.

64. R. T. Hanlon, J. W. Forsythe, and D. E. Joneschild, "Crypsis, conspicuouness, mimicry and polyphenism as antipredator defences of foraging octopuses on Indo-Pacific coral reefs, with a method of quantifying crypsis from video tapes," *Biological Journal of the Linnean Society*, vol. 66, no. 1, pp. 1–22, 1999.

65. L. F. Toledo and C. F. B. Haddad, "Reproductive biology of *Scinax fuscomarginatus* (Anura, Hylidae) in south-eastern Brazil," *Journal of Natural History*, vol. 39, no. 32, pp. 3029–3037, 2005.

66. P. C. Eterovick and I. Sazima, *Amphibians from Serra do Cipó*, Editora PUC Minas, Belo Horizonte, Brazil, 2004.

67. L. P. C. Morellato, R. R. Rogrigues, H. Leitão-Filho, and C. A. Joly, "Estudo comparativo da fenologia de espécies arbóreas de floresta de altitude e floresta mesófila semidescídua na Serra do Japi, Jundiaí, São Paulo," *Revista Brasileira de Biologia*, vol. 12, pp. 85–98, 1989.

68. W. H. Wente and J. B. Phillips, "Fixed green and brown color morphs and a novel color-changing morph of the pacific tree frog *Hyla regilla*," *The American Naturalist*, vol. 162, no. 4, pp. 461–473, 2003.

69. J. Bajger, "Diversity of defensive responses in populations of fire toads (*Bombina bombina and Bombina variegata*)," *Herpetologica*, vol. 36, pp. 133–137, 1980.

70. E. D. Lindquist and T. E. Hetherington, "Field studies on visual and acoustic signaling in the "earless" Panamanian golden frog, *Atelopus zeteki*," *Journal of Herpetology*, vol. 30, no. 3, pp. 347–354, 1996.

71. L. F. Toledo, L. D. Guimarães, L. P. Lima, R. P. Bastos, and C. F. B. Haddad, "Notes on courtship, egg-laying site, and defensive behavior of *Epipedobates flavopictus* (Anura: Dendrobatidae) from two mountain ranges of central and southeastern Brazil," *Phyllomedusa*, vol. 3, no. 2, pp. 145–147, 2004.
72. R. W. McDiarmid and S. Gorzula, "Aspects of the reproductive ecology and behavior of the Tepui toads, genus *Oreophrynella* (Anura, Bufonidae)," *Copeia*, vol. 2, pp. 445–451, 1989.
73. J. Mallet and M. Joron, "Evolution of diversity in warning color and mimicry: polymorphisms, shifting balance, and speciation," *Annual Review of Ecology and Systematics*, vol. 30, pp. 201–233, 1999.
74. K. Summers and M. E. Clough, "The evolution of coloration and toxicity in the poison frog family (Dendrobatidae)," *Proceedings of the National Academy of Sciences of the United States of America*, vol. 98, no. 11, pp. 6227–6232, 2001.
75. J. C. Santos, L. A. Coloma, and D. C. Cannatella, "Multiple, recurring origins of aposematism and diet specialization in poison frogs," *Proceedings of the National Academy of Sciences of the United States of America*, vol. 100, no. 22, pp. 12792–12797, 2003.
76. M. Vences, J. Kosuch, R. Boistel et al., "Convergent evolution of aposematic coloration in Neotropical poison frogs: a molecular phylogenetic perspective," *Organisms Diversity & Evolution*, vol. 3, no. 3, pp. 215–226, 2003.
77. D. R. Frost, T. Grant, J. Faivovich et al., "The amphibian tree of life," *Bulletin of the American Museum of Natural History*, no. 297, pp. 1–291, 2006.
78. Y. Chiari, M. Vences, D. R. Vieites et al., "New evidence for parallel evolution of colour patterns in Malagasy poison frogs (*Mantella*)," *Molecular Ecology*, vol. 13, no. 12, pp. 3763–3774, 2004.
79. J. A. Endler, "Defense against predators," in *Predator-Prey Relationships: Perspectives and Approaches from the Study of Lower Vertebrates*, M. E. Feder and G. V. Lauder, Eds., pp. 109–134, The University of Chicago Press, London, UK, 1986.
80. J. W. Daly, "The chemistry of poisons in amphibian skin," *Proceedings of the National Academy of Sciences of the United States of America*, vol. 92, no. 1, pp. 9–13, 1995.

10

Intraspecific Color Variation among Lizards in Distinct Island Environments Enhances Local Camouflage

Kate L. A. Marshall[1], Kate E. Philpot[2], Isabel Damas-Moreira[3] and Martin Stevens [2]

[1] Behavioural Ecology Group, Department of Zoology, University of Cambridge, Cambridge, United Kingdom,
[2] Centre for Ecology and Conservation, College of Life and Environmental Sciences, University of Exeter, Penryn Campus, Penryn, Cornwall, United Kingdom,
[3] CIBIO Research Centre in Biodiversity and Genetic Resources, InBIO, Universidade do Porto, Campus Agrário de Vairão, Rua Padre Armando Quintas, Vairão, Vila do Conde, Portugal, University of Sussex, United Kingdom,

ABSTRACT

Within-species colour variation is widespread among animals. Understanding how this arises can elucidate evolutionary mechanisms, such as those underlying reproductive isolation and speciation. Here, we investigated whether five island populations of Aegean wall lizards (*Podarcis erhardii*) have more effective camouflage against their own (local) island substrates than against other (non-local) island substrates to avian predators, and

Citation: Intraspecific Colour Variation among Lizards in Distinct Island Environments Enhances Local Camouflage. Marshall KLA, Philpot KE, Damas-Moreira I, Stevens M (2015) PLoS ONE 10(9): e0135241. doi:10.1371/journal.pone.0135241

whether this was linked to island differences in substrate appearance. We also investigated whether degree of local substrate matching varied among island populations and between sexes. In most populations, both sexes were better matched against local backgrounds than against non-local backgrounds, particularly in terms of luminance (perceived lightness), which usually occurred when local and non-local backgrounds were different in appearance. This was found even between island populations that historically had a land connection and in populations that have been isolated relatively recently, suggesting that isolation in these distinct island environments has been sufficient to cause enhanced local background matching, sometimes on a rapid evolutionary time-scale. However, heightened local matching was poorer in populations inhabiting more variable and unstable environments with a prolonged history of volcanic activity. Overall, these results show that lizard coloration is tuned to provide camouflage in local environments, either due to genetic adaptation or changes during development. Yet, the occurrence and extent of selection for local matching may depend on specific conditions associated with local ecology and biogeographic history. These results emphasize how anti-predator adaptations to different environments can drive divergence within a species, which may contribute to reproductive isolation among populations and lead to ecological speciation.

INTRODUCTION

Intraspecific variation of colour patterns is a long studied phenomenon in animals, and has profoundly contributed to our understanding of evolutionary processes. Classic experiments have shown that natural selection can drive variation in coloration among populations to optimize survival against predators in varying local environments ([1–3]). It is now understood that adaptation to local environments can cause reproductive isolation between divergent populations and potentially lead to ecological speciation (reviewed in [4–6]).

Various studies on mice and lizards have shown that divergent populations undergo selection to better match local backgrounds for camouflage, and that this involves mutations of key genes and is influenced by levels of gene flow (e.g. [7–11]). However, until recently, relatively few studies have based their conclusions on the visual systems of potential predators, such as hunting birds (but see [12, 13]). As the avian predator visual system and perception of background matching camouflage differs from that of humans (for example, birds can see ultraviolet light and probably a greater range of colours; [14,

15]), it is important to determine whether predators perceive divergence in camouflage among local environments (e.g. [16, 17]). As local adaptation for camouflage should involve selection for coloration that matches associated backgrounds, it can be inferred only if local background environments are different. This is because coloration could become variable through other processes, such as genetic drift, even among similar environments. Variation may also arise owing to phenotypic or developmental plasticity, whereby individuals can change colour during their lifetimes (e.g. [18–21]). Consequently, it is important to determine how predator-perceived camouflage and visual backgrounds vary among populations in order to understand local adaptation and its role in intraspecific divergence.

The Aegean wall lizard (*Podarcis erhardii;* [22]) is a valuable model species for this type of study because different island populations exhibit clearly visible colour variation (to humans) among ostensibly distinct island environments (see Fig 1). These island populations have been historically categorised as different subspecies due to their varying appearance ([23]; but see more recent molecular analyses in [24, 25]) and show genetic diversity between isolated islands with no recent common history ([26]). The variable environments of the islands are widely thought to have driven the apparent phenotypic and genetic diversification among the inhabitant *P. erhardii* populations (via the equivalent of a sustained bottleneck) dating from their time of separation ([25–28]), although this has not been formally tested.

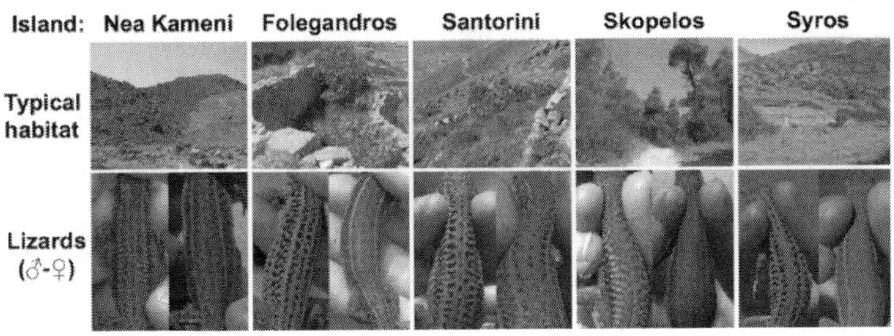

Figure 1. **Example images of Aegean wall lizards (*Podarcis erhardii*) and their typical island habitats.**

Images depict typical natural habitat and dorsal coloration of males (left image) and females (right image) in each focal island population (Nea Kameni, Folegandros, Santorini, Skopelos and Syros). Images were obtained in the field by the authors.

Studying *P. erhardii* in this island system differs from a range of previous studies focusing on the adaptive significance of colour polymorphisms among sympatric lizard populations (e.g. [7, 9, 13, 29]). In particular, examining colour variation between isolated island populations can potentially reveal how distinct ecological factors, as well as time of isolation and evolutionary history (e.g. historical geographical connections), may drive differential local adaptation of camouflage and thus further clarify the selective conditions under which it can occur. Specifically, island populations of *P. erhardii* vary in habitat type (e.g. geological characteristics and habitat composition), predation risk (i.e., number of resident avian predators), time of isolation (ranging from >200,000 to 400 years), and potential amounts of gene flow from other island populations depending on historical geographic connections and amounts of boat traffic (see Table 1; [25–28, 30–32]). For example, the Nea Kameni islet has been formed from intermittent volcanic eruptions over the last 400 years ([33]) and inhabitant *P. erhardii* rest on homogeneous black lava rock backgrounds, whereas the land bridge island of Syros has been isolated for over 13,000 years and lizards rest on various metamorphic rock backgrounds (Table 1; [27, 34]).

Table 1. Differences in the environments and evolutionary histories of the five focal islands inhabited by populations of Aegean wall lizards (*Podarcis erhardii*).

Island and subspecies [a]	Time of isolation (years) [b]	Historical geographical connections [b]	No. avian predator species [d]	Main habitat type	Main rock type [e]
Folegandros *P.e. naxensis*	11,650	Syros, Santorini	2	Rocky shrubland	Metamorphic (marble, schist)
Nea Kameni *P.e. naxensis*	400 (65) [c]	None; formed from volcanic eruptions	5	Lava dome	Igneous (black dacite)
Santorini *P.e. naxensis*	> 200,000	Folegandros, Syros	5	Rocky shrubland	Various (e.g. rhyolite, pumice, limestone, schist)
Skopelos *P.e. ruthveni*	5,900	None; separate island group	4	Pine forests, some fields	Sedimentary (dolomites, breccias, limestone)
Syros *P.e. mykonensis*	12,800	Folegandros, Santorini	5	Rocky shrubland	Metamorphic (schist, gneiss, marble)

[a] Subspecies identified following [23–25].

[b] Estimated date of separation from adjacent landmasses (except Nea Kameni; [27])

[c]The last volcanic eruption in 1950 probably eradicated the Nea Kameni population, so the current population age is approx. 65y [26].

[d] From [31]. Major avian predator species considered are: *Buteo buteo*, *Buteo rufinus*, *Falco tinnunculus*, *Falco eleonorae*, *Tyto alba*, *Corvus corax* and *Corvus corone*.

[e] From [34, 35].

A recent study showed that *P. erhardii* island populations are tuned to their local island environments to be conspicuous to conspecifics while also being camouflaged against avian predators ([36]). Specifically, *P. erhardii* coloration was found to be less perceptible by avian predators than by conspecifics on all three focal islands. Moreover, dorsal regions (upper and lower backs) exposed to aerially hunting birds were relatively more camouflaged against local rock backgrounds compared to less exposed lateral (flank) regions, and this was more evident on some islands compared to others. In addition, on all islands, sexually competing males showed inferior camouflage on their upper backs compared to that of females. Given the apparent environmental variation among the islands, these findings suggest that *P. erhardii* have adapted to their local environments to enhance dorsal camouflage against avian predators, and also that degree of local adaptation varies among islands, and between sexes and upper and lower backs. However, this study did not test whether the backgrounds and lizard populations were variable among islands, or whether lizards were more camouflaged against their local island backgrounds compared to that of different islands. Therefore, in the current study we directly test these hypotheses, and also consider how different ecological selection pressures (e.g. predation risk) and evolutionary history (e.g. time of island isolation) has influenced local selection on dorsal camouflage in each population (Table 1).

It is likely that island populations of *P. erhardii* have adapted to be camouflaged in their local environments for several reasons. The populations are under considerable though variable threat from avian predators (i.e., raptors and corvids), particularly as they tend to bask on rocks in open environments where they are often visible to hunting birds, as in other lizards (Table 1; [31, 37–39]; pers. observations). Therefore, it is likely that selection has favoured colour matching against local rock backgrounds to reduce detection by avian predators. However, the variable threat from avian predators among the islands may affect degree of local adaptation for camouflage, as shown in past work on other lizards (e.g. [13, 29, 40]). Additionally, various studies have shown that there is relatively little gene flow (i.e., via over-water dispersal) between most island populations ([26–28, 41]), which is similar to other lizards showing phenotypic diversification among recently isolated islands where dispersal across seawater is also extremely rare ([42]). Thus, most of the current *P. erhardii* populations are highly likely to be descendants of the original colonists that have adapted to their native environments since they became isolated from neighbouring landmasses. This is substantiated by past work showing genetic adaptation of lizard coloration to environments that have existed for only 6,000 years (e.g. [10, 29]), which is a comparable age to most island environments considered in the current study (Table 1; [27]).

We investigated whether *P. erhardii* populations are more camouflaged on their local island than on non-local islands to avian predators, and whether this was associated with lizard and environmental (substrate) differences in appearance between local and non-local islands. We predicted that there would be differences in coloration between *P. erhardii* island populations and between their rock background environments, and that lizards would show better camouflage against local backgrounds than against non-local backgrounds to avian predators. We also predicted that there would be variation in degree of camouflage among islands with different evolutionary histories and environments, and between sexes and body regions.

MATERIALS AND METHODS

Study Species and Sites

The Aegean wall lizard (*Podarcis erhardii*) is a diurnal, small lacertid distributed across most of the South Balkans and widespread throughout many Aegean islands where it is found in all ecosystems ([43, 44]). We conducted field research during the activity season (April-July; [45]) between 2012–2014 on five Aegean islands with three different inhabitant subspecies (classified based on appearance; [23]): *P.e. naxensis* (Folegandros [36°37' N, 24°54' E], Santorini [36°25' N, 25°26' E] and the neighbouring islet Nea Kameni [36°24' N, 25°24' E]; *P.e. mykonensis* (Syros [37°27' N, 24°54' E]); and *P.e. ruthveni* (Skopelos [39°7' N, 23°43' E]). We chose to sample lizards from these islands because they have ostensibly varying coloration and environments, substantial but variable risk from avian predators, and different evolutionary histories (see Fig 1; Table 1).

Ethics Statement

P. erhardii is listed as a species of 'least concern' under the IUCN Red List classification, so it is not an endangered or protected species ([46]). We conducted field research with permission from the Greek Ministry of Environment (permit numbers: 166648/356 and 107222/707). These permits were granted for all study sites, and all land used for fieldwork was publicly accessible. The permits also gave us permission to sample lizard coloration using all methods as described. Sampling methods were non-invasive, involving photography of lizards in situ. However, in one study site (Nea Kameni islet) lizards were captured for photography, because this islet has a low population density and we had restricted access with a limited amount of time to obtain an appropriate sample size. Lizards were caught using a noose consisting of a slipknot attached to a telescopic pole, which is widely used by herpetologists as a non-harmful capture method (e.g. [47]). Nonetheless, we endeavoured to release the lizard from the noose as quickly as possible after capture, which was usually within five seconds and never more than 10 seconds. All researchers had prior experience using this capture method and with handling lizards. After capture, lizards were immediately photographed, marked with a small amount of non-toxic (water-based), inconspicuous (dark green) paint (Montana, Heidelberg, Germany) on the vertical surface of their hind leg so as not to alter dorsal camouflage, and then released at the original (marked) point of capture. During capture, we ensured that we handled lizards in a way that minimized the likelihood of tail autotomy and aimed to release them as soon as possible, which was typically within five minutes. Lizards were never harmed and all showed normal behaviour after release. We captured the minimum number of lizards necessary to rigorously test our hypotheses ($N = 32$; 19 males, 13 females). All protocols conformed to the policies and requirements of the ethics committee of our institution.

In situ Photography

We used digital imaging to sample lizards and their corresponding background coloration, because it has various advantages over alternative spectrometry methods (see [36, 48]). As in ([36]), during in situ photography we took images of stationary lizards and their corresponding natural backgrounds with a Fujifilm IS Pro ultraviolet (UV)-sensitive digital camera with a quartz

CoastalOpt UV lens (Coastal Optical Systems), fitted with a UV and infrared (IR) blocking filter for photographs in the human-visible spectrum (Baader UV/IR Cut filter; transmitting between 400 and 700 nm), and with a UV pass filter (Baader U filter; transmitting between 300 and 400nm) for UV images. After the photographed lizard had fled, we took human-visible and UV images of a Spectralon™ grey reflectance standard (Labsphere, Congleton, UK), which reflects light equally at 40 per cent between 300 and 750 nm. Following this 'sequential method' as used in past work ([49, 50]), images of the standard were taken at the same distance, with the same camera settings, and in the same location and light conditions, as the photographed lizard. This enabled us to undertake the required standardization for ambient light conditions (i.e. image linearization and RGB-equalization; see below) (see [48]). Light conditions rarely changed in the typical 2–3 minutes between photographing the lizard and the standard. However, on the few occasions when this did occur (e.g. clouds passed over the sun), we re-photographed the standard once the lighting had returned to normal. We endeavoured to photograph most lizards (and hence standards) in full sunlight. In the few cases where shadows partly obscured the image, our selections avoided shadowed areas and only measured parts of the lizards, backgrounds and standards that were in sunlight.

We recorded photographed lizards' locations using a Garmin eTrex GPS device (Schauffhausen, Switzerland) and marked it with coloured tape to indicate sex and life-stage ([43]). We confirmed these estimations were 99% reliable by comparing estimated (from photographs) and observed sex and life-stages from captured lizards (see [36]). We avoided pseudoreplication by never repeating photography of a lizard of the same sex within the same home range (i.e. within 10 m) (see [36, 51]).

On the Nea Kameni islet we captured lizards for photography. Capture location was marked with coloured tape for subsequent release. To minimise stress and to avoid any colour fading through decreases in body temperature, which can sometimes arise during long-lasting stressful situations in lizards (e.g. [52, 53]), lizards were photographed within five minutes after capture. The grey reflectance standard was photographed and lizards were marked as described previously. Observations and other mark re-capture studies showed that marks lasted for at least three weeks (K. Marshall and I. Damas-Moreira, unpublished data). After allowing the paint to dry, lizards were released at the marked capture location. Capture attempts were not made within 10m of the same location again. To ensure that lizards would be subsequently compared to their natural corresponding backgrounds, each (marked) background that each lizard had been resting on at the point of capture was photographed with the grey standard in the image.

Image Analysis and Visual Modelling

We followed methods previously described in ([36]). Human-visible and UV images of lizards and their backgrounds were linearized with respect to light intensity and transformed to reflectance (RGB-equalized; see [48]). Any images that were overexposed and/or could not be RGB-equalized were discarded from the analysis. We then used a mapping process based on the spectral sensitivity of our camera's sensors (derived prior to photography; see [36]) to convert the images to correspond to avian predicted photon catch cone values ([48]). This is highly accurate compared to spectrometry-derived estimates of animal photon catch values (see [54–56]). We converted the aligned images from camera colour space to the relative photon catches of an avian predator's longwave (LW), mediumwave (MW), shortwave (SW) and UV sensitive cone photoreceptors using the spectral sensitivity of a peafowl (*Pavo cristatus*; [14]). We also mapped the images to peafowl double cone photoreceptors ([57]) to represent achromatic (luminance) avian perception ([58, 59]). The peafowl visual system is often used as a representative of the violet-sensitive (VS) class of colour vision in birds ([60, 61]). The VS-system is typical of the predatory birds that hunt *Podarcis* lizards and other lacertids in Europe (i.e. raptors and corvids; [31, 37, 62]), which has recently been supported by molecular studies tracing the evolution of ultraviolet vision in birds ([63]). Although some birds with ultraviolet-sensitive (UVS) vision might also prey on *P. erhardii* (e.g. gulls and some Turdidae (thrush) species that are present in our study sites; [63–66]), we considered VS-sensitive raptors and corvids as the more important avian predators of adult lizards in Europe ([31, 37, 62, 63, 67]). Indeed, we often observed hooded crows (*Corvus cornix*) and a variety of raptors (e.g. buzzards, *Buteo buteo*; common kestrels, *Falco tinnunculus;* Eleonora's falcon, *Falco eleonorae*) hunting in our study sites, while gulls and thrushes were rarely seen. Calibrations were performed in MATLAB v. R2011b (The MathWorks, Inc., MA, USA) using self-written programs. Both calibrations were restricted to the 300–700 nm range, which encompasses most of the visual spectrum of diurnal birds ([61]).

LW, MW, SW and UV photon catches of each lizard and their corresponding background (i.e., the background we photographed the lizard resting on) were extracted from the calibrated images in ImageJ using the selection tool. Selection of a particular area in the image generated an average photon catch value for that patch. Background selections were limited to rock backgrounds because *Podarcis* lizards most frequently bask on rocks (given their good thermal quality) where they are potentially visible to aerially hunting predators in open environments (e.g. [38, 39, 68]; pers. observations). We randomly selected areas of rock that were adjacent to the photographed lizard, so that the selection touched but did not overlap it, and avoided areas of lichen and moss. Lizards sometimes rested on rock backgrounds that consisted of different patches of colour patterns, although we could usually identify the predominant (largest) patch from which we made our background selection. However, in some images we could not identify a predominant background patch. In these cases, we made up to three selections of different patches and averaged the photon catches across patches as a representative substrate colour.

Lizard selections were made from two dorsal body regions (posterior lower and anterior upper backs). The average photon catch value extracted by these selections included dorsal patterning, which is particularly extensive in males (see Fig 1). We assumed that this would depict how lizards appear to birds hunting from a distance, especially given that the lizards are small (i.e. ≈7cm snout-to-vent length; [43]). However, we acknowledge that at close range patterning might be more distinguishable by birds and serve another function (e.g. disruptive camouflage; [69]). Separate selections of lower and upper backs were taken due to observed colour differences between these regions, and selection of these body regions was standardized across all images. Specifically, upper back selections were taken under the base of the head and lower backs selections were taken above the base of the tail (see [36]). A total of 295 adult lizards and their corresponding rock backgrounds were sampled from the five focal island populations (Folegandros = 100; 52 males, 48 females, Syros = 49; 34 males, 15 females; Santorini = 58; 34 males, 24 females; Skopelos = 56; 29 males, 27 females; and Nea Kameni = 32; 19 males, 13 females).

Island Differences in Lizards and Backgrounds

We first aimed to ascertain any differences in colour and luminance among the island lizard populations and their rock background environments. We calculated a variety of metrics to use in our analysis. *Saturation* (the amount of a certain colour compared to white light) refers to the Euclidian distance a colour patch is in a tetrahedral colour space from the achromatic (grey) point, with larger values representing more saturated colours (see [16]). We also quantified the type of colour (*hue*). This was based on a ratio of the different colour channels present (i.e., LW, MW, SW and UV) to represent relative photoreceptor stimulation by the colour patch spanning different parts of the light spectrum and receptor types ([55, 70–72]). This followed an approach developed in past work, which aims to quantify colour so that it broadly reflects how colour is encoded by an animal's visual system (see [56, 71, 72]). Specifically, we used a ratio based on the concept that colour is encoded using opponent (antagonistic) colour channels or neural pathways, such as the 'red-green' and 'blue-yellow' pathways in trichromatic primates ([59, 73–75]). Since it is unknown which specific opponent colour channels birds have, we identified biologically relevant putative colour channels by performing a principle component analysis (PCA) on a covariance matrix of the standardized cone photon catches derived from our avian visual model ([71]). This enabled us to determine one or two colour channels that best described the variation in colours existing among lizards ([56, 72]). Two PCs were extracted and together explained almost all of the variance in lizard and background coloration (lower backs = 95.2%; upper backs = 93.6%; backgrounds = 97.8%). PC1 showed highest eigenvalue loadings on the LW and SW channels, and PC2 showed highest loadings on the MW and UV channels. Therefore, we used the following standardized ratio to identify perceived differences in the identified opponent colour channels ((LW+SW)-(UV+MW)/UV+SW+MW+LW); see [69, 71]). Consequently, this ratio would reflect differences between perceived 'purple' and 'green-UV' coloration. Finally, for a measure of *luminance* (i.e. perceived lightness), we used standardized double cone values derived from the avian visual model ([57–59]).

Degree of Island Differences in Lizards and Backgrounds

In addition to calculating the *hue*, *saturation* and *luminance* metrics separately, we sought more insight into the divergence between islands by determining the extent of the perceived differences in lizards and their rock backgrounds, and to understand perceptual differences in colour accounting for both *hue* and *saturation* together. To do this, we measured the degree of contrast between each island population (i.e., lizards vs. lizards) to identify how similar lizards were to members of their local population and to members of non-local populations. We also measured the degree of contrast between the measured backgrounds (i.e., backgrounds vs. backgrounds) to identify how similar rock backgrounds were to those of the local island and to those of non-local islands. We quantified degree of colour (chromatic) contrast according to the widely used log form of a receptor noise model ([76]), which predicts visual discrimination abilities in observers. We also quantified luminance contrasts using a version of the model based on achromatic differences using peafowl double cones ([77]). To account for receptor noise, we used a Weber fraction value of 0.05 for the most frequent cone type based on data in other vertebrates ([76, 78]). We used relative proportions of cone types in the peafowl retina to calculate avian predator-perceived chromatic contrast (LW = 0.92, MW = 1.00, SW = 0.81, UV = 0.54; [14]). The degree of chromatic and achromatic contrast generated from these models is expressed as a measure of contrast called "just-noticeable-difference" (JND), which predicts discrimination behaviour in observers. Generally, JND values increasing above 3.00 indicate increasingly improved discrimination ([77]).

We compared each lizard with every lizard belonging to all island populations, including to its local population and to all four non-local populations. We performed the same comparisons with the backgrounds to determine whether differences in visual backgrounds were large enough to be perceptibly different to avian predators, and thus cause changes in lizard appearance to match them on each island. To control for sexual dichromatism in lizards' dorsal coloration and variation in coloration between upper and lower backs ([36]), lizards were compared with only members of the same sex and only equivalent body regions were compared. After obtaining JNDs for all comparisons, we calculated the average JND (i.e., degree of contrast) for each lizard and each background to its local island and to each of the four non-local islands.

Local Adaptation of Camouflage

Lastly, we determined to what extent the lizard populations matched different (local and non-local) rock backgrounds for background-matching camouflage. To ascertain any differences in camouflage among islands, we used the receptor noise model described above ([76]) to compare each lizard with all local island backgrounds and with all non-local backgrounds from the four different islands. We calculated the average JND for each lizard to ascertain how well they matched local and non-local island backgrounds.

Predictions and Statistical Analyses

Normality tests and residuals analysis showed a normal distribution of *hue*, but the *saturation*, *luminance* and JND data were not normally distributed with positively skewed distributions. Therefore, we transformed this data to normality using a logarithmic transformation and used the transformed data in all statistical analyses. However, to illustrate and describe our results, we report raw (back-transformed) JND data in figures. All statistical analyses were conducted in SPSS® (v20).

First, we predicted that coloration and luminance of lizards and their rock backgrounds would differ among islands and be more similar to those on local islands as opposed to those on non-local islands, particularly between islands that have never shared a historical geographical connection and that have different subspecies (see Table 1). We also expected lizards to vary between sexes and body regions due to sexual dichromatism ([36]). To address these predictions, we first conducted a series of general linear models (GLMs) on the colour (*hue* and *saturation*) and *luminance* metrics. To test for differences among island backgrounds, we carried out univariate GLMs with island (Folegandros, Nea Kameni, Santorini, Skopelos and Syros) included as the between-subjects factor. To test for differences in lizard populations, we conducted mixed GLMs with island and sex as between-subjects factors and body region (upper and lower backs) as the within-subjects factor. We also conducted mixed GLMs on the chromatic and achromatic lizards vs. lizards JND data (tests 1 and 2) and backgrounds vs. backgrounds JND data (tests 3 and 4). In tests 1 and 2, island and sex were included as between-subjects factors and body region (upper and lower backs) and 'compared island' (i.e., which island the compared lizard belonged to) were included as the within-subjects factors. The same variables were included in tests 3 and 4, with the exception of body region and sex.

Second, we predicted that lizards would match their local backgrounds better than backgrounds on other islands for camouflage against avian predators, and that this would be accompanied by environmental (rock background) differences between islands. In addition, we predicted that this effect would be weaker in populations that have been isolated for relatively shorter evolutionary periods and have ecological characteristics that reduce the need for camouflage, such as lower levels of risk from avian predators (see Table 1). We also predicted that females would show better local adaptation for camouflage compared to sexually competing males, and that lower backs would show enhanced local camouflage compared to sexually dichromatic upper backs. To address these predictions, we conducted two mixed GLMs, one testing chromatic contrast (test 5) and the other analysing achromatic contrast (test 6) against backgrounds (JND). In both GLMs, we included sex and island lizard population as between-subjects factors. The 'compared island' backgrounds (i.e. which island background the lizard was compared to) and body region were included as within-subjects factors. In both tests 5 and 6 we specifically looked for a factor interaction between island lizard population and 'compared island' background to determine if degree of camouflage (JND) depended on whether lizards were compared with their local island background or with a different (non-local) island background.

In all tests we also report any significant effects of sex and body region on lizard differences between islands and degree of local adaptation of camouflage. We report the size of all effects in ETA^2 (η^2), which can be interpreted as the proportion of variance in the dependent variable that is attributable to each effect. To determine the sources of variation in any significant effects, planned pairwise comparisons were conducted by re-running the GLMs with only the factors of interest included that were relevant to our predictions. Factors with non-significant effects were removed from the analysis before re-running the GLM to obtain the best model. The number of comparisons was limited to the number of "spare" experimental degrees of freedom (n-1), because these are more powerful than conservative, multiple unplanned post hoc comparisons ([79]).

RESULTS

Island Divergence in Lizards and Backgrounds

As predicted, lizard coloration (*hue* and *saturation*) and *luminance* varied between some but not all island populations, and rock backgrounds were also variable between most islands (see Figure A and Tables A and B). Differences between islands were most evident in terms of *luminance* and patterns of variation were generally similar in lizards and backgrounds (see Figure A). For example, Nea Kameni lizards were darker than all other populations, and Syros lizards were lighter than all populations except Santorini (see Figure A and Table A; $F_{4, 290} = 28.346$, $P < 0.001$, $\eta^2 = 0.281$). Accordingly, Nea Kameni rock backgrounds were also darker than those on all other islands, and Syros rocks were lighter than those on Skopelos and Nea Kameni (see Figure A and Table B; $F_{4, 290} = 37.755$, $P < 0.001$, $\eta^2 = 0.342$).

Differences in colour (*hue* and *saturation*) between lizard populations were less evident and usually found in only one sex and mainly in upper backs (Table A; island*body region: $F_{4, 285} = 4.997$, $P = 0.001$, $\eta^2 = 0.066$; island*sex: $F_{5, 285} = 3.316$, $P = 0.006$, $\eta^2 = 0.005$). For example, upper back coloration was more saturated in Nea Kameni lizards than in the Syros population (Figure A; $F_{1, 79} = 6.140$, $P = 0.015$, $\eta^2 = 0.072$) and differences in perceived purple and UV-green coloration (*hue*) were lower in Nea Kameni males compared with Santorini and Folegandros males (Table A; vs. Folegandros: $F_{1, 69} = 7.978$, $P = 0.006$, $\eta^2 = 0.104$; vs. Santorini: $F_{1, 51} = 6.478$, $P = 0.014$, $\eta^2 = 0.113$). A similar trend in variation was found in the rock backgrounds of these populations (Figure A and Table B).

Degree of island differences in lizards and backgrounds

Mauchly's Test of Sphericity showed that the assumption of sphericity had been violated in the 'compared island' factor (test 1: $\chi^2_{(9)} = 251.639$, $P < 0.001$; test 2: $\chi^2_{(9)} = 666.118$, $P < 0.001$; test 3: $\chi^2_{(9)} = 345.430$, $P < 0.001$; test 4: $\chi^2_{(9)} = 669.401$, $P < 0.001$; test 5: $\chi^2_{(9)} = 373.429$, $P < 0.001$; test 6: $\chi^2_{(9)} = 925.522$, $P < 0.001$). Therefore, the Greenhouse-Geisser correction was used when reporting results for this factor.

As predicted, lizard dorsal regions were more similar to that of members of their own population than to those of other populations (Fig 2). In both chromatic and achromatic avian visual models, this effect varied among islands and depended on which island populations were compared (*chromatic*: $F_{11.690, 832.911} = 15.476$, $P < 0.001$, $\eta2 = 0.178$; *achromatic*: $F_{7.760, 552.911} = 20.849$, $P < 0.001$, $\eta2 = 0.226$; see supporting information for more details [Table A and Figure A). Moreover, this overall effect varied with sex and with body region in both models (*chromatic*: sex, $F_{14.612, 832.911} = 11.553$, $P < 0.001$, $\eta2 = 0.169$; body region, $F_{13.404, 955.027} = 6.608$, $P < 0.001$, $\eta2 = 0.085$; *achromatic*: sex: $F_{7.760, 552.911} = 4.205$, $P < 0.001$, $\eta2 = 0.056$; body region: $F_{11.846, 844.019} = 8.365$, $P < 0.001$, $\eta2 = 0.029$). As predicted, in most populations, avian predators perceived upper backs to be more variable than lower backs, and males to be more variable than females, regarding both colour and luminance when compared to other island populations (Table A).

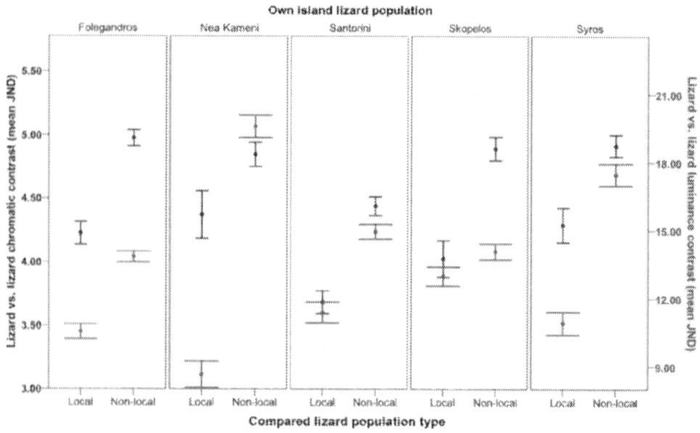

Figure 2. Degree of lizard contrast between island populations.

Degree of contrast refers to how different Aegean wall lizards (*Podarcis erhardii*) are to their local island population relative to how different they are to other (non-local) island populations overall (Folegandros, Nea Kameni, Santorini, Skopelos and Syros) in terms of chromatic contrast (left axis; black data points) and luminance contrast (right axis; red data points) (mean JND). JND values increasing >3.00 depict populations that are progressively distinguishable by avian predators. Error bars represent +/- 1 S.E.

As in the lizards, rock backgrounds were more similar to those on local islands compared to those on different islands, as predicted (Fig 3). Again, this effect varied among islands and depended on which islands were compared, in both visual models (Table A and Figure B; chromatic: $F_{10.075, 730.407} = 10.164$, $P < 0.001$, $\eta 2 = 0.123$; achromatic: $F_{7.055, 511.458} = 30.001$, $P < 0.001$, $\eta 2 = 0.293$).

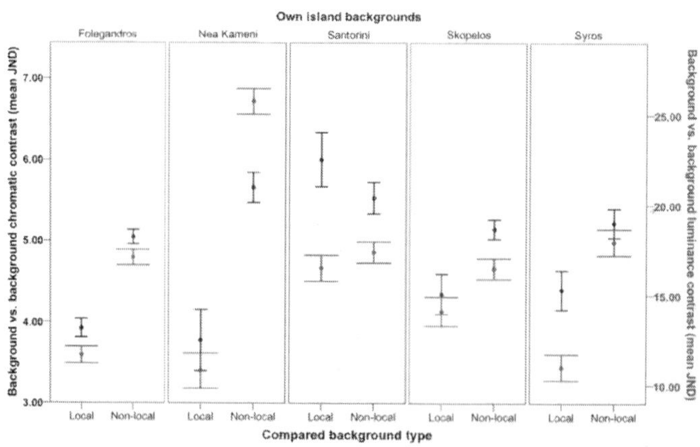

Figure 3. Degree of rock background contrast between islands.

Degree of contrast refers to how different Aegean wall lizard (*Podarcis erhardii*) rock backgrounds are to that of their local island population relative to how different they are to the backgrounds of other (non-local) island populations overall (Folegandros, Nea Kameni, Santorini, Skopelos and Syros) in terms of chromatic contrast (left axis; black data points) and luminance contrast (right axis; red data points) (mean JND). JND values increasing >3.00 depict rock backgrounds that are progressively distinguishable by avian predators. Error bars represent +/- 1 S.E.

Local Adaptation of Camouflage

Tests 5 and 6 reported a highly significant interaction effect on JNDs (i.e., lizard-background comparisons) between the lizard island populations and the island backgrounds with which they were compared ('compared island' background), which explained most of the variance in the achromatic model (Fig 4; Figure A in S3 File; *chromatic*: $F_{10.372,\ 738.991} = 5.003$, $P < 0.001$, $\eta 2 = 0.066$; *achromatic*: $F_{6.366,\ 453.585} = 30.813$, $P < 0.001$, $\eta 2 = 0.302$). This overall effect varied with sex and body region in both models (Figures B and C in S3 File; *chromatic*: sex, $F_{10.372,\ 738.991} = 3.227$, $P < 0.001$, $\eta 2 = 0.043$; body region, $F_{9.954,\ 709.190} = 2.358$, $P = 0.010$, $\eta 2 = 0.032$; *achromatic*: sex, $F_{6.366,\ 453.585} = 2.992$, $P = 0.006$, $\eta 2 = 0.040$; body region, $F_{9.397,\ 669.527} = 3.438$, $P < 0.001$, $\eta 2 = 0.046$). These results revealed that lizards' dorsal regions were typically better matched against local rock backgrounds than against other (non-local) island rock backgrounds (see Fig 4). However, as predicted, this effect depended on the island population and which other island backgrounds the lizards were compared to, as well as on sex and body region in some populations (Table 2; see Table A and Figures B and C in S3 File).

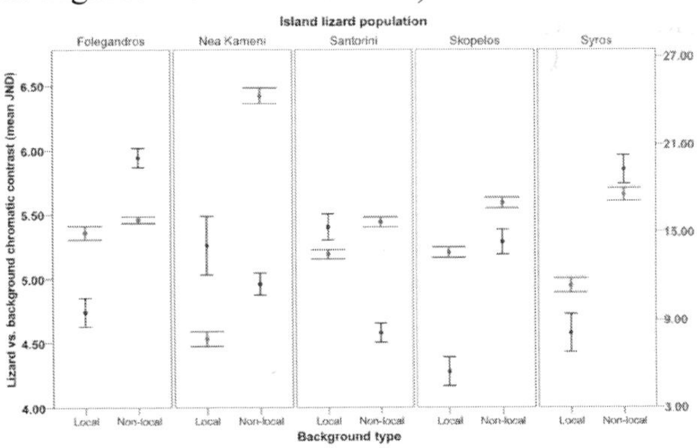

Figure 4. Local vs. non-local camouflage of Aegean wall lizards (*Podarcis erhardii*).

Degree of dorsal camouflage is shown against local island backgrounds and against different (non-local) island backgrounds overall (Folegandros, Nea Kameni, Santorini, Skopelos and Syros). Degree of camouflage is shown in terms of chromatic contrast (left axis; black data points) and luminance contrast (right axis; red data points) of lizards' dorsal regions against the background (mean JND). JND values increasing >3.00 depict lizards that are progressively distinguishable from the background by avian predators. Error bars represent +/- 1 S.E.

Table 2. Occurrence of enhanced local camouflage in Aegean wall lizards (*Podarcis erhardii*).

Enhanced local matching was verified when lizards' dorsal regions matched their local island backgrounds better than other non-local island backgrounds in terms of chromatic and luminance background matching (JND) to avian predators (see main text and table footnotes for symbol definitions). Lizard and background differences refer to significant differences in degree of contrast (JND) and in colour (*hue* and *saturation*) and *luminance* (see main text) between local and non-local islands.

Local island and lizard subspecies †	Non-local island	Lizard differences? Yes (Y) No (N)	Background differences? Yes (Y) No (N)	Enhanced local camouflage (chromatic)	Enhanced local camouflage (luminance)
Folegandros *P.e.naxensis*	Nea Kameni	Y	Y	✓	✓
	Santorini	Y	Y	✓	✓♀
	Skopelos	Y	Y	✓♂	❖
	Syros	Y	Y	✓	✓
Nea Kameni *P.e.naxensis*	Folegandros	Y	Y	❖	✓
	Santorini	Y	Y	✓	✓
	Skopelos	Y	Y	❖♀	✓
	Syros	Y	Y	–	✓
Santorini *P.e.naxensis*	Folegandros	Y	Y	❖	❖♀
	Nea Kameni	Y	Y	❖♀	✓
	Skopelos	Y	N (lum) Y (col)	❖♀	✓♂
	Syros	Y	Y	❖♂	✓♂
Skopelos *P.e.ruthveni*	Folegandros	Y	Y	–	✓
	Nea Kameni	Y	Y	✓	✓
	Santorini	Y	Y	✓	✓
	Syros	Y	Y	✓	✓

Syros *P.e.mykonensis*	Folegandros	Y	Y	–	✓♂
	Nea Kameni	Y	Y	✓	✓
	Santorini	N (lum) Y (col)	Y	✓	✓
	Skopelos	Y	Y	✓	✓♂

✓ Enhanced camouflage against local backgrounds vs. non-local backgrounds (JND) ($P < 0.05$)

❖ Inferior camouflage against local backgrounds vs. non-local backgrounds (JND) ($P < 0.05$)

– No difference in camouflage against local vs. non-local backgrounds (JND) ($P > 0.05$)

♂/♀ Difference in camouflage only in males/females

† Subspecies identified following [23–25].

Sex differences in local adaptation

Although Folegandros females typically showed better local chromatic camouflage compared to males in some island comparisons (vs. Santorini: $F_{1, 98} = 7.602$, $P = 0.007$, $\eta 2 = 0.072$; vs. Syros: $F_{1, 98} = 34.316$, $P < 0.001$, $\eta 2 = 0.259$), in other comparisons, better camouflage against local backgrounds was found in only males and in only females (Table 2). Moreover, Santorini and Syros males were sometimes the only sex to show enhanced achromatic local matching.

Body Region Differences in Local Adaptation

As predicted, Folegandros lizards' lower backs were generally more camouflaged in colour against local backgrounds compared to their upper backs (Figure B; vs. Santorini: $F_{1, 98} = 18.347$, $P < 0.001$, $\eta 2 = 0.158$; vs. Syros: $F_{1, 98} = 5.860$, $P = 0.017$, $\eta 2 = 0.056$; vs. Nea Kameni: $F_{1, 98} = 5.414$, $P = 0.022$, $\eta 2 = 0.052$). Similarly, Skopelos lizards' lower backs sometimes showed relatively better luminance matching against local backgrounds than their upper backs (Figure C; vs. Nea Kameni: $F_{1, 54} = 5.741$, $P = 0.020$, $\eta 2 = 0.096$). In contrast, in one case, only the upper backs of Folegandros lizards showed enhanced achromatic camouflage against local backgrounds (Table A and Figure C; vs Nea Kameni: $F_{1, 98} = 17.716$, $P < 0.001$, $\eta 2 = 0.153$).

DISCUSSION

We investigated whether dorsal coloration among island populations of Aegean wall lizards (*Podarcis erhardii*) differs in order to match local rock backgrounds for camouflage against avian predators. We found that lizards were better camouflaged against local island backgrounds compared to different island backgrounds to reduce detection by avian predators, with few exceptions (see Table 2; Fig 4; Figure A in S3 File). This was typically accompanied by island differences in lizards and the backgrounds they rest on, which were sufficient to be perceived by a typical avian predator visual system (see Table 2; Figs Figs22 and and3;3; S1 and S2 Files).

We found that better matching to local backgrounds occurred in most *P. erhardii* populations (Fig 4), and this arose when lizards were compared to most other non-local island backgrounds, in terms of luminance and/or chromatic matching (Table 2). These findings suggest that selection has typically driven better matching against local backgrounds to enhance camouflage against avian predators, in line with our predictions and with previous studies on lizards and mice (e.g. [12, 13, 80, 81]). Contrary to our predictions, in most populations, both sexes and dorsal body regions showed equally enhanced camouflage against local backgrounds (with some exceptions; Table 2; Figures B and C in S3 File). This is probably because selection favours camouflage on dorsal regions exposed to aerially hunting birds to offset the relatively heightened conspicuousness in sexually competing males and to further improve female camouflage (e.g. [13, 36, 82]). In addition, we did not find a clear association between predator risk (i.e. number of resident avian predator species; Table 1) and degree of local camouflage. For example, Santorini lizards exhibited relatively poor chromatic camouflage despite being under high risk, while Folegandros lizards were comparatively well camouflaged in their lower risk habitats (see Tables Tables11 and and2).2). This unexpected result may be because the number of resident avian predator species does not sufficiently reflect degree of predator pressure on each island. A more rigorous test would need a precise count of resident avian predators (e.g. number of breeding pairs) along with a measurement of relative survival (e.g. survival experiments with artificial models) on each island.

Local achromatic matching was particularly enhanced, with 18 out of 20 island comparisons showing better achromatic matching against local backgrounds (Table 2). This is likely to aid lizard camouflage by preventing discrimination of visual textures and small targets by hunting birds, particularly during motion (escape) ([58, 83, 84]). These results were consistent with frequent differences in lizard luminance (perceived lightness) between islands, and were also substantiated by corresponding patterns of variation between lizards and backgrounds. For example, Nea Kameni lizards and their rock backgrounds were both darker than those of most other islands, and those on Syros were typically lighter (see S1 File). Moreover, in many comparisons, lizards and their backgrounds were more similar in luminance to that of their own population as opposed to non-local populations (Figs (Figs22 and and3;3; see S2 File). Accordingly, these island populations consistently showed better luminance camouflage to avian predators against local backgrounds than against all non-local backgrounds (Table 2; Fig 4; S3 File). We found only one puzzling result. Syros lizards showed better luminance matching against local backgrounds than against non-local Santorini backgrounds, despite the lizards from these populations being no different in luminance (Table 2). This might be explained by subtle differences in lightness between these populations that were not identified by our analysis, together with the finding that they rest on backgrounds that strongly differ in luminance (see S1 and S2 Files).

Relatively heightened chromatic matching against local backgrounds was found in 11 out of 20 island comparisons (Table 2), which probably reduces the risk of detection by birds ([59, 85]). This consistently occurred when lizards and their backgrounds were more similar in colour to that of their own population than to that of non-local populations (Table 2; S2 File). These results suggest that heightened local chromatic camouflage against avian predators has been caused by adaptive convergence of coloration in each island population to match local backgrounds. Overall, these results suggest that the isolation of certain *P. erhardii* island populations in different environments has been generally sufficient to cause local adaptation for camouflage, as indicated by previous work ([36]).

These assumptions are supported by various indications that selection should favour dorsal matching to local backgrounds in *P. erhardii*. Past studies have shown that background matching dorsal camouflage in lizards reduces detection by avian predators in local environments (e.g. [12, 13, 36]). In addition, survival experiments with artificial models of *P. erhardii* and of other lizards have shown that more conspicuous models are attacked more frequently by avian predators compared to more camouflaged models ([86, 87]). Thus, dorsal matching to local rock backgrounds is likely to counteract considerable risk from avian predators, especially because *P. erhardii*, like other lizards, typically bask and display on rocks in open environments where they are highly visible to hunting birds ([31, 37]; pers. observations). Moreover, there is known genetic diversity and little gene flow (via over-water dispersal) between most Aegean island populations ([26, 27, 41]), similarly to other island lizards ([42]). This means that many of the current populations are likely to be descendants of the original colonists, which are highly likely to have adapted to match their native rock backgrounds since their time of isolation. This is supported by previous findings of adaptation to match local backgrounds on a comparably short evolutionary time-scale (i.e., 6,000 years; 10; Table 1). Moreover, our results show better local camouflage relative to non-local camouflage in comparisons between islands that were historically geographically connected and that have the same inhabitant subspecies, again suggesting that isolation of these island populations in their local environment has been sufficient to drive diversification for camouflage (e.g. Folegandros vs. Santorini; [23, 27]; Tables Tables11 and and2;2; Figure A in S3 File). However, we cannot rule out the influence of phenotypic plasticity given that (non-rapid) colour change appears to occur in *P. erhardii* over different seasonal periods to optimize camouflage (K. Marshall, unpublished data). Indeed, other lizards show similar colour change abilities, such as over different seasonal periods, photoperiods and developmental stages, which have sometimes been shown to enhance camouflage (e.g. [18, 19, 88, 89]).

Although past studies have reported local adaptation of camouflage in sympatric lizard populations (e.g. [10, 12, 13, 80]), here we studied separate island populations with distinct environments and evolutionary histories. This system has been useful in identifying certain factors that may affect local selection for camouflage. In particular, in some island populations, enhanced local background matching (relative to matching against non-local island backgrounds) was not found, or was found only in either chromatic or luminance matching, even when local and non-local lizards and background environments were variable (i.e. in 9 comparisons; Table 2). Often in such cases, the local island had been isolated for a relatively short period of time with potentially high levels of re-colonization from boat traffic (i.e., Nea Kameni; [26, 27]) and had experienced relatively low levels of risk from fewer resident avian predator species (i.e., Folegandros; [31]) (see Tables Tables11 and and2).2). In addition, the prolonged (> 200,000 year) period of volcanic eruptions on Santorini and the resulting geological complexity of the island (Table 1; [33, 34]) may have caused the inferior local selection for chromatic camouflage in the inhabitant lizards (Table 2; Fig 4). In particular, volcanic eruptions may have eradicated past populations on Santorini and/or changed the rock backgrounds to which they had become adapted, and produced a more heterogeneous environment to prevent more specialized camouflage ([90, 91]). These findings therefore indicate that, as predicted, local selection for camouflage in certain populations may be influenced to some degree by traits specific to their local island, such as predation risk and biogeographic history. Further survival experiments and genetic analyses would corroborate these ideas.

Nonetheless, enhanced local camouflage was found in relatively young and novel environments with a history of re-colonization, indicating rapid adaptation or phenotypic changes to enhance camouflage. For example, Nea Kameni lizards showed extremely strong achromatic camouflage in the form of dark (almost black) coloration against their black lava backgrounds (Table 2; Fig 4), much like other melanic lizard morphs have adapted to match darker substrates (e.g. [29]). This was despite a potentially high level of gene flow from the nearby Santorini population via unusually large amounts of boat traffic ([26]). The possibility that Nea Kameni lizards have become darker than their recolonizing population to match their darker backgrounds is supported by findings in other island lizards showing that phenotypic divergence has occurred despite high levels of gene flow between populations ([92]). Again, genetic analyses would substantiate these assumptions.

Our key findings show that Aegean wall lizard populations match their local backgrounds better than non-local backgrounds to enhance camouflage against avian predators, similarly to past work (e.g. [12, 13]), and that this appears to depend on ecological conditions and evolutionary history of the local environment. The exact mechanism underlying our results requires corroborative work, including genetic analyses and survival experiments, to investigate the relative roles of genetic adaptation, predation risk and phenotypic plasticity. Additionally, further work exploring how selection may enhance sexual signals against local backgrounds could elucidate the potential for reproductive isolation between the divergent island populations and subsequent speciation.

ACKNOWLEDGMENTS

We are grateful to N.B. Davies and to two anonymous referees for comments on previous versions. We thank the Greek Ministry of Environment and P. Pafilis for granting permission to conduct this research. We are grateful to S. Finlay, M. Moore and A. Török for excellent assistance in the field. K.L.A.M. thanks Magdalene College and the Department of Zoology, Cambridge.

FUNDING STATEMENT

This work was supported by a Biotechnology and Biological Sciences Research Council studentship (www.BBSRC.com), the British Herpetological Society (www.thebhs.org), the Cambridge Philosophical Society (http://www.cambridgephilosophicalsociety.org) and Magdalene College, Cambridge (to KLAM), and a Biotechnology and Biological Sciences Research Council and David Philips Research Fellowship (grant number BB/G022887/1) to MS. The funders had no role in study design, data collection and analysis, decision to publish, or preparation of the manuscript.

REFERENCES

1. Clarke B. Natural selection in mixed populations of two polymorphic snails. Heredity. 1962; 17: 319–45.
2. Kettlewell HBD. A resume of investigations on the evolution of melanism in the Lepidoptera. P Roy Soc B. 1956; 145(920): 297–303.
3. Kettlewell HBD. The phenomenon of industrial melanism in Lepidoptera. Annu Rev Entomol. 1961; 6: 245–62.
4. Rundle HD, Nosil P. Ecological speciation. Ecol Lett. 2005; 8: 336–52.
5. Schluter D. Ecology and the origin of species. Trends Ecol Evol. 2001; 16: 372–80.
6. Stevens M. Sensory Ecology, Behaviour and Evolution. Oxford: Oxford University Press; 2013.
7. Hoekstra HE, Hirschmann RJ, Bundey RA, Insel PA, Crossland JP. A single amino acid mutation contributes to adaptive beach mouse color pattern. Science. 2006; 313: 101–4.
8. Manceau M, Domingues VS, Mallarino R, Hoekstra HE. The developmental role of agouti in color pattern evolution. Science. 2011; 331: 1062–5. 10.1126/science.1200684
9. Nachman MW, Hoekstra HE, D'Agostino SL. The genetic basis of adaptive melanism in pocket mice. P Natl Acad Sci USA. 2003; 100(9): 5268–73.
10. Rosenblum EB, Römpler H, Schöneberg T, Hoekstra HE. Molecular and functional basis of phenotypic convergence in white lizards at White Sands. P Natl Acad Sci USA. 2010; 107: 2113–7.
11. Steiner CC, Weber JN, Hoekstra HE. Adaptive variation in beach mice produced by two interacting pigmentation genes. PloS Biol. 2007; 6: e36.
12. McLean CA, Moussalli A, Stuart-Fox DM. Local adaptation and divergence in colour signal conspicuousness between monomorphic and polymorphic lineages in a lizard. J Evol Biol. 2014; 27: 2654–64. 10.1111/jeb.12521
13. Stuart-Fox DM, Moussalli A, Johnston GR, Owens IPF. Evolution of color variation in dragon lizards: Quantitative tests of the role of crypsis and local adaptation. Evolution. 2004; 58: 1549–59.
14. Hart NS. Vision in the peafol (Aves: *Pavo cristatus*). J Exp Biol. 2002; 205: 3925–35.

15. Lind O, Mitkus M, Olsson P, Kelber A. Ultraviolet sensitivity and colour vision in raptor foraging. J Exp Biol. 2013; 216: 1819–26. 10.1242/jeb.082834
16. Endler JA, Mielke PW. Comparing entire colour patterns as birds see them. Biol J Linn Soc Lond. 2005; 86(4): 405–31.
17. Stevens M. Predator perception and the interrelation between different forms of protective coloration. P Roy Soc B. 2007; 274(1617): 1457–64.
18. Fan M, Stuart-Fox DM, Cadena V. Cyclic colour change in the bearded dragon *Pogona vitticeps* under different photoperiods. PloS One. 2014; 9: e111504 10.1371/journal.pone.0111504
19. Hawlena D, Boochnik R, Abramsky Z, Bouskila A. Blue tail and striped body: why do lizards change their infant costume when growing up? Beh Ecol. 2006; 17: 889–96.
20. Hemmi JM, Marshall J, Pix W, Vorobyev M, Zeil J. The variable colours of the fiddler crab *Uca vomeris* and their relation to background and predation. J Exp Biol. 2006; 209: 4140–53.
21. Stevens M, Rong CP, Todd P. Colour change and camouflage in the horned ghost crab *Ocypode ceratophthalmus* . Biol J Linn Soc Lond. 2013; 109: 257–70.
22. Jv Bedriaga. Die Amphibien und Reptilien Griechenlands. Bull Soc Imp Nat Moscou 1882; 56: 43–103, 278–344.
23. Chondropoulos BP. A checklist of the Greek reptiles. I. The lizards. Amphibia-Reptilia. 1986; 7(3): 217–35.
24. Poulakakis N, Goulielmos G, Antoniou A, Zouros E, Mylonas M. Isolation and characterization of polymorphic microsatellite markers in the wall lizard *Podarcis erhardii* (Squamata: Lacertidae). Mol Ecol Notes. 2005; 5: 549–51.
25. Poulakakis N, Lymberakis P, Antoniou A, Chalkia D, Zouros E, Mylonas M, et al. Molecular phylogeny and biogeography of the wall-lizard *Podarcis erhardii* (Squamata: Lacertidae). Mol Phylogenet Evol. 2003; 28(1): 38–46.
26. Hurston H, Voith L, Bonanno J, Foufopoulos J, Pafilis P, Valakos E, et al. Effects of fragmentation on genetic diversity in island populations of the Aegean wall lizard *Podarcis erhardii* (Lacertidae, Reptilia). Mol

Phylogenet Evol. 2009; 52: 395–405. 10.1016/j.ympev.2009.03.028

27. Foufopoulos J, Ives AR. Reptile extinctions on land-bridge islands: Life-history attributes and vulnerability to extinction. Am Nat. 1999; 153: 1–25.
28. Poulakakis N, Kapli P, Lymberakis P, Trichas A, Vardinoyiannis K, Sfenthourakis S, et al. A review of phylogeographic analyses of animal taxa from the Aegean and surrounding regions. J Zool Syst Evol Res. 2014; 53: 18–32.
29. Rosenblum EB. Convergent evolution and divergent selection: Lizards at the White Sands ecotone. Am Nat. 2006; 167: 1–15.
30. Brock KM, Bednekoff PA, Pafilis P, Foufopoulos J. Evolution of antipredator behavior in an island lizard species, *Podarcis erhardii* (Reptilia: Lacertidae): The sum of all fears? Evolution. 2014; 69: 216–31. 10.1111/evo.12555
31. Handrinos G, Akriotis T. The Birds of Greece. London: Christopher Helm Ltd; 1997.
32. Perissoratis C, Conispoliatis N. The impacts of sea-level changes during latest Pleistocene and Holocene times on the morphology of the Ionian and Aegean seas (SE Alpine Europe). Mar Geol. 2003; 196: 145–56.
33. Kalogeropoulos S, Paritsis S. Geological and geochemical evolution of the Santorini Volcano: a review In: Hardy DA, editor. Proceedings of the Third International Congress; Santorini, Greece. London: The Thera Foundation; 1990. p. 164–71.
34. Higgins MD, Higgins R. A geological companion to Greece and the Aegean. Ithaca, NY/London: Cornell University Press/Duckworth Publishers; 1996.
35. Photiades A, Keay S. Geological and geochronological data for Sikinos and Folegandros metamorphic units (Cyclades, Greece): Their tectono-stratigraphic significance. Bulletin of the Geological Society of Greece. 2003; 35: 35–45.
36. Marshall KLA, Stevens M. Wall lizards display conspicuous signals to conspecifics and reduce detection by avian predators. Beh Ecol. 2014; 25(6): 1325–37.
37. Castilla AM, Gosá A, Galán P, Pérez-Mellado V. Green tails in lizards of the Genus *Podarcis*: Do they influence the intensity of predation? Herpetologica. 1999; 55: 530–7.

38. Huyghe K, Vanhooydonck B, Herrel A, Tadić Z, Van Damme R. Morphology, performance, behavior and ecology of three color morphs in males of the lizard *Podarcis melisellensis* . Integr Comp Biol. 2007; 47: 211–20. 10.1093/icb/icm043

39. Van Damme R, Bauwens D, Castilla AM, Verheyen RP. Comparative thermal ecology of the sympatric lizards *Podarcis tiliguerta* and *Podarcis sicula* . Acta Oecol. 1990; 11: 503–12.

40. Macedonia JM, Brandt Y, Clark DL. Sexual dichromatism and differential conspicuousness in two populations of the common collared lizard (*Crotaphytus collaris*) from Utah and New Mexico, USA. Biol J Linn Soc. 2002; 77: 67–85.

41. Valakos E, Maragou P, Mylonas M. Geographical distribution: *Podarcis erhardii* . Herp Rev. 1999; 30: 52–3.

42. Jordan MA, Snell HL. Historical fragmentation of islands and genetic drift in populations of Galápagos lava lizards (*Microlophus albemarlensis* complex). Mol Ecol. 2008; 17: 1224–37. 10.1111/j.1365-294X.2007.03658.x

43. Arnold N, Ovenden D. Field Guide to Reptiles and Amphibians of Britain and Europe. London: Harper Collins; 2002.

44. Valakos E. The feeding ecology of *Podarcis erhardii* (Reptilia: Lacertidae) in a main insular ecosystem. J Herpetol. 1986; 1: 118–21.

45. Grimm A, Ramírez AMP, Moulherat S, Reynaud J, Henle K. Life-history trait database of European reptile species. Nat Cons. 2014; 9: 45–67.

46. *Podarcis erhardii* [database on the Internet]. IUCN Red List of Threatened Species Version 2013.2. 2009 [cited 25 November 2013].

47. García-Muñoz E, Sillero N. Two new types of noose for capturing herps. Acta Herpetologica. 2010; 5, 259–263.

48. Stevens M, Párraga CA, Cuthill IC, Partridge JC, Troscianko TS. Using digital photography to study animal coloration. Biol J Linn Soc Lond. 2007; 90(2): 211–37.

49. Bergman TJ, Beehner JC. A simple method for measuring colour in wild animals: validation and use on chest patch colour in geladas (*Theropithecus gelada*). Biol J Linnean Soc. 2008; 94(2): 231–40.

50. Stevens M, Stoddard MC, Higham JP. Studying primate color: towards visual system-dependent methods. Int J Primatol. 2009; 30: 893–917.

51. Verwaijen D, Van Damme R. Wide home ranges for widely foraging

lizards. Zoology. 2008; 111(1): 37–47.

52. Cooper WE, Greenberg N. Reptilian Coloration and Behavior In: Crews D, Gans C, editors. Biology of the Reptilian: Hormones, Brain, and Behaviour. New York: Academic Press; 1992. p. 298–422.
53. Fitze PS, Cote J, San-Jose LM, Meylan S, Isaksson C, Andersson S, et al. Carotenoid-based colours reflect the stress response in the common lizard. PloS One. 2009; 4: e5111 10.1371/journal.pone.0005111
54. Pike TW. Using digital cameras to investigate animal colouration: estimating sensor sensitivity functions. Behav Ecol Sociobiol. 2011; 65: 849–58.
55. Stevens M, Cuthill IC. Hidden messages: are ultraviolet signals a special channel in avian communication? BioScience. 2007; 57(6): 501–7.
56. Stevens M, Lown AE, Wood LE. Color change and camouflage in juvenile shore crabs *Carcinus maenas* . Front Ecol Evol. 2014; 2: 14.
57. Hart NS, Partridge JC, Cuthill IC, Bennett ATD. Visual pigments, oil droplets, ocular media and cone photoreceptor distribution in two species of passerine bird: the blue tit (Parus caeruleus L.) and the blackbird (Turdus merula L.). Journal of Comparative Physiology. 2000; 186: 375–87.
58. Jones CD, Osorio D. Discrimination of oriented visual textures by poultry chicks. Vision Research. 2004; 44: 83–9.
59. Osorio D, Vorobyev M. Photoreceptor spectral sensitivities in terrestrial animals: adaptations for luminance and colour vision. Proc R Soc B. 2005; 272: 1745–52.
60. Cuthill IC. Color Perception In: Hill GE, McGraw KJ, editors. Bird Coloration. Cambridge, MA: Harvard University Press; 2006. p. 3–40.
61. Hart NS, Hunt DM. Avian visual pigments: characteristics, spectral tuning, and evolution. Am Nat. 2007; 169(S1): S7–S26.
62. Martín J, López P. Avian predation on a large lizard (*Lacerta lepida*) found at low population densities in Mediterranean habitats: An analysis of bird diets. Copeia 1996; 1996: 722–6.
63. Ödeen A, Håstad O. 2013. The phylogenetic distribution of ultraviolet sensitivity in birds. BMC Evol Biol. 13:36 10.1186/1471-2148-13-36
64. Castilla AM, Labra A. Predation and spatial distribution of the lizard *Podarcis hispanica atrata*: an experimental approach. Acta Oecologica.

1997; 19: 107–114.

65. Collar NJ. Family Turdidae (thrushes) In: del Hoyo J, Elliot A, Sargatal J, editors. Handbook of the Birds of the World Vol. 10. Cuckoo-shrikes to thrushes. Barcelona: Lynx Edicions; 2005. p. 514–807.
66. Sazima I, D'Angelo GB. The pale-breasted thrush (*Turdud leucomelas*) preys on a gekkonid lizard and an anomalepidid snake. Revista Brasileira de Ornitologia. 2011; 19: 450452.
67. Martín J, López P. Amphibians and reptiles as prey of birds in southwestern Europe. Washington: Smithsonian Herpetological Information Service; 1990.
68. Castilla AM, Bauwens D. Thermal biology, microhabitat selection, and conservation of the insular lizard *Podarcis hispanica astrata* . Oecologia. 1990; 85: 366–74.
69. Stevens M, Merilaita S. Defining disruptive coloration and distinguishing its functions. Phil Trans B. 2009; 364: 481–488.
70. Komdeur J, Oorebeek M, Overveld Tv, Cuthill IC. Mutual ornamentation, age, and reproductive performance in the European starling. Beh Ecol. 2005; 16: 805–17.
71. Spottiswoode CN, Stevens M. How to evade a coevolving brood parasite: egg discrimination versus egg variability as host defences. Proc R Soc B. 2011; 278: 3566–73. 10.1098/rspb.2011.0401
72. Stevens M, Lown AE, Denton AM. Rockpool gobies change colour for camouflage. PloS One. 2014; 9: e110325 10.1371/journal.pone.0110325
73. Osorio D, Smith AC, Vorobyev M, Buchanan-Smith HM. Detection of fruit and the selection of primate visual pigments for color vision. Am Nat. 2004; 164: 696–708.
74. Osorio D, Vorobyev M. Colour vision as an adaptation to frugivory in primates. Proc R Soc B. 1996; 263: 593–9.
75. Rowe MH. Trichromatic color vision in primates. Physiology. 2002; 17: 93–8.
76. Vorobyev M, Osorio D. Receptor noise as a determinant of colour thresholds. Proc R Soc B. 1998; 265(1394): 351–8.
77. Siddiqi A, Cronin TW, Loew ER, Vorobyev M, Summers K. Interspecific and intraspecific views of color signals in the strawberry poison frog *Dendrobates pumilio* . J Exp Biol. 2004; 207: 2471–85.

78. Vorobyev M, Osorio D, Bennett ATD, Marshall NJ, Cuthill IC. Tetrachromacy, oil droplets and bird plumage colours. J Comp Physiol A. 1998; 183: 621–33.

79. Ruxton GD, Beauchamp G. Time for some a priori thinking about post hoc testing. Behav Ecol. 2008; 19(3): 690–3.

80. Rosenblum EB, Hoekstra HE, Nachman MW. Adaptive reptile color variation and the evolution of the MC1R gene. Evolution. 2004; 58(8): 1794–808.

81. Vignieri SN, Larson JG, Hoekstra HE. The selective advantage of crypsis in mice. Evolution. 2010; 64(7): 2153–8. 10.1111/j.1558-5646.2010.00976.x

82. Garcia JE, Rohr D, Dyer AG. Trade-off between camouflage and sexual dimorphism revealed by UV digital imaging: the case of Australian Mallee dragons (*Ctenophorus fordi*). J Exp Biol. 2013; 216: 4290–8. 10.1242/jeb.094045

83. Osorio D, Miklosi A, Gonda Z. Visual ecology and perception of coloration patterns by domestic chicks. Evol Ecol. 1999; 13: 673–89.

84. Stuart-Fox DM, Whiting MJ, Moussalli A. Camouflage and colour change: responses to bird and snake predators across multiple populations in a dwarf chameleon. Biol J Linn Soc Lond. 2006; 88: 437–46.

85. Kelber A, Vorobyev M, Osorio D. Animal colour vision–behavioural tests and physiological concepts. Biological Reviews. 2003; 78: 81–118.

86. Husak JF, Macedonia JM, Fox SF, Sauceda RC. Predation cost of conspicuous male coloration in collared lizards (*Crotaphytus collaris*): An experimental test using clay-covered model lizards. Ethology. 2006; 112(6): 572–80.

87. Stuart-Fox DM, Moussalli A, Marshall NJ, Owens IPF. Conspicuous males suffer higher predation risk: visual modelling and experimental evidence from lizards. Anim Behav. 2003; 66(3): 541–50.

88. Hamilton PS, Gaalema DE, Sullivan BK. Short-term changes in dorsal reflectance for background matching in Ornate Tree lizards (*Urosaurus ornatus*). Amphibia-Reptilia. 2008; 29: 473–7.

89. Hernández-Gallegos O, Domínguez-Vega H. Seasonal changes in the dorsal coloration in the lizard *Aspidoscelis costata costata* (Squamata: Teiidae). Rev Biol Trop. 2012; 60: 405–12.

90. Houston AI, Stevens M, Cuthill IC. Animal camouflage: compromise of specialize in a 2 patch-type envrionment? Beh Ecol. 2007; 18: 769–75.
91. Merilaita S, Tuomi J, Jormalainen V. Optimization of cryptic coloration in heterogeneous habitats. Biol J Linn Soc Lond. 1999; 67: 151–61.
92. Jordan MA, Snell HL, Snell HM, Jordan WC. Phenotypic divergence despite high levels of gene flow in Galápagos lava lizards (*Microlophus albemarlensis*). Mol Ecol. 2005; 14: 859–67.

11

A Virtual Motion Camouflage Approach for Cooperative Trajectory Planning of Multiple UCAVs

Xueqiang Gu, Lincheng Shen, Jing Chen, Yu Zhang, and Wanpeng Zhang

College of Mechatronic Engineering and Automation, National University of Defense Technology, Changsha, Hunan 410073, China

ABSTRACT

This paper investigates cooperative trajectory planning of multiple unmanned combat aerial vehicles (multi-UCAV) in performing autonomous cooperative air-to-ground target attack missions. By integrating an approximate allowable attack region model, several constraint models, and a multicriterion objective function, the problem is formulated as a cooperative trajectory optimal control problem (CTOCP). Then, a virtual motion camouflage (VMC) for cooperative trajectory planning of multi-UCAV, combining with the differential flatness theory, Gauss pseudospectral method (GPM), and nonlinear programming, is designed to solve the CTOCP. In particular, the notion of the virtual time is introduced to the VMC problem formulation to handle the temporal cooperative constraints.

Citation: A Virtual Motion Camouflage Approach for Cooperative Trajectory Planning of Multiple UCAVs. Xueqiang Gu, Lincheng Shen, Jing Chen, Yu Zhang, and Wanpeng Zhang. Hindawi Publishing Corporation Mathematical Problems in Engineering Volume 2014, Article ID 748974, 15 pages http://dx.doi.org/10.1155/2014/748974

The simulation experiments validate that the CTOCP can be effectively solved by the cooperative trajectory planning algorithm based on VMC which integrates the spatial and temporal constraints on the trajectory level, and the comparative experiments illustrate that VMC based algorithm is more efficient than GPM based direct collocation method in tackling the CTOCP.

INTRODUCTION

Nowadays, it is an active research area to perform autonomous cooperative air-to-ground target attack (CA/GTA) missions using multiple unmanned combat aerial vehicles (multi-UCAV) [1]. However, compared with single UCAV planning and coordinated formation control problems [2], new technical challenges in the CA/GTA missions are emerging. The cooperative trajectory planning is one of the key challenging technologies, due to its high dimensionality, severe equality and inequality constraints involved, and the requirement of spatial-temporal cooperation of multi-UCAV.

Recently, various algorithms have been developed to solve this cooperative trajectory planning problem [3, 4], including artificial neural network methods [5], sample-based planning methods [6], maneuver automation (MA) [7], and optimal control methods. There is no doubt that the optimal control theory is the most natural framework for this type of problem with dynamic constraints [8]. However, the rapid solution to optimal control problems (OCPs) for complicated nonlinear systems, such as UCAVs, is a challenging task [9]. Analytical solutions are seldom available or even possible. As a result, one usually resorts to numerical techniques [3]. The techniques can be classified into two general types, namely, indirect and direct methods. Indirect methods [10] solve the OCPs by formulating the first-order optimality conditions, applying Pontryagin's Maximum Principle and nonlinear programming (NLP) to tackle the resulting two-point boundary value problem numerically. Direct methods [11], on the other hand, are devoted to reduce the OCPs to finite-dimensional NLP problems by discretization and parameterization of subsets of the state and control vectors and then are solved by developed optimizers. As one of direct methods, Gauss pseudospectral method (GPM) is applied widely and efficiently in the trajectory optimization field [12–14], due to its advantages of fewer parameters and higher precision in the calculation procedure. However, the method to achieve the trajectory planning for a single vehicle or cooperative multivehicle required a high computational load.

To reduce the computational complexity, the inverse dynamics methods [15, 16] and differential flatness theory (DFT) based methods [17–20] are introduced. Compared to pseudospectral methods, these methods can use any models and any performance indexes and transform the OCP into a low dimensional NLP problem. However, they are still time-consuming. Particularly, when the number of the vehicles is too large, the computational cost will be unacceptable. For this purpose, the virtual motion camouflage (VMC) algorithm [21–23] is introduced into this work. Inspired by the biological motion known as the motion camouflage, the VMC algorithm can dramatically reduce the problem dimension and the computational cost. Some numerical simulations [22–24] suggest that its computational speed could be much faster than the pseudospectral methods. Motivated by these advantages, this paper employs virtual motion camouflage approach to develop cooperative trajectory planning algorithms.

For the cooperative trajectory planning of multi-UCAV, the objective is to generate dynamically feasible trajectories for UCAVs, which can guide them to the goals in the shortest time or distance, without collisions with obstacles or each other. To date, a number of theories and techniques have been developed to accomplish the cooperative tasks, including several collision avoidance techniques and time adjustment strategies [17, 25–30]. Bollino and Lewis [25] addressed the optimal path planning of UAVs in obstacle-rich environments and proposed a collision-free path planning for multi-UAV using optimal control techniques and pseudospectral methods. Kuwata and How [26] presented a cooperative distributed robust trajectory optimization approach, using RH-MILP with independent dynamics but coupled objectives and hard constraints. To improve the convergence rate, the virtual motion camouflage method was applied to the cooperative electronic combat air vehicles (ECAVs) [27] and unmanned aerial vehicle (UAV) formation control [28]. Reference [27] described how an interesting bioinspired motion strategy can be used to design real-time trajectories for cooperative electronic combat air vehicles. The constant speed motion camouflage law was developed to derive the feasible condition of a constant speed ECAVs and PT coherent mission, and its feasibility conditions were found. Reference [28] proposed a divide and conquer hierarchical approach in three levels to solve the UAV formation flight trajectory plan problem considering dynamics, state, and control variable inequality and equality constraints. These approaches mentioned above could generate collision-free trajectories, but the temporal cooperation was ignored. In contrast, Lian [17] introduced a differential flatness theory (DFT) based method to optimally formulate the cooperative path planning for multiagent dynamical systems considering spatial and temporal constraints. McLain and Beard [29] proposed the coordination variables (CV) and coordination function (CF) based strategy to achieve cooperative timing among teams of vehicles by coordinating the velocity and path length of each vehicle. But these approaches failed to deal with the dynamics constraints of vehicles, and the generated paths were not always flyable and smooth. Kaminer et al. [30] presented a general framework for coordinated control problem of multiple autonomous agents. On the basis of decoupling space and time in the problem formulation, it reduced the number of optimization parameters and made it easy to implement optimization in real time. However, the time coordination lays in the path following and the design of control laws in path following algorithms was much more difficult. Although the previous investigations have described several valuable strategies in the cooperative path planning, these methods cannot tackle the point-to-region cooperative trajectory planning for CA/GTA missions directly, which needs to integrate both the spatial and temporal constraints on the trajectory level.

To address the problems mentioned previously, a novel cooperative trajectory planning method for multi-UCAV in performing the CA/GTA missions is presented in this paper. Firstly, some constraints including individual and cooperative constraints are modeled. Particularly, an approximate allowable attack region is built for the critical terminal constraints. Then, after the multicriterion objective function is constructed, the cooperative trajectory planning problem is formulated as a cooperative trajectory optimal control problem (CTOCP). Owing to the temporal constraints, a notion of the virtual time is introduced and the VMC problem is reformulated in the virtual time domain. Inspired by VMC, DFT, and GPM algorithms, a new cooperative trajectory planning algorithm in an optimal control framework is proposed. The proposed approach is demonstrated by two typical CA/GTA examples, and the comparative experiments between GPM and VMC are carried out. The results show that the proposed approach is feasible, effective, and efficient.

The rest of the paper is organized as follows. In Section 2, the integrated model is set up. In Section 3, the problem is formulated as a CTOCP followed by a multicriteria objective function. Section 4 develops a novel cooperative trajectory planning algorithm based on virtual motion camouflage in virtual time domain, which can solve the problem efficiently. Section 5 presents several numerical examples, and finally the conclusions and future works are outlined in Section 6.

MODELING

Aircraft Model

The kinematic and dynamic model of vehicles is needed in the cooperative trajectory planning of multi-UCAV. In this work, a team of N_V homogeneous UCAVs is considered, each of which is described by the kinematic and dynamic model according to a full-blown three-degree of freedom (3-DOF) model as follows [31]:

$$\dot{x} = V\cos\gamma\cos\psi, \qquad \dot{y} = V\cos\gamma\sin\psi, \qquad \dot{z} = V\sin\gamma,$$

$$\dot{V} = g\left(n_x - \sin\gamma\right), \qquad \dot{\gamma} = \frac{g}{V}\left(n_z\cos\mu - \cos\gamma\right),$$

$$\dot{\psi} = \frac{g}{V\cos\gamma}n_z\sin\mu, \tag{1}$$

where x, y, and z are the aircraft coordinates, that is, longitude, latitude, and height, V is the aircraft velocity, γ is the flight-path angle, ψ is the heading angle, μ is the roll angle, g is the gravity acceleration, and n_x and n_z are the longitudinal and normal components of the load factor, respectively.

For the optimal control problem, the state vector x and control vector u can be defined as

$$\mathbf{x} = \left\{ (x, y, z, V, \gamma, \psi) \mid \begin{matrix} z_{\min} \le z \le z_{\max} \\ V_{\min} \le V \le V_{\max} \\ \gamma_{\min} \le \gamma \le \gamma_{\max} \\ \psi_{\min} \le \psi \le \psi_{\max} \end{matrix} \right\} \in \mathbf{X} \subset \mathbb{R}^6,$$

$$\mathbf{u} = \left\{ (\mu, n_x, n_z) \mid \begin{matrix} \mu_{\min} \le \mu \le \mu_{\max} \\ n_{x,\min} \le n_x \le n_{x,\max} \\ n_{z,\min} \le n_z \le n_{z,\max} \end{matrix} \right\} \in \mathbf{U} \subset \mathbb{R}^3. \tag{2}$$

Allowable Attack Region Model

For the point-to-region trajectory planning problem, the allowable attack regions (AARs) of targets are defined as the areas where weapon delivery operations can be effectively performed by UCAVs. Therefore, in order to plan accurate and optimal attack trajectories, the AARs and the delivery parameters need to be obtained. The AARs of the ith target TARi, denoted as (TARi), are such sets of all feasible release states that TARi can be effectively attacked whenever the aircraft belongs to those state sets. When nV ($n\mathrm{V} \geq 2$) UCAVs are assigned to attack the same target TARi cooperatively, there will be nV AARs in (TARi); that is, $R(\mathrm{TAR}_i) = \bigcup_j^{n_V} \mathrm{AAR}_j$ can be formulated as an abstract 6-dimensional space [32]:

$$\mathrm{AAR}_j = \{x, y, z, V, \gamma, \psi\} \subset \mathbb{R}^6. \tag{3}$$

Obviously, AAR_j is a high-dimensional nonlinear space, which is difficult to handle. By presetting an appropriate weapon release speed $\widehat{V}_r$ and the flight-path angle $\widehat{\gamma}_r$ based on estimating the weapon impact effects and destruction requirements to the target and predetermining release heading $\widehat{\psi}_r$, AAR_j can be reduced to a 3-dimensional space as

$$\mathrm{LAR}^j_{\mathrm{sp}}\left(\mathrm{TAR}_i, \left[\widehat{V}_r, \widehat{\gamma}_r\right]\right) = \left\{\left(x, y, z, \widehat{V}_r, \widehat{\gamma}_r, \widehat{\psi}_r\right) \mid \begin{array}{l} (x, y, z) \in \mathrm{LAR}^j_{\mathrm{sp}}\left(\mathrm{TAR}_i, \left[\widehat{V}_r, \widehat{\gamma}_r, \widehat{\psi}_r\right]\right) \\ \widehat{\psi}_r = \text{azimuth of } \mathrm{TAR}_i \text{ relative to position } (x, y, z) \end{array}\right\}. \tag{4}$$

In the point-to-region trajectory planning, weapon delivery points (WDPt) as terminals of trajectories need to be included in the AARs, that is, meeting terminal constraints [33]. The formula can be denoted as $\mathrm{WDPt}_j \in \mathrm{AAR}_j$; that is

$$\left|x^j_A - x^j_f\right| \leq \Delta x, \qquad \left|y^j_A - y^j_f\right| \leq \Delta y,$$
$$\left|z^j_A - z^j_f\right| \leq \Delta z, \tag{5}$$

where (x^j_A, y^j_A, z^j_A) are the coordinates of the center of the AAR_j, (x^j_f, y^j_f, z^j_f) are the coordinates of WDPt_j, and $(\Delta x, \Delta y, \Delta z)$ are the thresholds of errors.

No-Fly Zone Model

In the battlefield environment, the no-fly zone (NFZ) is an area where vehicles are not permitted to fly over, due to the presence of military restrictions (e.g., armed enemies and the missile killing range), the physical obstacles (e.g., mountains and buildings within the natural and urban environment, resp.), and civil restrictions mainly due to safety reasons (e.g., densely populated areas and severe weather condition zones).

It is complex and unnecessary to describe a NFZ's exact shape and size. In this work, the p-norm method [34] is used to mathematically model the shapes of the NFZs, given by

$$h(x, y, z) = \left\| \left(\frac{x - x_c}{k_1 + b_s}, \frac{y - y_c}{k_2 + b_s}, \frac{z - z_c}{k_3 + b_s} \right) \right\|_p^p - k_4^p, \tag{6}$$

$$\|(x, y, z)\|_p = \left(|x|^p + |y|^p + |z|^p \right)^{1/p}, \quad p \in \mathbf{N}, \tag{7}$$

where (x_c, y_c, z_c) indicates the location of the geometric center of the NFZ, h(x, y, z) is the distance between a point (x, y, z) and the boundary of the NFZ, while $k_1 \sim k_4$ are the constant parameters chosen to define the size of the NFZ, and bs represents the width of a safe buffer which can effectively expand the NFZ boundary by an amount that accounts for the size of the vehicle and improve the robustness of plans. By varying parameters (p, $k_1 \sim k_4$), one can easily model a number of generic shapes (see Figure 1). And any NFZ can be approximately modeled by fitting one of those shapes around

Accordingly, the above NFZs can be expressed by the following path constraints in the OCP formulation as

$$h_n (x, y, z) \geq 0, \quad n = 1, 2, \ldots, N_{\text{NFZ}}, \tag{8}$$

where h_n represents the nth NFZ and N_{NFZ} is the total of NFZs. However, (8) is not well scaled. If a large p is chosen or the vehicle is far from the NFZ, h(x, y, z) can produce very large numbers; hence, for computational efficiency, the path constraint is scaled by the natural logarithm function as follows:

$$h_n (x, y, z) = \ln \left(\frac{1}{(k_4^n)^{p_n}} \left(\left(\frac{x - x_c^n}{k_1^n + b_s} \right)^{p_n} + \left(\frac{y - y_c^n}{k_2^n + b_s} \right)^{p_n} + \left(\frac{z - z_c^n}{k_3^n + b_s} \right)^{p_n} \right) \right) \geq 0. \tag{9}$$

Cooperative Constraint Model

In order to make the UCAVs arrive in the AARs of targets safely in an expected time sequence, cooperative trajectories should satisfy the following constraints.

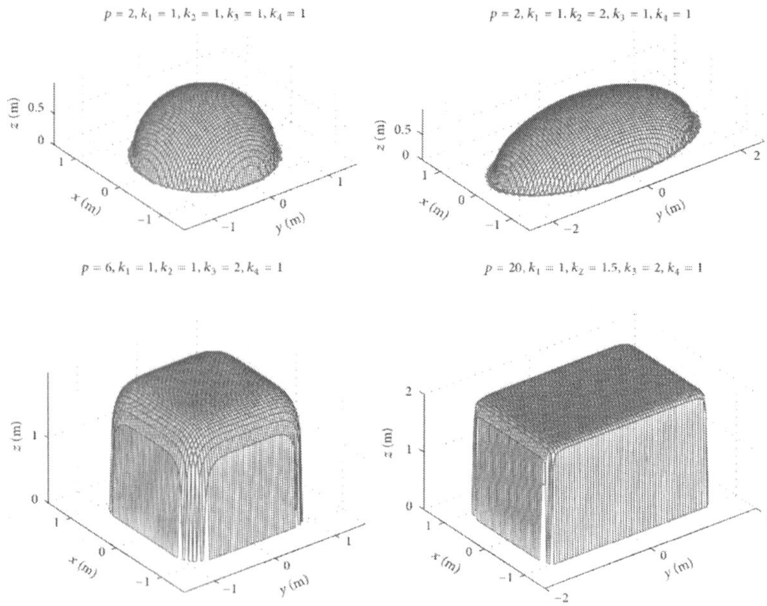

Figure 1: Several shapes generated from (6).

Spatial Constraints

During the mission, multi-UCAV should maintain a safe distance to guarantee them not to collide with each other. The model can be denoted as

$$\left\|\rho^{j}(t_i)-\rho^{k}(t_i)\right\|_2 \geq \max\left(d^{j}_{\text{safe}}, d^{k}_{\text{safe}}\right),$$
$$\forall j \neq k, \quad j,k = 1,2,\ldots,N_v, \tag{10}$$

where $\rho^{j}(t_i) = \{x^{j}(t_i), y^{j}(t_i), z^{j}(t_i)\}$ is the spatial position of each UCAV$_j$ at the time t_i, d^{j}_{safe} is the predetermined minimum safety radius of UCAV$_j$ (based on the vehicle size, maneuver agility, etc.), and N_V is the total of UCAVs.

Temporal Constraints

The temporal constraints include the simultaneous arrival constraint and sequencing constraint, which are described as

$$t_f^j + \Delta_{jk} - t_f^k \leq 0, \quad \forall j \neq k, \quad j,k = 1,2,\ldots,N_v, \tag{11}$$

where t_f^j is the terminal time of each UCAV$_j$ and Δ_{jk} is the arrival interval between UCAV$_j$ and UCAV$_k$. For the simultaneous arrival constraint, Δ=0, and for the sequencing constraint, Δ is the plus fixed quantity. Equation (11) is of the general form $f_T(t_f^1, \ldots, t_f^{N_v}) \leq 0.$

PROBLEM FORMULATION

Objective Function

The objective function of the entire team is a sum of each individual cost, which is a multicriterion objective function, and each criterion could compete with each other as follows:

$$J = \min \sum_{j=1}^{N_v} J^j = \min \sum_{j=1}^{N_v} \left(w_t^j J_t^j + w_p^j J_{\mathrm{prd}}^j + w_r^j J_{\mathrm{rob}}^j \right), \tag{12}$$

and the criterions can be, respectively, defined as

$$J_t^j = \int_t dt = t_f^j - t_0^j, \tag{13}$$

$$J_{\mathrm{prd}}^j = \int_t \mathrm{PRD}^j(t) = \int_t \left(1 - \prod_{r=1}^{n_r} \left(1 - P_d^j(t,r) \right) \right), \tag{14}$$

$$J_{\mathrm{rob}}^j = \int_t R^j(t) = \int_t \sum_{n=1}^{N_{\mathrm{NFZ}}} w_r^{n,j} \left(e^{e^{-h_n}} - 1 \right), \tag{15}$$

where t_0^j and t_f^j denote the initial and terminal time of UCAV$_j$. The first term, J_t^j;, denotes the total fight time of UCAV$_j$. The second term, J_{prd}^j, , denotes the detection probability of the nr-radar system to UCAV$_j$ at the time t, which describes the threat risk criterion of UCAV$_j$ as follows:

$$\text{PRD}^j(t) = 1 - \prod_{r=1}^{n_r} \left(1 - P_d^j(t,r)\right), \tag{16}$$

Where $P_d^j(t,r)$ is the model of the radar detection probability between the trajectory point of UCAV$_j$ at the time t and the rth radar [35]. The final term, J_{rob}^j, guarantees the robustness of the vehicle collision avoidance against the model imprecision, disturbing in the battlefield and operating errors that could otherwise cause vehicles to collide with obstacles or other vehicles. The robustness cost is defined as

$$R^j(t) = \sum_{n=1}^{N_{\text{NFZ}}} w_r^{n,j}\left(e^{e^{-h_n}} - 1\right), \tag{17}$$

where $w_r^{n,j}$ are weights that are incremented to increase the maneuver robustness when the trajectory of a vehicle passes in close proximity to an obstacle or another vehicle, which can be chosen by the method presented by Hurni et al. [36].

As can be seen, the individual objective function J^j is defined by the weighted sum of the three separate running cost terms with appropriate weighting factors w_t^j, w_p^j, and w_r^j. The three criteria (13)–(15) represent different physical meanings, and they are difficult to be compared directly. Hence the selection of these weighting factors is a skilled technique. In the experimentations, the multi-objective fuzzy optimization method proposed by Wang et al. [37] is employed. The method includes two main steps: (a) normalizing each single objective function and (b) solving the membership function about each criterion by using fuzzy distribution function.

Cooperative Trajectory Planning Problem Formulation

After establishing the above models, the cooperative trajectory planning problem is formulated in this section. The problem under consideration is a cooperative scheme, consisting of N_V homogeneous UCAVs. According to the system equations (1), the general form of the system for UCAV$_j$ is given by

$$\dot{\mathbf{x}}^j(t) = f^j\left[\mathbf{x}^j(t), \mathbf{u}^j(t), t\right], \quad j = 1, \ldots, N_v. \tag{18}$$

As stated above, to obtain optimal or suboptimal cooperative trajectories, the cooperative trajectory planning for the CA/GTA missions can be formulated as a cooperative trajectory optimal control problem (CTOCP).

Problem 1 (CTOCP)

Find the trajectories, which drive the system from given initial conditions to desired final conditions over time horizons $[t0,\ tf]$, while the cooperative objective function is minimized as

$$
\begin{aligned}
\min \quad J &= \sum_{j=1}^{N_v} J^j(\mathbf{x},\mathbf{u}) \\
&= \sum_{j=1}^{N_v} \left[w_t^j \left(t_f^j - t_0^j\right) + w_p^j \int_{t_0^j}^{t_f^j} \left(1 - P_d^j(t,r)\right) dt \right. \\
&\qquad \left. + w_r^j \int_{t_0^j}^{t_f^j} \sum_{n=1}^{N_{\mathrm{NFZ}}} w_r^{n,j} \left(e^{e^{-h_n}} - 1\right) dt \right],
\end{aligned}
\tag{19}
$$

subject to the system equation (18) and the boundary constraints (i.e., the initial and terminal states (5))

$$
\Phi\left[\mathbf{x}^j\left(t_0^j\right), \mathbf{u}^j\left(t_0^j\right), t_0^j, \mathbf{x}^j\left(t_f^j\right), \mathbf{u}^j\left(t_f^j\right), t_f^j\right] = 0,
\qquad \forall j = 1, 2, \ldots, N_v,
\tag{20}
$$

and several inequality and equality constraints, the individual and cooperative constraints, including the state and control vectors (2) and (9)–(11), are denoted as

$$
\begin{aligned}
&\delta^j\left[\mathbf{x}^j(t), \mathbf{u}^j(t), t\right] \le 0, \\
&\varphi^j\left[\mathbf{x}^j(t), \mathbf{u}^j(t), \mathbf{x}^k(t), \mathbf{u}^k(t), t\right] \le 0, \\
&f_T^j\left(t_f^j, t_f^k\right) \le 0, \quad \forall j \ne k, \ j,k = 1, 2, \ldots, N_v.
\end{aligned}
\tag{21}
$$

VIRTUAL MOTION CAMOUFLAGE BASED COOPERATIVE TRAJECTORY PLANNING

To tackle the cooperative trajectory planning for CA/GTA missions, an efficient virtual motion camouflage based planning method is proposed for numerically solving the CTOCP, which combines with the benefits of several other classical trajectory generation techniques, including DFT, GPM, and NLP. Correspondingly, this task contains three primary steps. The first is to determine outputs by VMC and DFT in the virtual time domain, such that the dynamics system can be mapped to a lower-dimensional output space. Then, GPM is used to discretize the outputs. Finally, the system is transformed into a NLP problem and the sequential quadratic programming (SQP) is used to solve the NLP to minimize the cost and subject to constraints in output space.

Virtual Motion Camouflage and Problem Formulation in the Output Space

Virtual Motion Camouflage

Inspired by mating hoverflies, Srinivasan and Davey [38] described a new form of the stealth strategy which is the so-called motion camouflage (MC). MC involves two moving objects, a prey and an aggressor, and occurs when the aggressor conceals its motion (i.e., maintains a constant bearing in the prey's coordinates), except the inevitable size change, as viewed by the moving prey. Here the MC concept is introduced into the optimal trajectory planning and is called virtual motion camouflage (VMC) [21–23] because the "prey" is a virtual one. In the VMC framework, the aggressor trajectory (i.e., the aggressor position vector) $\boldsymbol{\rho}_a(t) \in \mathbb{R}^{n_a}$ is confined by the virtual prey trajectory (VPT) $\boldsymbol{\rho}_p(t) \in \mathbb{R}^{n_a}$, the selected reference point $\boldsymbol{\rho}_{\text{ref}}(t) \in \mathbb{R}^{n_a}$, and the path control parameter (PCP) $V(t) \in \mathbb{R}$ as follows:

$$\boldsymbol{\rho}_a(t) = \boldsymbol{\rho}_{\text{ref}} + v(t)\left(\boldsymbol{\rho}_p(t) - \boldsymbol{\rho}_{\text{ref}}\right). \tag{22}$$

The reference point is considered fixed over time, so the derivatives of the trajectory can be described as

$$\dot{\boldsymbol{\rho}}_a(t) = \dot{v}(t)\left(\boldsymbol{\rho}_p(t) - \boldsymbol{\rho}_{\text{ref}}\right) + v(t)\dot{\boldsymbol{\rho}}_p(t),$$

$$\ddot{\boldsymbol{\rho}}_a(t) = \ddot{v}(t)\left(\boldsymbol{\rho}_p(t) - \boldsymbol{\rho}_{\text{ref}}\right) + v(t)\ddot{\boldsymbol{\rho}}_p(t) + 2\dot{v}(t)\dot{\boldsymbol{\rho}}_p(t). \tag{23}$$

From [23], it can be known that when the proposed VMC suboptimal trajectory design method is used, the complexity of the problem can be reduced by $(n_{sV} + m_{cV})$ times from the collocation method or n_{sV} times from the differential inclusion, where n_{sV} and m_{cV} are the numbers of the state variables and the control variables. In VMC framework, if the dimension of the position vector x(t) is one (i.e., $n_a = 1$), the dimension of the problem is the same as that of the differential inclusion method. Otherwise, when $n_a > 1$, the reference point can be fixed or regarded as a parameter to be optimized, whereas VPT can be predefined or predetermined by the initial and terminal conditions of the aggressor trajectory. Thus, the dimension of the problem is lowered from n_a to one, such that the solving speed can increase.

VMC Problem Formulation in Virtual Time Domain

Obviously, the time factor of trajectories is an argument of the state and control. To deal with temporal constraints and achieve the time cooperation of multi-UCAV, the time along the trajectories should be considered separately. Therefore, the independent intermediate variable (called virtual time here, $\bar{t} \in [0, 1]$) is introduced and described as

$$\bar{t} = \frac{(t - t_0)}{(t_f - t_0)}, \tag{24}$$

such that the trajectories can be generated in the virtual time domain from $\bar{t}_0 = 0$ to $\bar{t}_f = 1$.

For UCAVs, which are required to take off at the same time, it can be assumed that the initial time of all UCAVs is zero ($t_0 = 0$).Thus, the terminal time t_f can be written as $tf = t/\bar{t}$.That is, t_f denotes the ratio between the true time variable and the newly defined virtual time variable. To coordinate the arrival time of all UCAVs, the terminal time can be defined as an argument to be optimized in the dynamical system, designated as T_f. Then, the following relationship between the virtual time domain and true time domain can be obtained for an arbitrary variable χ as follows:

$$\chi'(\bar{t}) = \frac{d\chi(t)}{d\bar{t}} = \frac{T_f d\chi(t)}{dt} = T_f \dot{\chi}(t),$$

$$\chi''(\bar{t}) = \frac{d\chi'(\bar{t})}{d\bar{t}} = \frac{T_f d\chi'(\bar{t})}{dt} = \frac{T_f d\left(T_f \dot{\chi}(t)\right)}{dt} = T_f^2 \ddot{\chi}(t). \tag{25}$$

Particularly, the derivative of the speed variable can be denoted as

$$V(\bar{t}) = \sqrt{(x')^2 + (y')^2 + (z')^2}$$
$$= T_f \sqrt{\dot{x}^2 + \dot{y}^2 + \dot{z}^2} = T_f V(t), \tag{26}$$

where the superscript "$'$" represents the derivative with respect to the virtual time. According to (18) and (24), the system equations can be rearranged as

$$\mathbf{x}'^{j}(\bar{t}) = T_f \dot{\mathbf{x}}^j(t) = T_f f^j\left[\mathbf{x}^j(t), \mathbf{u}^j(t), t\right]$$
$$= g^j\left[\mathbf{x}^j(\bar{t}), \mathbf{u}^j(\bar{t}), T_f^j\right],$$
$$T_f'^{j} = 0, \quad j = 1, \ldots, N_v. \tag{27}$$

Hence, the Problem 1 can be reformulated in virtual time domain (VTD) as follows.

Problem 2 (CTOCP-VTD)

Minimize the cooperative objective function (19) of all UCAVs represented with respect to the new independent variable $\bar{t}$ as

$$\min \quad J = \sum_{j=1}^{N_v} J^j\left(\mathbf{x}, \mathbf{u}, T_f^j\right)$$
$$= \sum_{j=1}^{N_v}\left[w_t^j T_f^j + w_p^j \int_0^1 \mathrm{PRD}^j(\bar{t})\, d\bar{t}\right.$$
$$\left. + w_r^j \int_0^1 R^j(\bar{t})\, d\bar{t}\right], \tag{28}$$

subject to the system equations (27), and the boundary constraints (20), written as

$$\phi\left[\mathbf{x}^j(0), \mathbf{u}^j(0), \mathbf{x}^j(1), \mathbf{u}^j(1), T_f^j\right] = 0, \tag{29}$$

Meanwhile satisfying the inequality and equality constraints (21) and additional temporal constraints as follows:

$$\delta^j\left[\mathbf{x}^j(\bar{t}), \mathbf{u}^j(\bar{t}), T_f^j\right] \le 0,$$
$$\varphi^j\left[\mathbf{x}^j(\bar{t}), \mathbf{u}^j(\bar{t}), \mathbf{x}^k(\bar{t}), \mathbf{u}^k(\bar{t}), T_f^j, T_f^k\right] \le 0,$$
$$f_T^j\left(t_f^j, t_f^k\right) \le 0, \quad \forall j \ne k, \ j,k = 1,2,\ldots,N_v. \tag{30}$$

From (1), it can be clearly known that the dynamics system under consideration is a higher-dimensional space system. Due to the complexity in solving this type of the problem, the VMC based method is used to lower the system dimension and the state and control vectors can be related by the differential flatness theory based method [39]. In this system, the state vector x ∈ $\mathbb{R}^6$ can be separated into two parts: the trajectory position (e.g., spatial trajectory) $\rho_a = [x, y, z] \in \mathbb{R}^3$ and the corresponding state rate $\rho_r = [V, \psi, \gamma] \in \mathbb{R}^3$, by which the trajectory position in virtual time domain and the terminal time T_f can be defined as the flat output vector

$$\vartheta = \left[\rho_a(\bar{t}), T_f\right] = \left[x(\bar{t}), y(\bar{t}), z(\bar{t}), T_f\right]^T. \tag{31}$$

Thus, the original state vector x and control vector u can be recovered from the flat outputs and their derivatives as

$$\mathbf{x} = \alpha\left(\vartheta, \vartheta', \vartheta''\right) = \alpha\left(\rho_a, \rho_a', \rho_a'', T_f\right),$$
$$\mathbf{u} = \beta\left(\vartheta, \vartheta', \vartheta''\right) = \beta\left(\rho_a, \rho_a', \rho_a'', T_f\right). \tag{32}$$

According to (1), the relation of the state vector and control vector in the VTD is described as follows:

$$\begin{aligned}
V &= \sqrt{(x')^2 + (y')^2 + (z')^2},\\
\psi &= \arctan\left(\frac{y'}{x'}\right),\\
\gamma &= \arctan\left(\frac{z'}{\sqrt{(x')^2 + (y')^2}}\right),\\
n_x &= \frac{V'}{\left(T_f^2 g\right)} + \sin\gamma,\\
n_z &= \frac{\sqrt{\left(V\gamma' + T_f^2 g\cos\gamma\right)^2 + (V\psi'\cos\gamma)^2}}{T_f g},\\
\mu &= \arctan\left(\frac{V\psi'\cos\gamma}{\left(V\gamma' + T_f^2 g\cos\gamma\right)}\right).
\end{aligned} \tag{33}$$

In the VMC framework, the trajectory position ρa can be defined as the aggressor position vector. As described above, the state and the control variables of the dynamics system can be recovered from the flat outputs. Hence, according to (22), these variables are also functions of the PCP, the predefined virtual prey motion, the reference point, and their corresponding derivatives. Once the predefined virtual prey motion and the reference point are selected, the system of cooperative trajectory planning of UCAVs, including the objective function and the constraints, is mapped to a lowerdimensional output space (here, the dimension is one). Obviously, the dynamic constraints of this system (27) can be automatically satisfied. Therefore, the Problem 2 can be redefined as follows.

Problem 3 (CTOCP-VTD in the output space)

Consider

$$\begin{aligned} \min \quad J(v) &= \sum_{j=1}^{N_v} J^j(v^j) \\ &= \sum_{j=1}^{N_v} \Big[\left(v^j, v'^j, \ldots, T_f^j\right) \\ &\qquad + \int_0^1 f^j\left(v^j, v'^j, \ldots, \bar{t}\right) d\bar{t} \Big] \\ \text{s.t.} \quad & \phi^j\left[v^j(0), v^j(1), T_f^j\right] = 0, \\ & \delta^j\left(v^j, v'^j, \ldots,\right) \le 0, \\ & \varphi^j\left(v^j, v'^j, \ldots, v^k, v'^k, \ldots\right) \le 0, \\ & f_T^j\left(t_f^j, t_f^k\right) \le 0, \ \forall j \ne k, \ j,k = 1,2,\ldots,N_v. \end{aligned} \tag{34}$$

Collocation Using Gauss Pseudospectral Method

In order to solve the Problem 3 through a NLP algorithm, the PCP history $v(\bar{t})$ should be discretized into $c = 0, 1, \ldots, N_c, N_c + 1$ nodes, with $v_0 = v(\bar{t}_0)$ and $v_{N_c+1} = v(\bar{t}_{N_c+1})$. In this paper, GPM is selected as the discretization method [12], which is an orthogonal collection method where the collocation points are the Legendre-Gauss (LG) points. Similar to other pseudospectral methods, a finite basis of global interpolating polynomials is used to approximate the state and control space at a set of discretization nodes in the GPM and then the optimal control equations can be transformed into nonlinear algebra equations, such that the OCP can be solved by a NLP algorithm.

The standard interval considered here is denoted as $\tau \in [-1, 1]$. By using a linear transformation, the virtual time $\bar{t}$ can be expressed as a function of τ via

$$\bar{t} = \frac{\left[\left(\bar{t}_f - \bar{t}_0\right)\tau + \left(\bar{t}_f + \bar{t}_0\right)\right]}{2} = \frac{(\tau + 1)}{2}. \tag{35}$$

Let $\tau 1 < \cdots < \tau Nc$ be collocation points in $(-1, 1)$ and $\tau 0 = -1$, $\tau f = 1$. The time history of the approximated PCP at the Nc LG points is given by

$$v(\tau) \approx \bar{v}(\tau) = \sum_{c=0}^{N_c} v(\tau_c) L_c(\tau), \tag{36}$$

where V(τ_c) is an approximant at the node τ_c and the base function $L_c(\tau)$ is the Lagrange interpolating polynomial of degree N_c, expressed as

$$L_c(\tau) = \prod_{l=0,\, l \neq c}^{N_c} \frac{\tau - \tau_l}{\tau_c - \tau_l}, \quad c = 0, 1, \ldots, N_c, \tag{37}$$

satisfying the isolation property

$$L_c(\tau_l) = \delta_{cl} = \begin{cases} 1, & c = l, \\ 0, & c \neq l. \end{cases} \tag{38}$$

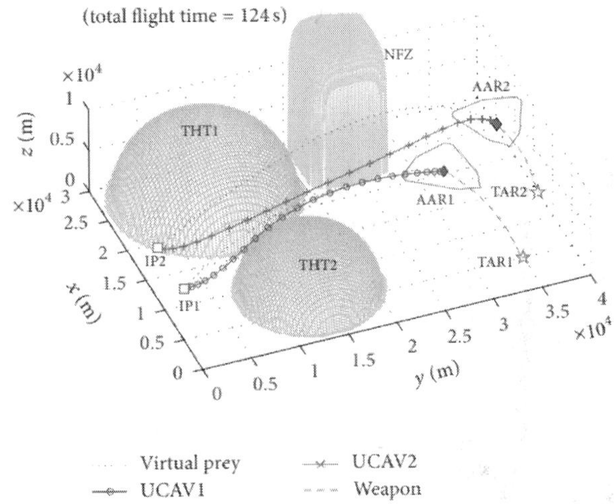

Figure 2: Two collision-free UCAV trajectories with arriving simultaneously

And the endpoint can be approximated according to the following formula:

$$v(\tau_f) \approx \bar{v}(\tau_f) = \bar{v}(\tau_0) + \frac{\bar{t}_f - \bar{t}_0}{2} \times \sum_{c=1}^{N_c} w_c f\left[\bar{v}(\tau_c), \tau_c\right], \tag{39}$$

where τ_c ($c = 1, \ldots, N_c$) are the LG points and the variables ω_c are the LG weights given by

$$w_c = \int_{-1}^{1} L_c(\tau)\, d\tau. \tag{40}$$

The derivative of the state can be approximated as the exact derivative of the interpolating polynomial. Evaluating the derivative at the LG points results in

$$v'(\tau_l) \approx \bar{v}'(\tau_l) = \sum_{c=0}^{N_c} v(\tau_c) L'_c(\tau_l)$$

$$= \sum_{c=0}^{N_c} v(\tau_c) D_{lc}, \quad l = 1, \dots, N_c, \tag{41}$$

where $\mathbf{D} = [D_{lc}] \in \mathbb{R}^{N_c \times (N_c+1)}$ is a differential matrix and can be offline determined as

$$D_{lc} = \sum_{k_1=0}^{N_c} \frac{\prod_{k_2=0, k_2 \neq c, k_1}^{N_c} \left(\tau_l - \tau_{k_2}\right)}{\prod_{k_2=0, k_2 \neq c}^{N_c} \left(\tau_c - \tau_{k_2}\right)},$$

$$l = 1, \dots N_c, \quad c = 0, \dots, N_c. \tag{42}$$

The discretized states and controls satisfying the vehicle dynamics can be computed by ensuring that equation

$$\sum_{c=0}^{N_c} v(\tau_c) D_{lc} - \frac{\bar{t}_f - \bar{t}_0}{2} f\left[v(\tau_l), \tau_l\right] = 0, \quad l = 1, \dots, N_c. \tag{43}$$

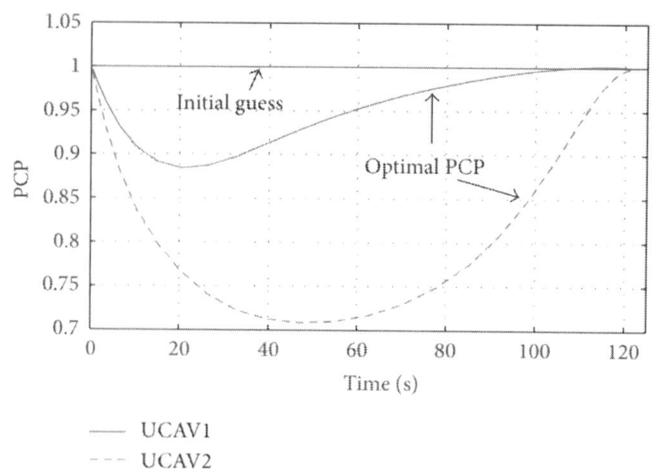

Figure 3: PCP histories of two UCAVs.

As described above, the continuous OCP can be transformed to the discretized NLP problem. The parameters to be optimized are the PCP nodes $\mathbf{v} = [v_c]_{c=0,1,\dots N_c+1}$. When the prey motion $\rho_{p,}$ is equal to the aggressor motion $\rho_{a,c}$ for $c = 0, N_c + 1$, one can assume that $v_0 = v_{N_c} = 1$, so the boundary conditions are no longer considered. And the discretized NLP problem can be denoted as follows.

Problem 4 (CTP-NLP)

Its standard form is denoted as

$$\min J(\mathbf{v}) = \sum_{j=1}^{N_v} J^j(\mathbf{v}^j) = \sum_{j=1}^{N_v} \left[\Phi^j \left[\mathbf{v}^j, T_f^j\right] + \frac{\bar{t}_f - \bar{t}_0}{2} \sum_{c=1}^{N_c} f^j\left(\mathbf{v}^j, T_f^j\right) \omega_c \right]$$

$$\text{s.t.} \quad \delta_l^j\left[\mathbf{v}^j\right] \le 0, \quad l = 1, \ldots, N_c,$$
$$\varphi^j\left[\mathbf{v}^j, \mathbf{v}^k\right] \le 0,$$
$$f_T^j\left(t_f^j, t_f^k\right) \le 0, \quad \forall j \ne k, \ j,k = 1,2,\ldots,N_v, \tag{44}$$

where $v^j = [v_c^j]_{c=1,\ldots N_c}$ is the vector form of the discretized PCP.

Then the resulting Problem 4 can be solved through welldeveloped algorithms, such as the SQP algorithm. In this paper, TOMLAB/SNOPT software toolbox is chosen due to its advantages on solution effectiveness for the large-scale NLP problems [40].

Convergence and Initial Guess

As analyzed in [41], the GPM exhibits global convergence properties in many applications. In addition, the convergence of the virtual motion camouflage has been proved by Xu et al. [21, 23], in which [23] showed that if the problem was only onedimensional, the VMC based trajectory design method could converge to the optimal solution, but when the problem was multi-dimensional and highly nonlinear, the result might be suboptimal in the full solution space, and [21] showed that the VMC framework together with the pseudospectral discretization method could converge to the optimal solution.

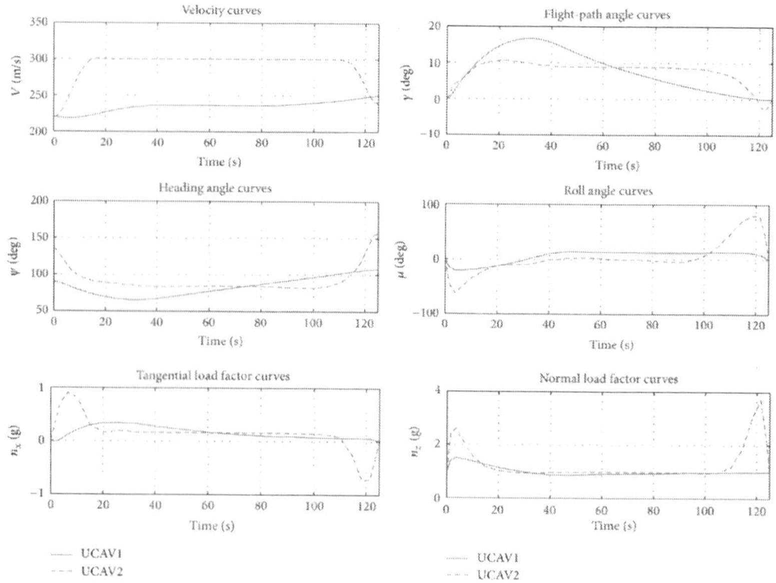

Figure 4: Example 1—state and control time histories of two UCAVs

By Introducing elastic programming concept and some advances in both mathematical programming techniques and pseudospectral methods, Ross and Gong [42] designed a guess-free trajectory optimization algorithm. These advances have the effect of eliminating the guessing problem in the trajectory optimization, which has been used in the current GPM algorithm. In the proposed VMC method, the motion of the virtual prey and a reference point are guessed. However, both of them have direct physical relation to the problem. Typically the virtual prey motion is selected according to the boundary conditions and a proper guess is not a difficult task [23]. In this paper, the trivial guess for the reference point can be the initial point (IP), and an initial guess of $V_c = 1$ is chosen for $c = 1, \ldots, N_c$. The virtual prey motion is defined as a 2-order curve, determined by the initial and terminal conditions.

NUMERICAL EXAMPLES

The basic ideas presented in this paper are illustrated in the following three examples. The specific vehicle platform used in simulations is the Storm shadow UCAV. Its relevant parameters are all taken from [43], summarized in Table 1

The experimental test environment is a rectangle area of 30 × 40 km^2, as shown in Figure 2, where the NFZ1 caused by the severe weather condition is modeled as a cuboid with infinite altitude, and the NFZs caused by enemy threats are modeled as two hemispheres, denoted as THT1 and THT2. According to Figure 1, the shape parameters can be set as follows: for the NFZ1, $k_1 = k_2 = k_3 = k_4 = 1$, $p=2$, and the length of a side is set as 8 km; for THT1 and THT2, $k_1 = k_2 = k_4 = 1$, $k_3 = 2$, $p = 6$, and the radius of the threats can be set as 7 km and 9 km, respectively. In order to avoid contact, a safe buffer b_s (b_s = 200 m) is added around the outside edge of each obstacle/threat boundary. All the results presented below are generated using TOMLAB/SNOPT software toolbox on a 2.4 GHz CPU and 2 G RAM computer running Windows XP and MATLAB R2009b. Table 2 summarizes the parameters used in the algorithm. To simplify the problem, the same weight coefficients for the objective function of each UCAV are set, and it is assumed that the target assignment is already completed before.

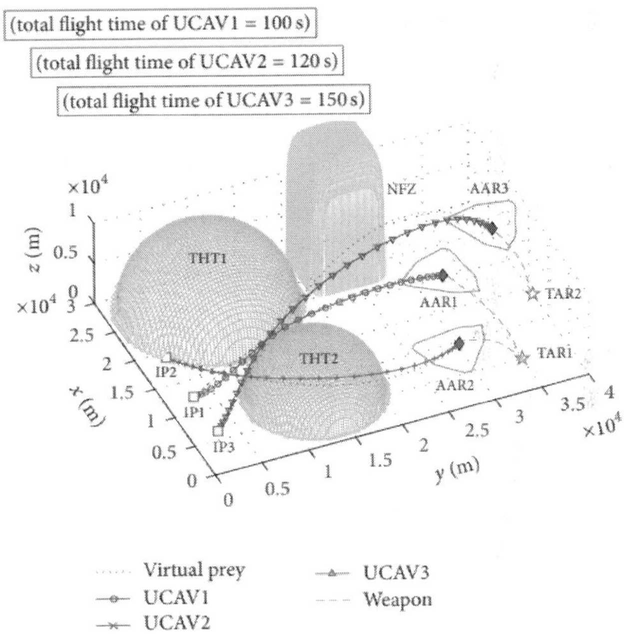

Figure 5: Three collision-free UCAV trajectories with arriving in sequence.

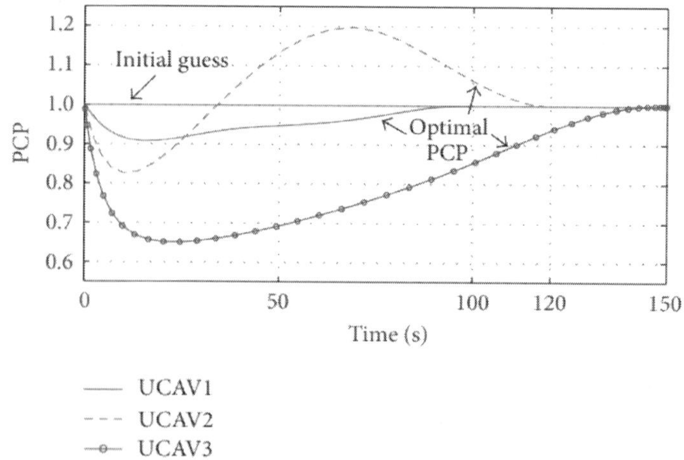

Figure 6: Path control parameter histories of three UCAVs.

Example 1: Cooperative Trajectories of Two UCAVs Arriving Simultaneously

In this example, two UCAVs cooperatively attack two stationary ground targets while avoiding a series of static obstacles/threats detected and collision with each other en route and meeting aircraft dynamical constraints, especially simultaneous arrival constraint. UCAV1 and UCAV2 start at each initial point, that is, IP1 (10, 2, 2) km and IP2 (17, 2, 2) km. Then they fly into the AARs of two targets, TAR1 (4, 34, 0) km and TAR2 (13, 40, 0) km, respectively. The initial remaining three states and control inputs of each UCAV are preset as $\rho^1_{r,0} = (V^1_0, \gamma^1_0, \psi^1_0) = (220\,\text{m/s}, 0^\circ, 90^\circ)$, $\rho^2_{r,0} = (V^2_0, \gamma^2_0, \psi^2_0) = (220\,\text{m/s}, 0^\circ, 140^\circ)$, and $\mathbf{u}^1_0 = \mathbf{u}^2_0 = (\mu_0, n_{x,0}, n_{z,0}) = (0^\circ, 0\,\text{g}, 1\,\text{g})$

as $\rho^1_{r,f} = (\bar{V}^1_f, \bar{\gamma}^1_f, \bar{\psi}^1_f) = (250\,\text{m/s}, 0^\circ, 110^\circ)$ and $\rho^2_{r,f} = (\bar{V}^2_f, \bar{\gamma}^2_f, \bar{\psi}^2_f) = (250\,\text{m/s}, 0^\circ, 160^\circ$. In VMC framework, the reference point ρ_{ref} for each UCAV is presumed as its IP, and the virtual prey path is predefined as a 2-order curve, determined by the initial and terminal conditions.

Figure 2 shows the overall collision-free attack trajectories of two UCAVs, whose total flight time is equal, equaling to 124 s. It means that two UCAVs arrive simultaneously. As can be seen from Figure 2, along the resulting smooth trajectories, the UCAVs can avoid collision with all obstacles, threats, and the other en route and then successfully fly into the AARs (fuchsin areas) to perform weapon delivery missions. In addition, the approximate weapon trajectories are drawn to illustrate the attack process. The PCP histories of two UCAVs are shown in Figure 3, which are smooth and lie in the range from 0 to 1. And the time histories of the UCAVs' states (V, γ, ψ) and control inputs (μ, n_x, n_z) are shown in Figure 4. It is clear that the constraints on these variables, especially the cooperative constraints, are all satisfied (see Tables 1 and 2), which means that the resulting trajectories are feasible and safe

Example 2: Cooperative Trajectories of Multi-UCAV Arriving in Sequence

In this example, three UCAVs attack two stationary ground targets cooperatively. The only additional requirement is that the UCAVs arrive at their AARs in sequence rather than simultaneously, and the intervals between UCAVs are set to 20 s and 30 s, respectively, denoted as Δ_{12} = 20 s and Δ_{23} = 30 s. The coordinates of the initial points IP1, IP2 and two targets TAR1, TAR2 are the same with those in Example 1, and the third initial point is IP3 (4, 2, 2) km. The result of the previously finished target assignment is that the UCAV1 and UCAV2 attack TAR1 and UCAV3 attacks the other one. The initial remaining three states and control inputs of the UCAV3 are preset as $\rho_{r,0}^{3} = (V_0^3, \gamma_0^3, \psi_0^3) = (220\,\mathrm{m/s}, 0^\circ, 45^\circ)$ and $\mathbf{u}_0^3 = (\mu_0, n_{x,0}, n_{z,0}) =$ (0° , 0 g, 1 g), and the terminal remaining three states of the UCAV3 are predetermined as $\rho_{r,f}^{3} = (\bar{V}_f^3, \bar{\gamma}_f^3, \bar{\psi}_f^3) =$ (220 m/s, 0° , 160°). In VMC framework, the reference point and the virtual prey path are predetermined, using the same strategy as the one in Example 1.

The overall collision-free attack trajectories of multi-UCAV are shown in Figure 5, and the total flight time of three UCAVs is 100 s, 120 s, and 150 s, respectively. It can be clearly demonstrated that UCAVs can avoid all obstacles or threats and successfully fly into the AARs in sequence to perform the weapon delivery. The PCP histories of three UCAVs are shown in Figure 6. And Figure 7 shows the distance between each pair of UCAVs. From it, one can find that the minimum distance between each UCAV is more than the minimal safety radius of d_{safe} (d_{safe} = 500 m); that is, the UCAVs can avoid collision with each other en route. Figure 8 shows the time histories of the UCAVs' states (V, γ, ψ) and control inputs (μ, n_x, n_z). Obviously, the resulting trajectories are feasible and safe, since all the constraints listed in Tables 1 and 2 are satisfied.

Table 1: State and control constraints of UCAVs.

Item	Minimum value	Maximum value
Flight altitude (m)	$z_{\min} = 200$	$z_{\max} = 8\,000$
Airspeed (m/s)	$V_{\min} = 60$	$V_{\max} = 300$
Flight-path angle (deg)	$\gamma_{\min} = -89$	$\gamma_{\max} = 89$
Roll angle (deg)	$\mu_{\min} = -80$	$\mu_{\max} = 80$
Tangential load factor (g)	$n_{x,\min} = -0.725$	$n_{x,\max} = 0.91$
Normal load factor (g)	$n_{z,\min} = -3.2$	$n_{z,\max} = 8$
Roll angle rate (deg/s)	$\dot{\mu}_{\min} = -30$	$\dot{\mu}_{\max} = 30$
Rate of change of tangential load factor (g/s)	$\dot{n}_{x,\min} = -0.2$	$\dot{n}_{x,\max} = 0.2$
Rate of change of normal load factor (g/s)	$\dot{n}_{z,\min} = -2$	$\dot{n}_{z,\max} = 2$

Table 2: Parameter values for the algorithm.

Item	Parameter	Value
Minimum safety radius (m)	d_{safe}	500
	w_t	0.3
Weighting factors of the objective function	w_p	0.4
	w_r	0.3
	w_r^1	0.5
Weighting factors of the robust subobjective	w_r^2	0.3
	w_r^3	0.2
Safe buffer of NFZs (m)	b_s	200

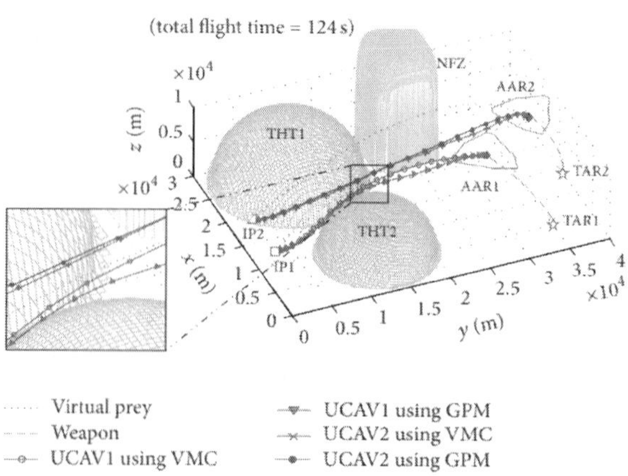

Figure 9: Comparison of two collision-free UCAV trajectories with arriving simultaneously

In order to validate the convergence rate of VMC based algorithm in cooperative trajectory planning for multi-UCAV further, the comparative experiments are carried out with different numbers of UCAVs (N_V is set as 1, 2, 3, 5, 8, and 10, resp.) and the same number of nodes ($N_c = 20$). The performance comparison of multi-UCAV is summarized in Table 3. As can be seen clearly, along with the increase of the number of UCAVs, the average planning time is rising at a slow rate, which is acceptable for the preplanning applications. Particularly, for one UCAV, the average planning time is 2.21 s. When the number of UCAVs increases to 10, the average planning time increases by only more than quadruple of that for one UCAV, which is 9.62 s.

Comparison between VMC Based Method and GPM Based Direct Collocation Method

To further evaluate the performance of the proposed algorithm, a recently developed optimal trajectory generation method, GPM based direct collocation method, is used for comparison, which is basically converting the original optimal control problem to a NLP problem via the GPM directly. In the following comparative experiments, Example 1 is simulated again using the two algorithms, respectively, performed on the same desktop with different numbers of nodes (N_c is set as 10, 15, 20, and 30, resp.). All the results are presented below.

Table 3: Performance comparison with the increase of the number of UCAVs.

Number of UCAVs (N_v)	Number of optimization variables	Average planning time (s)	Ratio of the planning time (%)
1	$(20 \times 1 + 1) \times 1 = 21$	2.21	—
2	$(20 \times 1 + 1) \times 2 = 42$	2.83	128.05
3	$(20 \times 1 + 1) \times 3 = 63$	3.32	150.23
5	$(20 \times 1 + 1) \times 5 = 105$	4.93	223.08
8	$(20 \times 1 + 1) \times 8 = 168$	7.70	348.42
10	$(20 \times 1 + 1) \times 10 = 210$	9.62	435.29

Figures 9 and 10 show the comparative solutions attained by using the GPM and VMC based methods, respectively. It can be seen that the trajectories and the state and control time histories of these two algorithms are similar. The detailed performance comparison is summarized in Table 4. With comparison, when the number of nodes increases, the number of optimization variables in GPM is much more than VMC. As a result, the computational speed of VMC is more than an order of magnitude faster than GPM, while the optimization performance has only small loss.

In order to compare the convergence rate and analyze the impacts of the number of UCAVs on the computation time between VMC and GPM further, the comparative experiment is run multiple times with different numbers of UCAVs (N_V is set as an integer from 1 to 10, resp.) and the same number of nodes ($N_c = 20$). Figure 11 shows the change of the average planning time required to solve the optimization problem as the number of UCAVs increases. It can be seen that the increase of the number of UCAVs results in an exponential increase of the required computation time for GPM. In contrast, the computation time for VMC based algorithm increases slowly. When the number of UCAVs increases to 10, the number of the optimization variables for GPM is up to 1810. However, for VMC, the number is only 210. Since the required computation time strongly depends on the number of optimization variables, the computational time required by VMC is more than an order of magnitude less than that required by GPM. Maybe it can be concluded that the larger the number of UCAVs is, the more obvious the advantage of VMC to GPM is.

CONCLUSIONS

This paper is devoted to explore the cooperative collision-free trajectory planning for multiple UCAVs in performing the CA/GTA missions. The main contributions of the paper are as follows. Firstly, the cooperative trajectory planning problem under consideration is mathematically formulated as a cooperative trajectory optimal control problem (CTOCP). In order to consider the weapon delivery constraints, an approximate allowable attack region model is proposed and integrated into the problem formulation. Secondly, a particular virtual motion camouflage approach combining with differential flatness theory, Gauss pseudospectral method, and nonlinear programming is used to develop the trajectory planner for solving CTOCP. The advantages of this method are that it can automatically satisfy all boundary conditions and use any models and any performance indexes, does not require numerical integration of differentiation of the state equations, and transforms the OCP into a NLP problem of very low dimension. Finally, in order to handle the temporal constraints, the notion of the virtual time is introduced into the virtual motion camouflage approach to realize the spatial-temporal cooperation.

The proposed approach is validated by numerical examples. The results show that the approach is able to generate both feasible and near-optimal attack trajectories with meeting the spatial and temporal constraints very efficiently. Moreover, the convergence rate and average planning time of the method and optimality of the generated trajectories are evaluated via a detailed comparison with GPM based direct collocation method. The results show that the computational speed of virtual motion camouflage approach is more than an order of magnitude faster than GPM, at small loss of optimality.

For the future work, we will analyze some uncertain factors in the true battlefield environment and carry out the research on the real-time cooperative trajectory planning. For the presence of a larger number of UCAVs, we will make further efforts to exploit more efficient trajectory planning algorithm and improve the time cooperative strategy. Moreover, we will try to study another important aspect about how to plan in the opposability battlefield environment.

ACKNOWLEDGMENTS

Great thanks are due to the editor for helpful comments and the reviewers for their valuable suggestions and comments on the revision of the paper. This work was supported by the Science and Technology on Electro-Optic Control Laboratory of China (no. 20125188006).

REFERENCES

1. R. M. Murray, "Recent research in cooperative control of multivehicle systems," *Journal of Dynamic Systems, Measurement and Control*, vol. 129, no. 5, pp. 571–583, 2007.
2. M. Fu and J. Jiao, "A hybrid approach for coordinated formation control of multiple surface vessels," *Mathematical Problems in Engineering*, vol. 2013, Article ID 794284, 8 pages, 2013.
3. J. T. Betts, "Survey of numerical methods for trajectory optimization," *Journal of Guidance, Control, and Dynamics*, vol. 21, no. 2, pp. 193–207, 1998.
4. C. Goerzen, Z. Kong, and B. Mettler, "A survey of motion planning algorithms from the perspective of autonomous UAV guidance," *Journal of Intelligent and Robotic Systems*, vol. 57, no. 1–4, pp. 65–100, 2010.
5. J. F. Horn, E. M. Schmidt, B. R. Geiger, and M. P. DeAngelo, "Neural network-based trajectory optimization for unmanned aerial vehicles," *Journal of Guidance, Control, and Dynamics*, vol. 35, no. 2, pp. 548–562, 2012.
6. S. Karaman and E. Frazzoli, "Sampling-based algorithms for optimal motion planning," *The International Journal of Robotics Research*, vol. 30, no. 7, pp. 846–894, 2011.
7. P. Narayan, P. Meyer, and D. Campbell, "Embedding human expert cognition into autonomous UAS trajectory planning," *IEEE Transaction on Cybernetics*, vol. 43, no. 2, pp. 530–543, 2013.
8. Q. Gong, L. R. Lewis, and I. M. Ross, "Pseudospectral motion planning for autonomous vehicles," *Journal of Guidance, Control, and Dynamics*, vol. 32, no. 3, pp. 1039–1045, 2009.
9. H. Liu, S. Chen, L. Shen et al., "An integrated multi-criterion hp-Adaptive pseudospectral method for direct optimal control problems solving," *Mathematical Problems in Engineering*, vol. 2012, Article ID 760890, 22 pages, 2012.
10. K. Yang, S. Sukkarieh, and Y. Kang, "Adaptive nonlinear model predictive path tracking control for a fixed-wing unmanned aerial vehicle," in *Proceedings of the AIAA Guidance, Navigation, and Control Conference and Exhibit*, Chicago, Ill, USA, August 2009.
11. O. A. Yakimenko, Y. Xu, and G. Basset, "Computing short-time aircraft maneuvers using direct methods," in *Proceedings of the AIAA Guidance, Navigation and Control Conference and Exhibit*, Honolulu,

Hawaii, USA, August 2008.

12. D. Benson, *A Gauss pseudospectral transcription for optimal control [Dissertation for the doctoral degree]*, Massachusettes Institute of Technology, Cambridge, Mass, USA, 2005.
13. M. Soler, A. Olivares, and E. Staffetti, "Hybrid optimal control approach to commercial aircraft trajectory planning," *Journal of Guidance, Control, and Dynamics*, vol. 33, no. 3, pp. 985–991, 2010.
14. J. Li, H. An, H. Zhu et al., "Geometric pseudospectral method on SE(3) for rigid-body dynamics with application to aircraft," *Mathematical Problems in Engineering*, vol. 2013, Article ID 595086, 16 pages, 2013.
15. A. Verma and J. L. Junkins, "Inverse dynamics approach for real-time determination of feasible aircraft reference trajectories," in *Proceedings of the 24th Atmospheric Flight Mechanics Conference*, 1999.
16. R. G. Drury, "Performance of NLP algorithms with inverse dynamics for near-real time trajectory generation," in *Proceedings of the AIAA Guidance, Navigation, and Control Conference*, Portland, Ore, USA, 2011.
17. F.-L. Lian, "Cooperative path planning of dynamical multi-agent systems using differential flatness approach," *International Journal of Control, Automation and Systems*, vol. 6, no. 3, pp. 401–412, 2008.
18. A. Chamseddine, Y. Zhang, C. A. Rabbath et al., "Trajectory planning and peplanning strategies applied to a quadrotor unmanned aerial vehicle," *Journal of Guidance, Control, and Dynamics*, vol. 35, no. 5, pp. 1667–1671, 2012.
19. X. Gu, Y. Zhang, S. Chen et al., "Cooperative trajectory planning for multi-UCAV performing air-to-ground target attack missions," *Advanced Materials Research*, vol. 718-720, pp. 1329–1334, 2013.
20. X. Gu, Y. Zhang, J. Chen et al., "Real-time cooperative trajectory planning using differential flatness approach and B-splines," *Applied Mechanics and Materials*, vol. 333-335, pp. 1338–1343, 2013.
21. Y. Xu, "Analytical solutions to spacecraft formation-flying guidance using virtual motion camouflage," *Journal of Guidance, Control, and Dynamics*, vol. 33, no. 5, pp. 1376–1386, 2010.
22. Y. Xu and G. Basset, "Sequential virtual motion camouflage method for nonlinear constrained optimal trajectory control," *Automatica*, vol.

48, no. 7, pp. 1273–1285, 2012.

23. Y. Xu, "Virtual motion camouflage and suboptimal trajectory design," in *Proceedings of the AIAA Guidance, Navigation, and Control Conference*, pp. 555–571, August 2007.
24. Y. Xu and G. Basset, "Pre and post optimality checking of the virtual motion camouflage based nonlinear constrained subspace optimal control," in *Proceedings of the AIAA Guidance, Navigation, and Control Conference and Exhibit*, Chicago, Ill, USA, August 2009.
25. K. P. Bollino and L. R. Lewis, "Collision-free multi-UAV optimal path planning and cooperative control for tactical applications," in *Proceedings of the AIAA Guidance, Navigation and Control Conference and Exhibit*, Honolulu, Hawaii, USA, August 2008.
26. Y. Kuwata and J. P. How Sr., "Cooperative distributed robust trajectory optimization using receding horizon MILP," *IEEE Transactions on Control Systems Technology*, vol. 19, no. 2, pp. 423–431, 2011.
27. Y. Xu and G. Basset, "Virtual motion camouflage based phantom track generation through cooperative electronic combat air vehicles," *Automatica*, vol. 46, no. 9, pp. 1454–1461, 2010.
28. Y. Xu, M. Xin, and J. Wang, "Unmanned aerial vehicle formation flight via a hierarchical cooperative control approach," in *Proceedings of the AIAA Guidance, Navigation, and Control Conference*, Portland, Ore, USA, 2011.
29. T. W. McLain and R. W. Beard, "Coordination variables, coordination functions, and cooperative-timing missions," *Journal of Guidance, Control, and Dynamics*, vol. 28, no. 1, pp. 150–161, 2005.
30. I. Kamlner, O. Yakimenko, V. Dobrokhodov et al., "Coordinated path following for time-critical missions of multiple UAVs via *L*1 adaptive output feedback controllers," in *Proceedings of the AIAA Guidance, Navigation, and Control Conference*, pp. 915–948, August 2007.
31. B. L. Stevens and F. L. Lewis, *Aircraft Control and Simulation*, John Wiley & Sons, New York, NY, USA, 2003.
32. Y. Zhang, J. Chen, and L. Shen, "Hybrid hierarchical trajectory planning for a fixed-wing UCAV performing air-to-surface multi-target attack," *Journal of Systems Engineering and Electronics*, vol. 23, no. 4, pp. 256–264, 2012.
33. X. Gu, Y. Zhang, N. Wang et al., "Robust multi-objective optimization based on NSGA-II for UCAV weapon delivery," in *Proceedings of the*

4th International Conference on Intelligent Human-Machine Systems and Cybernetics, 2012.

34. L. P. R. Lewis, *Rapid motion planning and autonomous obstacle avoidance for unmanned vehicles [M.S. thesis]*, Naval Postgraduate School, Monterey, Calif, USA, 2006.
35. E. Besada-Portas, L. de la Torre, J. M. de la Cruz, and B. de Andrés-Toro, "Evolutionary trajectory planner for multiple UAVs in realistic scenarios," *IEEE Transactions on Robotics*, vol. 26, no. 4, pp. 619–634, 2010.
36. M. A. Hurni, P. Sekhavat, M. Karpenko, and I. M. Ross, "A pseudospectral optimal motion planner for autonomous unmanned vehicles," in *Proceedings of the American Control Conference (ACC '10)*, pp. 1591–1598, Baltimore, Md, USA, July 2010.
37. Y. Wang, T. Wei, and X. Qu, "Study of multi-objective fuzzy optimization for path planning," *Chinese Journal of Aeronautics*, vol. 25, no. 1, pp. 51–56, 2012.
38. M. V. Srinivasan and M. Davey, "Strategies for active camouflage of motion," *Proceedings of the Royal Society B*, vol. 259, no. 1354, pp. 19–25, 1995.
39. M. Fliess, J. Levine, P. Martin, and P. Rouchon, "Flatness and defect of non-linear systems: introductory theory and examples," *International Journal of Control*, vol. 61, no. 6, pp. 1327–1361, 1995.
40. K. Holmström, A. O. Göran, and M. M. Edvall, *USer'S Guide For TOMLAB /SNOPT*, Department of Mathematics and Physics, Mälardalen University, Västerås, Sweden, 2008.
41. S. Chen, H. Liu, J. Chen et al., "Penetration trajectory planning based on radar tracking features for UAV," *Aircraft Engineering and Aerospace Technology*, vol. 85, no. 1, pp. 62–71, 2013.
42. I. M. Ross and Q. Gong, "Guess-free trajectory optimization," in *Proceedings of the AIAA/AAS Astrodynamics Specialist Conference and Exhibit*, Honolulu, Hawaii, USA, August 2008.
43. "Storm shadow UCAV performance," 2001, http://www.aerospaceweb.org/design/ucav/main.shtml.

INDEX

D

E

F